从入门到实战·微课视频

Java EE 框架整合开发入门到实战
——Spring+Spring MVC+MyBatis

微课版

◎ 陈恒 楼偶俊 张立杰 编著

清華大學出版社
北京

内容简介

本书详细讲解了Java EE中Spring、Spring MVC和MyBatis三大框架（SSM）的基础知识和实际应用。为了更好地帮助读者学习SSM框架，本书以大量案例介绍了SSM框架的基本思想、方法和技术。

全书共20章，分四部分介绍。第1～5章为第1部分，主要讲解Spring框架的相关知识，内容包括Spring入门、Spring IoC、Spring Bean、Spring AOP以及Spring的事务管理；第6～8章为第2部分，主要讲解MyBatis的相关知识，内容包括MyBatis开发入门、映射器以及动态SQL；第9～18章为第3部分，主要讲解Spring MVC的相关知识，内容包括Spring MVC入门、Controller、类型转换和格式化、数据绑定和表单标签库、拦截器、数据验证、国际化、统一异常处理、文件的上传和下载以及EL与JSTL；第19章、第20章为第4部分，主要讲解SSM框架整合的基本思想与实战开发，内容包括SSM框架整合以及基于SSM框架的电子商务平台的设计与实现。本书突出实用性、趣味性，内容组织合理、通俗易懂，使读者能够快速掌握SSM框架的基础知识、编程技巧以及完整的开发体系，从而为大型项目开发打下坚实的基础。

本书附有教学视频、源代码、课件、教学大纲、习题答案等配套资源，可以作为大学计算机及相关专业的教材或教学参考书，也可以作为Java技术的培训教材，同时适合广大Java EE应用开发人员阅读与使用。

图书在版编目（CIP）数据

Java EE框架整合开发入门到实战——Spring+Spring MVC+MyBatis：微课版 / 陈恒，楼偶俊，张立杰编著. —北京：清华大学出版社，2018 (2023. 1重印)

（从入门到实战·微课视频）

ISBN 978-7-302-50296-8

Ⅰ. ①J… Ⅱ. ①陈… ②楼… ③张… Ⅲ. ①JAVA语言-程序设计 Ⅳ. ①TP312.8

中国版本图书馆CIP数据核字（2018）第112440号

策划编辑： 魏江江
责任编辑： 王冰飞
封面设计： 刘 键
责任校对： 胡伟民
责任印制： 宋 林

出版发行： 清华大学出版社
网　　址： http://www.tup.com.cn, http://www.wqbook.com
地　　址： 北京清华大学学研大厦A座　　**邮　　编：** 100084
社 总 机： 010-83470000　　**邮　　购：** 010-62786544
投稿与读者服务： 010-62776969，c-service@tup.tsinghua.edu.cn
质 量 反 馈： 010-62772015，zhiliang@tup.tsinghua.edu.cn
印 装 者： 三河市君旺印务有限公司
经　　销： 全国新华书店
开　　本： 185mm×260mm　　**印　　张：** 23.25　　**字　　数：** 563千字
版　　次： 2018年9月第1版　　**印　　次：** 2023年1月第15次印刷
印　　数： 40501~43000
定　　价： 69.80元

产品编号：079720-01

前 言

本书适用于具有 Java 编程基础和一定 Java Web 相关知识的读者学习。

本书使用 Spring 5.0.2 + Spring MVC 5.0.2 + MyBatis 3.4.5 版本详细讲解了 SSM 三大框架的基础知识和使用方法。本书的重点不是简单地介绍三大框架的基础知识，而是精心设计了大量实例。读者通过本书可以快速地掌握 SSM 框架的实战应用，提高 Java EE 应用的开发能力。

全书共 20 章，各章的具体内容如下：

第 1 章主要讲解 Spring 框架入门的一些基础知识，包括 Spring 框架的体系结构、核心容器、开发环境以及入门程序等内容。

第 2 章主要介绍 Spring IoC 的基本概念、Spring IoC 容器以及依赖注入的类型等内容。

第 3 章对 Spring 中的 Bean 进行详细介绍，包括 Spring Bean 的配置、实例化、作用域、生命周期以及装配方式等内容。

第 4 章介绍 AOP 的相关知识，包括 AOP 的概念和术语、动态代理以及 AOP 的实现和 AspectJ 的开发等内容。

第 5 章主要介绍 Spring 框架所支持的事务管理，包括编程式事务管理和声明式事务管理。

第 6 章主要讲解 MyBatis 环境的构建、MyBatis 的工作原理以及与 Spring 框架的整合开发。

第 7 章对 MyBatis 的核心配置进行详细讲解，包括 MyBatis 配置文件、映射文件以及级联查询。

第 8 章主要讲解如何拼接 MyBatis 的动态 SQL 语句。

第 9 章主要讲解 MVC 的设计思想以及 Spring MVC 的工作原理。

第 10 章详细讲解基于注解的控制器，包括 Controller 注解和 RequestMapping 注解类型的使用，是 Spring MVC 框架的重点内容之一。

第 11 章介绍类型转换器和格式化转换器，包括内置的类型转换器和格式化转换器以及自定义类型转换器和格式化转换器等内容。

第 12 章讲解数据绑定和表单标签库，是 Spring MVC 框架的重点内容之一。

第 13 章主要介绍拦截器的概念、原理以及实际应用。

第 14 章详细讲解 Spring MVC 框架的输入验证体系，包括 Spring 验证和 JSR 303 验证等内容。

第 15 章介绍 Spring MVC 国际化的实现方法，包括 JSP 页面信息国际化以及错误消息国际化等内容。

第 16 章详细讲解如何使用 Spring MVC 框架进行异常的统一处理，是 Spring MVC 框架的重点内容之一。

第 17 章讲解如何使用 Spring MVC 框架进行文件的上传与下载。

第 18 章介绍 EL 与 JSTL 的基本用法。

第 19 章主要讲解 SSM 框架整合环境的构建，包括整合思路、整合所需 JAR 包以及整合应用测试等内容。

第 20 章以电子商务平台的设计与实现为综合案例，讲述如何使用 SSM（Spring+Spring MVC+MyBatis）框架整合开发一个 Web 应用。

为便于教学，本书提供源代码、课件、教学大纲、习题答案等配套资源，读者可以扫描封底课件二维码免费下载。扫描本书章节中的二维码，可以在线观看视频讲解。

由于编者水平有限，书中难免会有不足之处，敬请广大读者批评指正。

编　者
2018 年 5 月

目 录

源码下载

第1部分 Spring

第 2 部分 MyBatis

第 1 部分

Spring

第1章

Spring 入门

学习目的与要求

本章重点讲解 Spring 开发环境的构建。通过本章的学习，读者能够了解 Spring 的体系结构，掌握 Spring 开发环境的构建。

本章主要内容

- Spring 的体系结构;
- Spring 开发环境的构建;
- Spring 的核心容器;
- Spring 的入门程序。

Spring 是当前主流的 Java Web 开发框架，为企业级应用开发提供了丰富的功能，掌握 Spring 框架的使用是 Java 开发者必备的技能之一。本章将介绍如何使用 Eclipse 开发 Spring 入门程序，不过在此之前需要构建 Spring 的开发环境。

1.1 Spring 简介

1.1.1 Spring 的由来

Spring 是一个轻量级 Java 开发框架，最早由 Rod Johnson 创建，目的是为了解决企业级应用开发的业务逻辑层和其他各层的耦合问题。它是一个分层的 JavaSE/EE full-stack（一站式）轻量级开源框架，为开发 Java 应用程序提供全面的基础架构支持。Spring 负责基础架构，因此 Java 开发者可以专注于应用程序的开发。

1.1.2 Spring 的体系结构

Spring 框架至今已集成了 20 多个模块，这些模块分布在核心容器（Core Container）、数据访问/集成（Data Access/Integration）层、Web 层、AOP（Aspect Oriented Programming，面向切面的编程）模块、植入（Instrumentation）模块、消息传输（Messaging）和测试（Test）模块中，如图 1.1 所示。

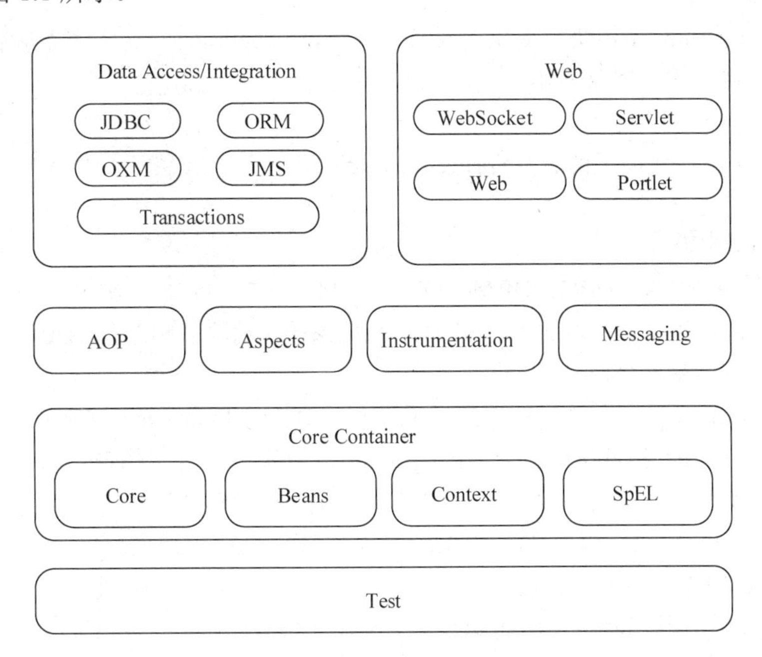

图 1.1　Spring 的体系结构

❶ 核心容器

Spring 的核心容器是其他模块建立的基础，由 Spring-core、Spring-beans、Spring-context、Spring-context-support 和 Spring-expression（Spring 表达式语言）等模块组成。

- Spring-core 模块：提供了框架的基本组成部分，包括控制反转（Inversion of Control，IoC）和依赖注入（Dependency Injection，DI）功能。
- Spring-beans 模块：提供了 BeanFactory，是工厂模式的一个经典实现，Spring 将管理对象称为 Bean。
- Spring-context 模块：建立在 Core 和 Beans 模块的基础之上，提供一个框架式的对象访问方式，是访问定义和配置的任何对象的媒介。ApplicationContext 接口是 Context 模块的焦点。
- Spring-context-support 模块：支持整合第三方库到 Spring 应用程序上下文，特别是用于高速缓存（EhCache、JCache）和任务调度（CommonJ、Quartz）的支持。
- Spring-expression 模块：提供了强大的表达式语言去支持运行时查询和操作对象图。这是对 JSP 2.1 规范中规定的统一表达式语言（Unified EL）的扩展。该语言支持设置和获取属性值、属性分配、方法调用、访问数组、集合和索引器的内容、逻辑和算术运算、变量命名以及从 Spring 的 IoC 容器中以名称检索对象。它还支持列表投影、选择以及常见的列表聚合。

❷ AOP 和 Instrumentation

- Spring-aop 模块：提供了一个符合 AOP 要求的面向切面的编程实现，允许定义方法拦截器和切入点，将代码按照功能进行分离，以便干净地解耦。

- Spring-aspects 模块：提供了与 AspectJ 的集成功能，AspectJ 是一个功能强大且成熟的 AOP 框架。
- Spring-instrument 模块：提供了类植入（Instrumentation）支持和类加载器的实现，可以在特定的应用服务器中使用。

❸ 消息

Spring 4.0 以后新增了消息（Spring-messaging）模块，该模块提供了对消息传递体系结构和协议的支持。

❹ 数据访问/集成

数据访问/集成层由 JDBC、ORM、OXM、JMS 和事务模块组成。

- Spring-jdbc 模块：提供了一个 JDBC 的抽象层，消除了烦琐的 JDBC 编码和数据库厂商特有的错误代码解析。
- Spring-orm 模块：为流行的对象关系映射（Object-Relational Mapping）API 提供集成层，包括 JPA 和 Hibernate。使用 Spring-orm 模块可以将这些 O/R 映射框架与 Spring 提供的所有其他功能结合使用，例如声明式事务管理功能。
- Spring-oxm 模块：提供了一个支持对象/XML 映射的抽象层实现，例如 JAXB、Castor、JiBX 和 XStream。
- Spring-jms 模块（Java Messaging Service）：指 Java 消息传递服务，包含用于生产和使用消息的功能。自 Spring 4.1 以后，提供了与 Spring-messaging 模块的集成。
- Spring-tx 模块（事务模块）：支持用于实现特殊接口和所有 POJO（普通 Java 对象）类的编程和声明式事务管理。

❺ Web

Web 层由 Spring-web、Spring-webmvc、Spring-websocket 和 Portlet 模块组成。

- Spring-web 模块：提供了基本的 Web 开发集成功能，例如多文件上传功能、使用 Servlet 监听器初始化一个 IoC 容器以及 Web 应用上下文。
- Spring-webmvc 模块：也称为 Web-Servlet 模块，包含用于 Web 应用程序的 Spring MVC 和 REST Web Services 实现。Spring MVC 框架提供了领域模型代码和 Web 表单之间的清晰分离，并与 Spring Framework 的所有其他功能集成，本书后续章节将会详细讲解 Spring MVC 框架。
- Spring-websocket 模块：Spring 4.0 以后新增的模块，它提供了 WebSocket 和 SockJS 的实现。
- Portlet 模块：类似于 Servlet 模块的功能，提供了 Portlet 环境下的 MVC 实现。

❻ 测试

Spring-test 模块支持使用 JUnit 或 TestNG 对 Spring 组件进行单元测试和集成测试。

1.2 Spring 开发环境的构建

视频讲解

在使用 Spring 框架开发 Web 应用前应先搭建 Web 应用的开发环境。

1.2.1　使用 Eclipse 开发 Java Web 应用

为了提高开发效率，通常需要安装 IDE（集成开发环境）工具。Eclipse 是一个可用于开发 Web 应用的 IDE 工具。登录“http://www.eclipse.org/ide”，选择 Java EE，根据操作系统的类型下载相应的 Eclipse。本书采用的是 eclipse-jee-oxygen-2-win32-x86_64.zip。

在使用 Eclipse 之前需要对 JDK、Web 服务器和 Eclipse 进行一些必要的配置，因此在安装 Eclipse 之前应先安装 JDK 和 Web 服务器。

❶ 安装 JDK

安装并配置 JDK（本书采用的 JDK 是 jdk-8u152-windows-x64.exe），按照提示安装完成 JDK 后，需要配置“环境变量”中的“系统变量”Java_Home 和 Path。在 Win10 系统下，系统变量示例如图 1.2 和图 1.3 所示。

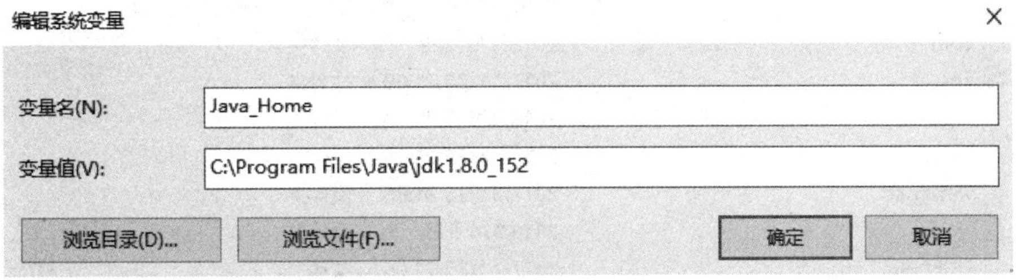

图 1.2　新建系统变量 Java_Home

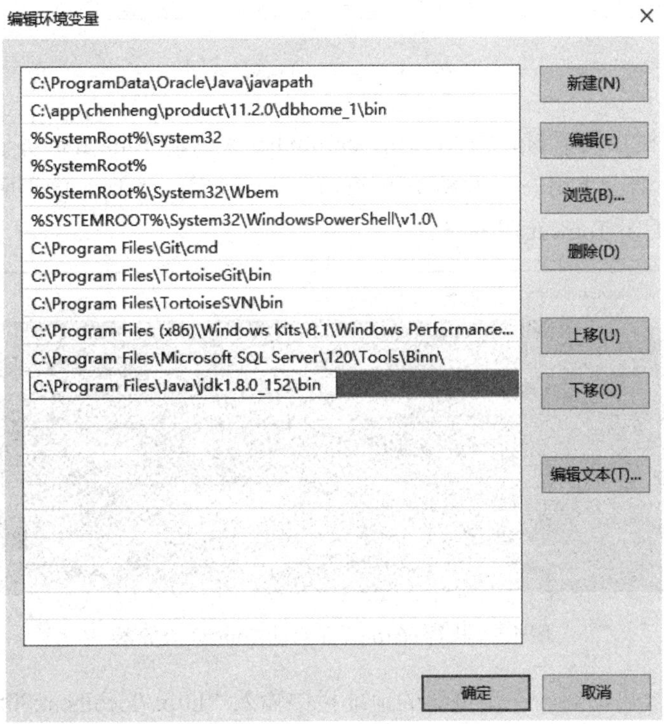

图 1.3　编辑系统变量 Path 的值

❷ Web 服务器

目前，比较常用的 Web 服务器包括 Tomcat、JRun、Resin、WebSphere、WebLogic 等，本书采用的是 Tomcat 9.0。

登录 Apache 软件基金会的官方网站"http://jakarta.Apache.org/tomcat"，下载 Tomcat 9.0 的免安装版（本书采用 apache-tomcat-9.0.2-windows-x64.zip）。登录网站后，首先在 Download 中选择 Tomcat 9，然后在 Binary Distributions 的 Core 中选择相应版本。

在安装 Tomcat 之前需要先安装 JDK 并配置系统环境变量 Java_Home。将下载的 apache-tomcat-9.0.2-windows-x64.zip 解压缩到某个目录下，例如解压缩到"E:\Java soft"，解压缩后将出现如图 1.4 所示的目录结构。

E:\Java soft\apache-tomcat-9.0.2

名称	修改日期	类型	大小
bin	2017/11/25 21:08	文件夹	
conf	2018/1/6 6:29	文件夹	
lib	2017/11/25 21:08	文件夹	
logs	2018/1/6 6:29	文件夹	
temp	2017/11/25 21:08	文件夹	
webapps	2017/11/25 21:08	文件夹	
work	2018/1/6 6:29	文件夹	
LICENSE	2017/11/25 21:08	文件	57 KB
NOTICE	2017/11/25 21:08	文件	2 KB
RELEASE-NOTES	2017/11/25 21:08	文件	7 KB
RUNNING.txt	2017/11/25 21:08	文本文档	17 KB

图 1.4　Tomcat 目录结构

执行 Tomcat 根目录下 bin 文件夹中的 startup.bat 来启动 Tomcat 服务器。执行 startup.bat 启动 Tomcat 服务器会占用一个 MS-DOS 窗口，出现如图 1.5 所示的界面，如果关闭当前 MS-DOS 窗口将关闭 Tomcat 服务器。

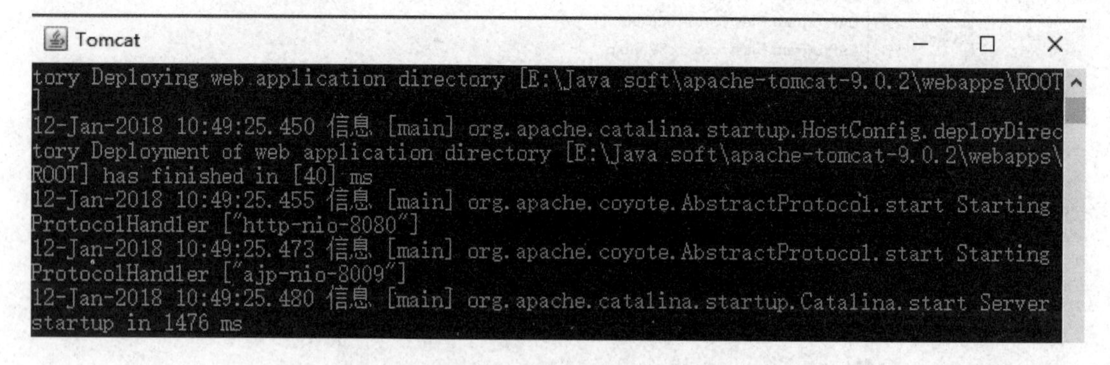

图 1.5　执行 startup.bat 启动 Tomcat 服务器

Tomcat 服务器启动后，在浏览器的地址栏中输入"http://localhost:8080"，将出现如图 1.6 所示的 Tomcat 测试页面。

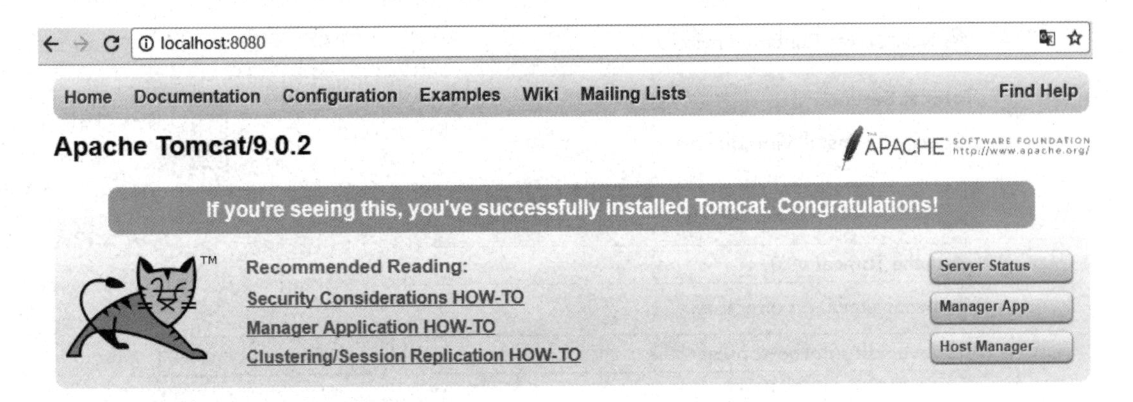

图 1.6 Tomcat 测试页面

❸ 安装 Eclipse

在 Eclipse 下载完成后解压缩到自己设置的路径下，即可完成安装。在 Eclipse 安装后，双击 Eclipse 安装目录下的 eclipse.exe 文件启动 Eclipse。

❹ 集成 Tomcat

启动 Eclipse，选择 Window→Preferences 命令，在弹出的对话框中选择 Server 下的 Runtime Environments，然后在弹出的对话框中单击 Add 按钮，弹出如图 1.7 所示的 New Server Runtime Environment 界面，在此可以配置各种版本的 Web 服务器。

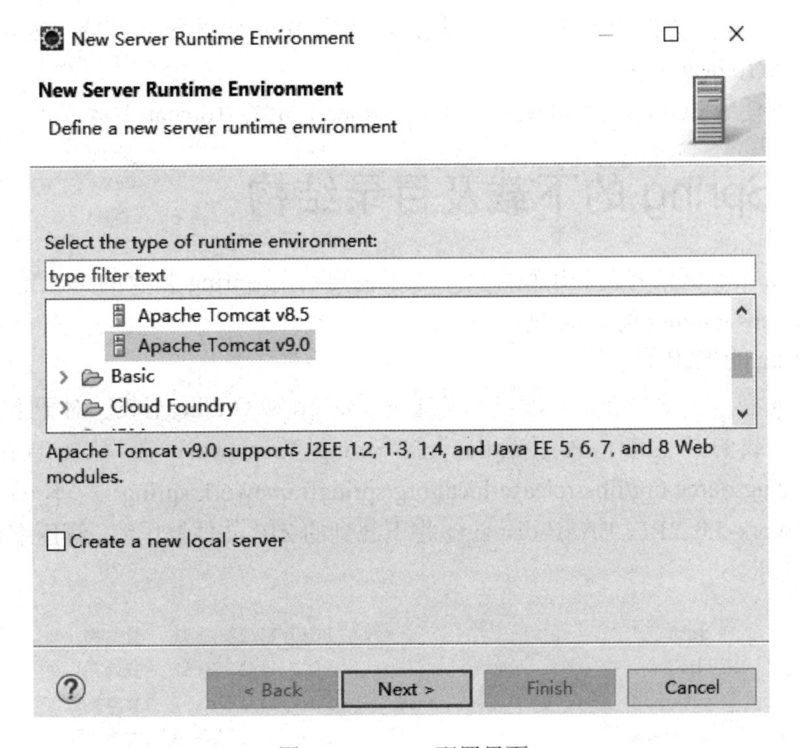

图 1.7 Tomcat 配置界面

在图 1.7 中选择 Apache Tomcat v9.0 服务器版本，单击 Next 按钮，进入如图 1.8 所示的界面。

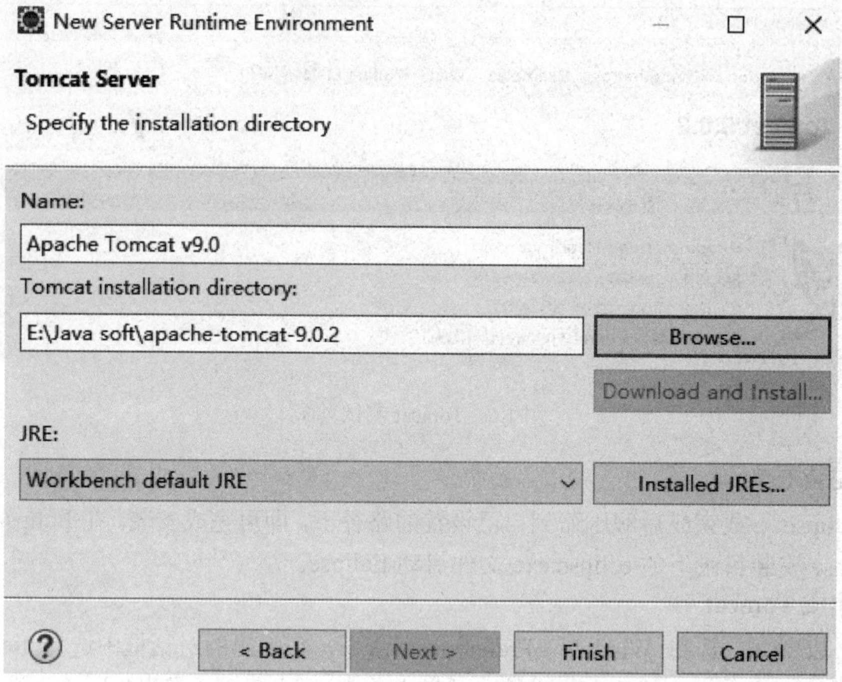

图 1.8 选择 Tomcat 目录

在图 1.8 所示的界面中单击 Browse 按钮，选择 Tomcat 的安装目录，然后单击 Finish 按钮即可完成 Tomcat 的配置。

至此可以使用 Eclipse 创建 Dynamic Web Project，并在 Tomcat 下运行。

1.2.2 Spring 的下载及目录结构

在使用 Spring 框架开发应用程序时，除了需要引用 Spring 自身的 JAR 包以外，还需要引用 commons.logging 的 JAR 包。

❶ Spring 的 JAR 包

在 Spring 官方网站升级后，建议都是通过 Maven 和 Gradle 下载。对于不使用 Maven 和 Gradle 下载的开发者，本书给出一个 Spring Framework jar 官方直接下载路径 http://repo.springsource.org/libs-release-local/org/springframework/spring/。本书采用的是 spring-framework-5.0.2.RELEASE-dist.zip。将下载到的 ZIP 文件解压缩，解压缩后的目录结构如图 1.9 所示。

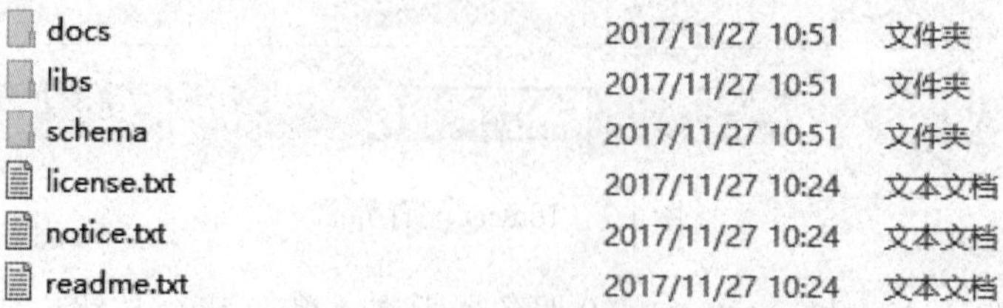

图 1.9 spring-framework-5.0.2 的目录结构

图 1.9 中，docs 目录包含 Spring 的 API 文档和开发规范；libs 目录包含开发 Spring 应用所需要的 JAR 包和源代码；schema 目录包含开发 Spring 应用所需要的 schema 文件，这些 schema 文件定义了 Spring 相关配置文件的约束。其中，libs 目录下有 3 类 JAR 文件：以 RELEASE.jar 结尾的文件是 Spring 框架 class 的 JAR 包，即开发 Spring 应用所需要的 JAR 包；以 RELEASE-javadoc.jar 结尾的文件是 Spring 框架 API 文档的压缩包；以 RELEASE-sources.jar 结尾的文件是 Spring 框架源文件的压缩包。在 libs 目录中有 4 个基础包，即 spring-core-5.0.2.RELEASE.jar、spring-beans-5.0.2.RELEASE.jar、spring-context-5.0.2.RELEASE.jar 和 spring-expression-5.0.2.RELEASE.jar，分别对应 Spring 核心容器的 4 个模块，即 Spring-core 模块、Spring-beans 模块、Spring-context 模块和 Spring-expression 模块。

❷ commons.logging 的 JAR 包

Spring 框架依赖于 Apache Commons Logging 组件，该组件的 JAR 包可以通过网址“http://commons.apache.org/proper/commons-logging/download_logging.cgi”下载，本书下载的是 commons-logging-1.2-bin.zip，解压缩后即可找到 commons-logging-1.2.jar。

对于 Spring 框架的初学者，在开发 Spring 应用时只需要将 Spring 的 4 个基础包和 commons-logging-1.2.jar 复制到 Web 应用的 WEB-INF/lib 目录下即可。如果用户不知道需要哪些 JAR 包，可以将 Spring 的 libs 目录中的 spring-XXX-5.0.2.RELEASE.jar 全部复制到 WEB-INF/lib 目录下。

1.3 使用 Eclipse 开发 Spring 入门程序

视频讲解

本节通过一个简单的入门程序向读者演示 Spring 框架的使用过程，具体如下：

❶ 使用 Eclipse 创建 Web 应用并导入 JAR 包

使用 Eclipse 创建一个名为 ch1 的 Web 应用，并将 Spring 的 4 个基础包和第三方依赖包 commons-logging-1.2.jar 复制到 ch1 的 WEB-INF/lib 目录中，如图 1.10 所示。

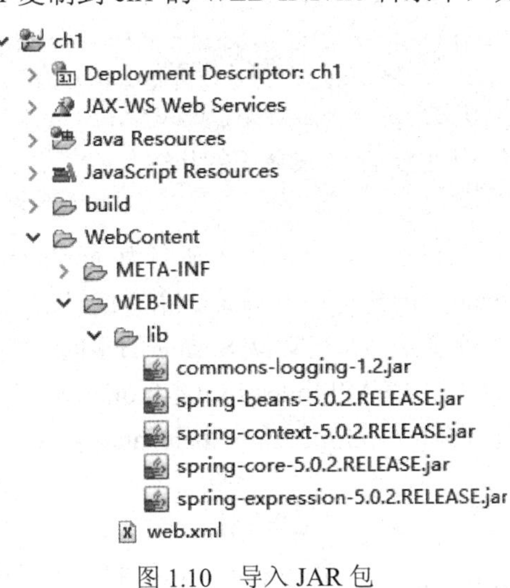

图 1.10 导入 JAR 包

注意：在讲解 Spring MVC 框架前本书的实例并没有真正运行 Web 应用，创建 Web 应用的目的是方便添加相关 JAR 包。

❷ **创建接口 TestDao**

Spring 解决的是业务逻辑层和其他各层的耦合问题，因此它将面向接口的编程思想贯穿整个系统应用。

在 src 目录下创建一个 dao 包，并在 dao 包中创建接口 TestDao，在接口中定义一个 sayHello 方法，代码如下：

```
package dao;
public interface TestDao {
    public void sayHello();
}
```

❸ **创建接口 TestDao 的实现类 TestDaoImpl**

在 dao 包下创建 TestDao 的实现类 TestDaoImpl，代码如下：

```
package dao;
public class TestDaoImpl implements TestDao{
    @Override
    public void sayHello() {
        System.out.println("Hello, Study hard!");
    }
}
```

❹ **创建配置文件 applicationContext.xml**

在 src 目录下创建 Spring 的配置文件 applicationContext.xml，并在该文件中使用实现类 TestDaoImpl 创建一个 id 为 test 的 Bean，代码如下：

```
<?xml version="1.0" encoding="UTF-8"?>
<beans xmlns="http://www.springframework.org/schema/beans"
    xmlns:xsi="http://www.w3.org/2001/XMLSchema-instance"
    xsi:schemaLocation="http://www.springframework.org/schema/beans
        http://www.springframework.org/schema/beans/spring-beans.xsd">
    <!-- 将指定类 TestDaoImpl 配置给 Spring，让 Spring 创建其实例 -->
    <bean id="test" class="dao.TestDaoImpl" />
</beans>
```

注：配置文件的名称可以自定义，但习惯上命名为 applicationContext.xml，有时候也命名为 beans.xml。有关 Bean 的创建将在本书第 3 章详细讲解，这里读者只需了解即可。另外，配置文件信息不需要读者手写，可以从 Spring 的帮助文档中复制（首先使用浏览器打开“\spring-framework-5.0.2.RELEASE\docs\spring-framework-reference\index.html，”在页面中单击超链接 Core，在 1.2.1 Configuration metadata 小节下即可找到配置文件的约束信息）。

❺ **创建测试类**

在 src 目录下创建一个 test 包，并在 test 包中创建 Test 类，代码如下：

```
package test;
import org.springframework.context.ApplicationContext;
import org.springframework.context.support.ClassPathXmlApplicationContext;
import dao.TestDao;
public class Test {
    public static void main(String[] args) {
        //初始化 Spring 容器 ApplicationContext，加载配置文件
        ApplicationContext appCon = new ClassPathXmlApplicationContext
       ("applicationContext.xml");
        //通过容器获取 test 实例
        TestDao tt=(TestDao)appCon.getBean("test");  //test 为配置文件中的 id
        tt.sayHello();
    }
}
```

在执行上述 main 方法后将在控制台输出“Hello, Study hard!”。在上述 main 方法中并没有使用 new 运算符创建 TestDaoImpl 类的对象，而是通过 Spring 容器来获取实现类对象，这就是 Spring IoC 的工作机制。本书将在第 2 章详细讲解 Spring IoC 的工作机制。

1.4　本章小结

本章首先简单介绍了 Spring 的体系结构；然后详细讲解了在 Eclipse 中如何构建 Spring 的开发环境；最后以 ch1 应用为例，简要介绍了 Spring 入门程序的开发流程。

习题 1

1．Spring 的核心容器由哪些模块组成？
2．如何找到 Spring 框架的官方 API？

第 2 章

Spring IoC

学习目的与要求

本章主要介绍 Spring IoC 的基本概念、Spring IoC 容器以及依赖注入的类型等内容。通过本章的学习，读者能够了解 Spring IoC 容器，掌握 Spring IoC 的基本概念以及依赖注入的类型。

本章主要内容

- Spring IoC 的基本概念;
- Spring IoC 容器;
- 依赖注入的类型。

IoC（控制反转）是 Spring 框架的基础，也是 Spring 框架的核心理念，本章将介绍 IoC 的基本概念、容器以及依赖注入的类型等内容。

2.1 Spring IoC 的基本概念

视频讲解

控制反转（Inversion of Control，IoC）是一个比较抽象的概念，是 Spring 框架的核心，用来消减计算机程序的耦合问题。依赖注入（Dependency Injection，DI）是 IoC 的另外一种说法，只是从不同的角度描述相同的概念。下面通过实际生活中的一个例子来解释 IoC 和 DI。

当人们需要一件东西时，第一反应就是找东西，例如想吃面包。在没有面包店和有面包店两种情况下，您会怎么做？在没有面包店时，最直观的做法可能是您按照自己的口味制作面包，也就是一个面包需要主动制作。然而时至今日，各种网店、实体店盛行，已经没有必要自己制作面包。想吃面包了，去网店或实体店把自己的口味告诉店家，一会就可以吃到面包了。注意，您并没有制作面包，而是由店家制作，但是完全符合您的口味。

上面只是列举了一个非常简单的例子，但包含了控制反转的思想，即把制作面包的主动权交给店家。下面通过面向对象编程思想继续探讨这两个概念。

当某个 Java 对象（调用者，例如您）需要调用另一个 Java 对象（被调用者，即被依赖对象，例如面包）时，在传统编程模式下，调用者通常会采用“new 被调用者”的代码方式来创建对象(例如您自己制作面包)。这种方式会增加调用者与被调用者之间的耦合性，不利于后期代码的升级与维护。

当 Spring 框架出现后，对象的实例不再由调用者来创建，而是由 Spring 容器（例如面包店）来创建。Spring 容器会负责控制程序之间的关系（例如面包店负责控制您与面包的关系），而不是由调用者的程序代码直接控制。这样，控制权由调用者转移到 Spring 容器，控制权发生了反转，这就是 Spring 的控制反转。

从 Spring 容器角度来看，Spring 容器负责将被依赖对象赋值给调用者的成员变量，相当于为调用者注入它所依赖的实例，这就是 Spring 的依赖注入。

综上所述，控制反转是一种通过描述（在 Spring 中可以是 XML 或注解）并通过第三方去产生或获取特定对象的方式。在 Spring 中实现控制反转的是 IoC 容器，其实现方法是依赖注入。

2.2 Spring IoC 容器

由 2.1 节得知，实现控制反转的是 Spring IoC 容器。Spring IoC 容器的设计主要是基于 BeanFactory 和 ApplicationContext 两个接口。

2.2.1 BeanFactory

BeanFactory 由 org.springframework.beans.factory.BeanFactory 接口定义，它提供了完整的 IoC 服务支持，是一个管理 Bean 的工厂，主要负责初始化各种 Bean。BeanFactory 接口有多个实现类，其中比较常用的是 org.springframework.beans.factory.xml.XmlBeanFactory，该类会根据 XML 配置文件中的定义来装配 Bean(有关 Bean 的知识将在本书第 3 章讲解)。

在创建 BeanFactory 实例时需要提供 XML 文件的绝对路径。例如可以将第 1 章 ch1 应用中 main 方法的代码修改如下：

```
public static void main(String[] args) {
    //初始化 Spring 容器，加载配置文件
    BeanFactory beanFac=new XmlBeanFactory(
            new FileSystemResource("D:\eclipse-workspace\ch1\src\
            applicationContext.xml")
    );
    //通过容器获取 test 实例
    TestDao tt=(TestDao)beanFac.getBean("test");
    tt.sayHello();
}
```

使用 BeanFactory 实例加载 Spring 配置文件在实际开发中并不多见，读者了解即可。

2.2.2 ApplicationContext

ApplicationContext 是 BeanFactory 的子接口，也称为应用上下文，由 org.springframework.context.ApplicationContext 接口定义。ApplicationContext 接口除了包含 BeanFactory 的所有功能以外，还添加了对国际化、资源访问、事件传播等内容的支持。

创建 ApplicationContext 接口实例通常有以下 3 种方法：

❶ 通过 ClassPathXmlApplicationContext 创建

ClassPathXmlApplicationContext 将从类路径目录（src 根目录）中寻找指定的 XML 配置文件，例如第 1 章 ch1 应用中 main 方法的代码：

```
public static void main(String[] args) {
    //初始化 Spring 容器 ApplicationContext，加载配置文件
    ApplicationContext appCon=new ClassPathXmlApplicationContext
    ("applicationContext.xml");
    //通过容器获取 test 实例
    TestDao tt=(TestDao)appCon.getBean("test");
    tt.sayHello();
}
```

❷ 通过 FileSystemXmlApplicationContext 创建

FileSystemXmlApplicationContext 将从指定文件的绝对路径中寻找 XML 配置文件，找到并装载完成 ApplicationContext 的实例化工作。例如，可以将第 1 章 ch1 应用中 main 方法的代码修改如下：

```
public static void main(String[] args) {
        //初始化 Spring 容器 ApplicationContext，加载配置文件
        ApplicationContext appCon=
      new  FileSystemXmlApplicationContext("D:\eclipse-workspace\ch1\src\
      applicationContext.xml");
        //通过容器获取 test 实例
        TestDao tt=(TestDao)appCon.getBean("test");
        tt.sayHello();
    }
```

采用绝对路径的加载方式将导致程序的灵活性变差，一般不推荐使用。因此，通常在 Spring 的 Java 应用中采取通过 ClassPathXmlApplicationContext 类来实例化 ApplicationContext 容器的方式，而在 Web 应用中，ApplicationContext 容器的实例化工作将交给 Web 服务器完成。

❸ 通过 Web 服务器实例化 ApplicationContext 容器

在 Web 服务器实例化 ApplicationContext 容器时，一般使用基于 org.springframework.web.context.ContextLoaderListener 的实现方式（需要将 spring-web-5.0.2.RELEASE.jar 复制到 WEB-INF/lib 目录中），此方法只需在 web.xml 中添加如下代码：

```
<context-param>
    <!-- 加载 src 目录下的 applicationContext.xml 文件 -->
    <param-name>contextConfigLocation</param-name>
    <param-value>
        classpath:applicationContext.xml
    </param-value>
  </context-param>
  <!-- 指定以 ContextLoaderListener 方式启动 Spring 容器 -->
  <listener>
    <listener-class>
        org.springframework.web.context.ContextLoaderListener
    </listener-class>
</listener>
```

2.3 依赖注入的类型

视频讲解

在 Spring 中实现 IoC 容器的方法是依赖注入，依赖注入的作用是在使用 Spring 框架创建对象时动态地将其所依赖的对象(例如属性值)注入 Bean 组件中。Spring 框架的依赖注入通常有两种实现方式，一种是使用构造方法注入，另一种是使用属性的 setter 方法注入。

2.3.1 使用构造方法注入

Spring 框架可以采用 Java 的反射机制，通过构造方法完成依赖注入。下面通过 Web 应用 ch2 讲解构造方法注入的实现过程，ch2 应用的目录结构如图 2.1 所示。

图 2.1 ch2 的目录结构

❶ **创建 dao 包**

在 ch2 应用中创建 dao 包，并在该包中创建 TestDIDao 接口和接口实现类 TestDIDaoImpl。创建 dao 包的目的是在 service 中使用构造方法依赖注入 TestDIDao 接口对象。

TestDIDao 接口的代码如下：

```
package dao;
public interface TestDIDao {
    public void sayHello();
}
```

TestDIDaoImpl 实现类的代码如下：

```
package dao;
public class TestDIDaoImpl implements TestDIDao{
    @Override
    public void sayHello() {
        System.out.println("TestDIDao say: Hello, Study hard!");
    }
}
```

❷ **创建 service 包**

在 ch2 应用中创建 service 包，并在该包中创建 TestDIService 接口和接口实现类 TestDIServiceImpl。在 TestDIServiceImpl 中使用构造方法依赖注入 TestDIDao 接口对象。

TestDIService 接口的代码如下：

```
package service;
public interface TestDIService {
    public void sayHello();
}
```

TestDIServiceImpl 实现类的代码如下：

```
package service;
import dao.TestDIDao;
public class TestDIServiceImpl implements TestDIService{
    private TestDIDao testDIDao;
    //构造方法，用于实现依赖注入接口对象 testDIDao
    public TestDIServiceImpl(TestDIDao testDIDao) {
        super();
        this.testDIDao = testDIDao;
    }
    @Override
    public void sayHello() {
        //调用 testDIDao 中的 sayHello 方法
        testDIDao.sayHello();
```

```
        System.out.println("TestDIService 构造方法注入 say: Hello, Study hard!");
    }
}
```

❸ **编写配置文件**

在 src 根目录下创建 Spring 配置文件 applicationContext.xml。在配置文件中首先将 dao.TestDIDaoImpl 类托管给 Spring，让 Spring 创建其对象，然后将 service.TestDIServiceImpl 类托管给 Spring，让 Spring 创建其对象，同时给构造方法传递实参。配置文件的具体代码如下：

```
<?xml version="1.0" encoding="UTF-8"?>
<beans xmlns="http://www.springframework.org/schema/beans"
    xmlns:xsi="http://www.w3.org/2001/XMLSchema-instance"
    xsi:schemaLocation="http://www.springframework.org/schema/beans
        http://www.springframework.org/schema/beans/spring-beans.xsd">
    <!-- 将指定类 TestDIDaoImpl 配置给 Spring，让 Spring 创建其实例 -->
    <bean id="myTestDIDao" class="dao.TestDIDaoImpl" />
    <!-- 使用构造方法注入 -->
    <bean id="testDIService" class="service.TestDIServiceImpl">
        <!-- 将 myTestDIDao 注入到 TestDIServiceImpl 类的属性 testDIDao 上-->
        <constructor-arg index="0" ref="myTestDIDao"/>
    </bean>
</beans>
```

在配置文件中，constructor-arg 元素用于定义类构造方法的参数，index 用于定义参数的位置，ref 指定某个实例的引用，如果参数是常量值，ref 由 value 代替。

❹ **创建 test 包**

在 ch2 应用中创建 test 包，并在该包中创建测试类 TestDI，具体代码如下：

```
package test;
import org.springframework.context.ApplicationContext;
import org.springframework.context.support.ClassPathXmlApplicationContext;
import service.TestDIService;
public class TestDI {
    public static void main(String[] args) {
        //初始化 Spring 容器 ApplicationContext，加载配置文件
        ApplicationContext appCon = new ClassPathXmlApplicationContext
        ("applicationContext.xml");
        //通过容器获取 testDIService 实例，测试构造方法注入
        TestDIService ts = (TestDIService)appCon.getBean("testDIService");
        ts.sayHello();
    }
}
```

2.3.2 使用属性的 setter 方法注入

使用 setter 方法注入是 Spring 框架中最主流的注入方式，它利用 Java Bean 规范所定义的 setter 方法来完成注入，灵活且可读性高。对于 setter 方法注入，Spring 框架也是使用 Java 的反射机制实现的。下面接着 2.3.1 节的 ch2 应用讲解使用属性的 setter 方法注入的实现过程。

❶ 创建接口实现类 TestDIServiceImpl1

在 service 包中创建接口实现类 TestDIServiceImpl1，在 TestDIServiceImpl1 中使用属性的 setter 方法依赖注入 TestDIDao 接口对象，具体代码如下：

```
package service;
import dao.TestDIDao;
public class TestDIServiceImpl1 implements TestDIService{
    private TestDIDao testDIDao;
    //添加 testDIDao 属性的 setter 方法，用于实现依赖注入
    public void setTestDIDao(TestDIDao testDIDao) {
        this.testDIDao=testDIDao;
    }
    @Override
    public void sayHello() {
        //调用 testDIDao 中的 sayHello 方法
        testDIDao.sayHello();
        System.out.println("TestDIService setter 方法注入 say: Hello, Study hard!");
    }
}
```

❷ 将 TestDIServiceImpl1 类托管给 Spring

将 TestDIServiceImpl1 类托管给 Spring，让 Spring 创建其对象，同时调用 TestDIServiceImpl1 类的 setter 方法完成依赖注入。在配置文件中添加如下代码：

```
<!-- 使用 setter 方法注入 -->
<bean id="testDIService1" class="service.TestDIServiceImpl1">
<!-- 调用 TestDIServiceImpl1 类的 setter 方法，将 myTestDIDao 注入到 TestDIServiceImpl1 类的属性 testDIDao 上-->
    <property name="testDIDao" ref="myTestDIDao"/>
</bean>
```

❸ 在 test 中测试 setter 方法注入

在主类中添加如下代码测试 setter 方法注入：

```
//通过容器获取 testDIService 实例，测试 setter 方法注入
TestDIService ts1=(TestDIService)appCon.getBean("testDIService1");
ts1.sayHello();
```

2.4　本章小结

本章首先通过生活中的实例讲解了 IoC 的基本概念；然后详细介绍了 IoC 容器的实现方式；最后通过实例演示了依赖注入的两种实现方式，即使用构造方法注入和使用属性的 setter 方法注入。

习题 2

1．举例说明 IoC 容器的实现方式有哪些？

2．在 Spring 框架中，什么是控制反转？什么是依赖注入？使用控制反转与依赖注入有什么优点？

3．Spring 框架采用 Java 的（　　）机制进行依赖注入。

A．反射　　B．异常　　C．事件　　D．多态

第 3 章

Spring Bean

学习目的与要求

本章主要介绍 Spring Bean 的配置、实例化、作用域、生命周期以及装配方式等内容。通过本章的学习，读者能够了解 Spring Bean 的生命周期，掌握 Spring Bean 的配置、实例化、作用域以及装配方式等内容。

本章主要内容

- Bean 的配置;
- Bean 的实例化;
- Bean 的作用域;
- Bean 的生命周期;
- Bean 的装配方式。

在 Spring 的应用中，Spring IoC 容器可以创建、装配和配置应用组件对象，这里的组件对象称为 Bean。本章将重点介绍如何将 Bean 装配注入到 Spring IoC 容器中。

3.1 Bean 的配置

Spring 可以看作一个大型工厂，用于生产和管理 Spring 容器中的 Bean。如果要使用这个工厂生产和管理 Bean，需要开发者将 Bean 配置在 Spring 的配置文件中。Spring 框架支持 XML 和 Properties 两种格式的配置文件，在实际开发中常用 XML 格式的配置文件。

从前面的学习得知 XML 配置文件的根元素是<beans>，<beans>中包含了多个<bean>子元素，每个<bean>元素定义一个 Bean，并描述 Bean 如何被装配到 Spring 容器中。<bean>元素的常用属性及其子元素如表 3.1 所示。

表 3.1 <bean>元素的常用属性及其子元素

属性或子元素名称	描　述
id	Bean 在 BeanFactory 中的唯一标识，在代码中通过 BeanFactory 获取 Bean 实例时需要以此作为索引名称
class	Bean 的具体实现类，使用类的名（例如 dao.TestDIDaoImpl）
scope	指定 Bean 实例的作用域，具体属性值及含义参见 3.3 节“Bean 的作用域”
<constructor-arg>	<bean>元素的子元素，使用构造方法注入，指定构造方法的参数。该元素的 index 属性指定参数的序号，ref 属性指定对 BeanFactory 中其他 Bean 的引用关系，type 属性指定参数类型，value 属性指定参数的常量值
<property>	<bean>元素的子元素，用于设置一个属性。该元素的 name 属性指定 Bean 实例中相应的属性名称，value 属性指定 Bean 的属性值，ref 属性指定属性对 BeanFactory 中其他 Bean 的引用关系
<list>	<property>元素的子元素，用于封装 List 或数组类型的依赖注入，具体用法参见 3.5 节“Bean 的装配方式”
<map>	<property>元素的子元素，用于封装 Map 类型的依赖注入，具体用法参见 3.5 节“Bean 的装配方式”
<set>	<property>元素的子元素，用于封装 Set 类型的依赖注入，具体用法参见 3.5 节“Bean 的装配方式”
<entry>	<map>元素的子元素，用于设置一个键值对，具体用法参见 3.5 节“Bean 的装配方式”

Bean 的配置示例代码如下：

```
<?xml version="1.0" encoding="UTF-8"?>
<beans xmlns="http://www.springframework.org/schema/beans"
   xmlns:xsi="http://www.w3.org/2001/XMLSchema-instance"
   xsi:schemaLocation="http://www.springframework.org/schema/beans
      http://www.springframework.org/schema/beans/spring-beans.xsd">
   <!-- 使用 id 属性定义 myTestDIDao，其对应的实现类为 dao.TestDIDaoImpl-->
   <bean id="myTestDIDao" class="dao.TestDIDaoImpl" />
   <!-- 使用构造方法注入 -->
   <bean id="testDIService" class="service.TestDIServiceImpl">
      <!-- 给构造方法传引用类型的参数值 myTestDIDao -->
      <constructor-arg index="0" ref="myTestDIDao"/>
   </bean>
</beans>
```

3.2 Bean 的实例化

视频讲解

在面向对象编程中，如果想使用某个对象，需要事先实例化该对象。同样，在 Spring 框架中，如果想使用 Spring 容器中的 Bean，也需要实例化 Bean。Spring 框架实例化 Bean 有 3 种方式，即构造方法实例化、静态工厂实例化和实例工厂实例化（其中，最常用的方法是构造方法实例化）。

3.2.1 构造方法实例化

在 Spring 框架中，Spring 容器可以调用 Bean 对应类中的无参数构造方法来实例化 Bean，这种方式称为构造方法实例化。下面通过 ch3 应用来演示构造方法实例化的过程。

❶ 创建 Web 应用 ch3

创建一个名为 ch3 的 Web 应用，并导入 Spring 支持和依赖的 JAR 包，ch3 的目录结构如图 3.1 所示。

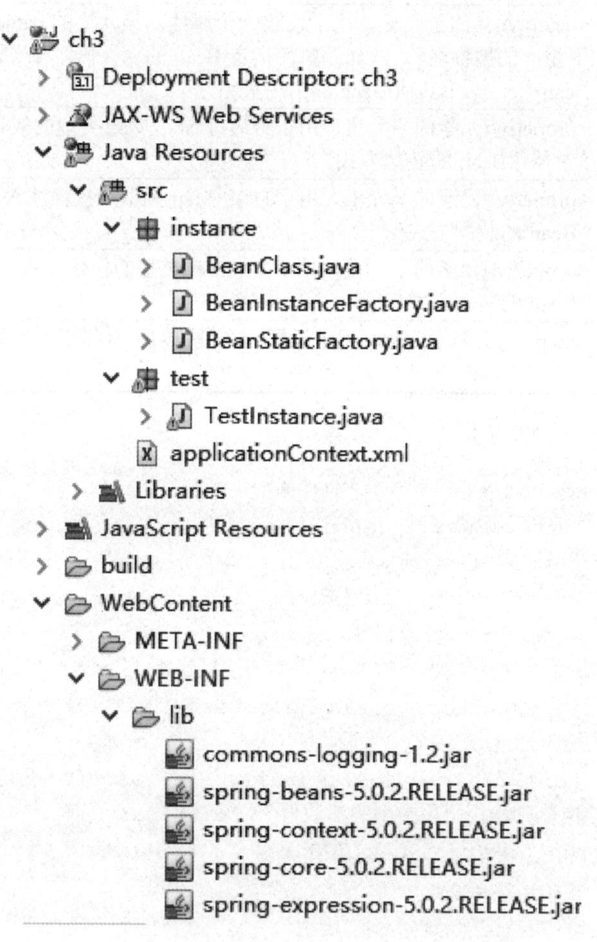

图 3.1　ch3 的目录结构

❷ 创建 BeanClass 类

在 ch3 的 src 目录下创建 instance 包，并在该包中创建 BeanClass 类，代码如下：

```
package instance;
public class BeanClass {
    public String message;
    public BeanClass() {
        message = "构造方法实例化 Bean";
```

```
    }
    public BeanClass(String s) {
        message = s;
    }
}
```

❸ 创建配置文件

在 ch3 的 src 目录下创建 Spring 的配置文件 applicationContext.xml，在配置文件中定义一个 id 为 constructorInstance 的 Bean，代码如下：

```
<?xml version="1.0" encoding="UTF-8"?>
<beans xmlns="http://www.springframework.org/schema/beans"
    xmlns:xsi="http://www.w3.org/2001/XMLSchema-instance"
    xsi:schemaLocation="http://www.springframework.org/schema/beans
        http://www.springframework.org/schema/beans/spring-beans.xsd">
    <!-- 构造方法实例化 Bean -->
    <bean id="constructorInstance" class="instance.BeanClass"/>
</beans>
```

❹ 创建测试类

在 ch3 的 src 目录下创建 test 包，并在该包下创建测试类 TestInstance，代码如下：

```
package test;
import org.springframework.context.ApplicationContext;
import org.springframework.context.support.ClassPathXmlApplicationContext;
import instance.BeanClass;
public class TestInstance {
    public static void main(String[] args) {
        //初始化 Spring 容器 ApplicationContext，加载配置文件
        ApplicationContext appCon = new ClassPathXmlApplicationContext
        ("applicationContext.xml");
        //测试构造方法实例化 Bean
        BeanClass b1 = (BeanClass)appCon.getBean("constructorInstance");
        System.out.println(b1+ b1.message);
    }
}
```

运行上述测试类，控制台的输出结果如图 3.2 所示。

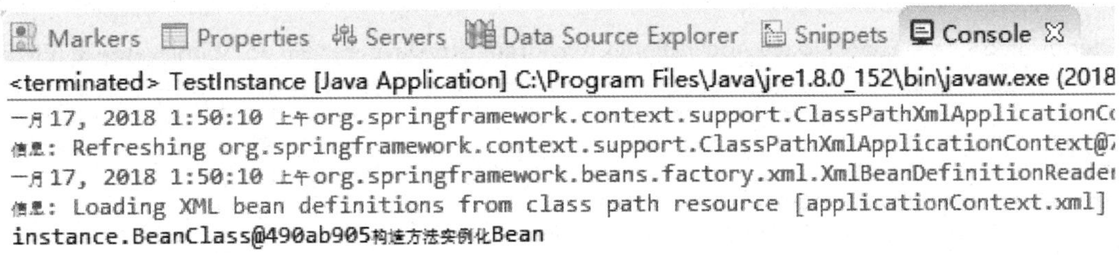

```
Markers  Properties  Servers  Data Source Explorer  Snippets  Console ☒
<terminated> TestInstance [Java Application] C:\Program Files\Java\jre1.8.0_152\bin\javaw.exe (2018
一月 17, 2018 1:50:10 上午 org.springframework.context.support.ClassPathXmlApplicationC
信息: Refreshing org.springframework.context.support.ClassPathXmlApplicationContext@7
一月 17, 2018 1:50:10 上午 org.springframework.beans.factory.xml.XmlBeanDefinitionReade
信息: Loading XML bean definitions from class path resource [applicationContext.xml]
instance.BeanClass@490ab905构造方法实例化Bean
```

图 3.2　构造方法实例化 Bean 的运行结果

3.2.2 静态工厂实例化

在使用静态工厂实例化 Bean 时要求开发者在工厂类中创建一个静态方法来创建 Bean 的实例。在配置 Bean 时，class 属性指定静态工厂类，同时还需要使用 factory-method 属性指定工厂类中的静态方法。下面通过 ch3 应用测试静态工厂实例化。

❶ 创建工厂类 BeanStaticFactory

在 instance 包中创建工厂类 BeanStaticFactory，该类中有一个静态方法来实例化对象，具体代码如下：

```
package instance;
public class BeanStaticFactory {
    private static BeanClass beanInstance = new BeanClass("调用静态工厂方法
    实例化 Bean");
    public static BeanClass createInstance() {
        return beanInstance;
    }
}
```

❷ 编辑配置文件

在配置文件 applicationContext.xml 中添加如下配置代码：

```
<!-- 静态工厂方法实例化 Bean，createInstance 为静态工厂类 BeanStaticFactory 中的
静态方法 -->
<bean id="staticFactoryInstance" class="instance.BeanStaticFactory" factory-
method="createInstance"/>
```

❸ 添加测试代码

在测试类 TestInstance 中添加如下代码：

```
//测试静态工厂方法实例化 Bean
BeanClass b2 = (BeanClass)appCon.getBean("staticFactoryInstance");
System.out.println(b2+b2.message);
```

此时，测试类的运行结果如图 3.3 所示。

```
Markers  Properties  Servers  Data Source Explorer  Snippets  Console
<terminated> TestInstance [Java Application] C:\Program Files\Java\jre1.8.0_152\bin\javaw.exe (2018
一月 17, 2018 1:50:10 上午 org.springframework.context.support.ClassPathXmlApplicationCo
信息: Refreshing org.springframework.context.support.ClassPathXmlApplicationContext@7
一月 17, 2018 1:50:10 上午 org.springframework.beans.factory.xml.XmlBeanDefinitionReader
信息: Loading XML bean definitions from class path resource [applicationContext.xml]
instance.BeanClass@490ab905构造方法实例化Bean
instance.BeanClass@56ac3a89调用静态工厂方法实例化Bean
```

图 3.3　实例化 Bean 的运行结果

3.2.3 实例工厂实例化

在使用实例工厂实例化 Bean 时要求开发者在工厂类中创建一个实例方法来创建 Bean 的实例。在配置 Bean 时需要使用 factory-bean 属性指定配置的实例工厂，同时还需要使用 factory-method 属性指定实例工厂中的实例方法。下面通过 ch3 应用测试实例工厂实例化。

❶ 创建工厂类 BeanInstanceFactory

在 instance 包中创建工厂类 BeanInstanceFactory，该类中有一个实例方法来实例化对象，具体代码如下：

```
package instance;
public class BeanInstanceFactory {
    public BeanClass createBeanClassInstance() {
        return new BeanClass("调用实例工厂方法实例化 Bean");
    }
}
```

❷ 编辑配置文件

在配置文件 applicationContext.xml 中添加如下配置代码：

```
<!-- 配置工厂 -->
    <bean id="myFactory" class="instance.BeanInstanceFactory"/>
<!-- 使用 factory-bean 属性指定配置工厂，使用 factory-method 属性指定使用工厂中的哪
个方法实例化 Bean -->
    <bean id="instanceFactoryInstance" factory-bean="myFactory" factory-method=
    "createBeanClassInstance"/>
```

❸ 添加测试代码

在测试类 TestInstance 中添加如下代码：

```
//测试实例工厂方法实例化 Bean
BeanClass b3=(BeanClass)appCon.getBean("instanceFactoryInstance");
System.out.println(b3 + b3.message);
```

此时，测试类的运行结果如图 3.4 所示。

```
Markers  Properties  Servers  Data Source Explorer  Snippets  Console
<terminated> TestInstance [Java Application] C:\Program Files\Java\jre1.8.0_152\bin\javaw.exe (2018
一月 17, 2018 2:01:16 上午 org.springframework.context.support.ClassPathXmlApplicationC
信息: Refreshing org.springframework.context.support.ClassPathXmlApplicationContext@
一月 17, 2018 2:01:16 上午 org.springframework.beans.factory.xml.XmlBeanDefinitionReade
信息: Loading XML bean definitions from class path resource [applicationContext.xml]
instance.BeanClass@490ab905构造方法实例化Bean
instance.BeanClass@56ac3a89调用静态工厂方法实例化Bean
instance.BeanClass@27c20538调用实例工厂方法实例化Bean
```

图 3.4 实例化 Bean 的运行结果

3.3 Bean 的作用域

视频讲解

在 Spring 中不仅可以完成 Bean 的实例化，还可以为 Bean 指定作用域。在 Spring 5.0 中为 Bean 的实例定义了如表 3.2 所示的作用域。

表 3.2 Bean 的作用域

作用域名称	描 述
singleton	默认的作用域，使用singleton定义的Bean在Spring容器中只有一个Bean实例
prototype	Spring容器每次获取prototype定义的Bean，容器都将创建一个新的Bean实例
request	在一次HTTP请求中容器将返回一个Bean实例，不同的HTTP请求返回不同的Bean实例。仅在Web Spring应用程序上下文中使用
session	在一个HTTP Session中，容器将返回同一个Bean实例。仅在Web Spring应用程序上下文中使用
application	为每个ServletContext对象创建一个实例，即同一个应用共享一个Bean实例。仅在Web Spring应用程序上下文中使用
websocket	为每个WebSocket对象创建一个Bean实例。仅在Web Spring应用程序上下文中使用

在表 3.2 所示的 6 种作用域中，singleton 和 prototype 是最常用的两种，后面 4 种作用域仅在 Web Spring 应用程序上下文中使用，在本节将会对 singleton 和 prototype 进行详细的讲解。

3.3.1 singleton 作用域

当将 bean 的 scope 设置为 singleton 时，Spring IoC 容器仅生成和管理一个 Bean 实例。在使用 id 或 name 获取 Bean 实例时，IoC 容器将返回共享的 Bean 实例。

由于 singleton 是 scope 的默认方式，因此有两种方式将 bean 的 scope 设置为 singleton。配置文件示例代码如下：

```
<bean id="constructorInstance" class="instance.BeanClass"/>
```

或

```
<bean id="constructorInstance" class="instance.BeanClass" scope="singleton"/>
```

测试 singleton 作用域，代码如下：

```
package test;
import org.springframework.context.ApplicationContext;
import org.springframework.context.support.ClassPathXmlApplicationContext;
import instance.BeanClass;
public class TestInstance {
    public static void main(String[] args) {
        //初始化 Spring 容器 ApplicationContext，加载配置文件
```

```
        ApplicationContext appCon = new ClassPathXmlApplicationContext
        ("applicationContext.xml");
        //测试构造方法实例化 Bean
        BeanClass b1 = (BeanClass)appCon.getBean("constructorInstance");
        System.out.println(b1);
        BeanClass b2 = (BeanClass)appCon.getBean("constructorInstance");
        System.out.println(b2);

    }
}
```

上述测试代码的运行结果如图 3.5 所示。

```
Markers  Properties  Servers  Data Source Explorer  Snippets  Console
<terminated> TestInstance [Java Application] C:\Program Files\Java\jre1.8.0_152\bin\javaw.exe (2018
一月 18, 2018 5:51:23 上午 org.springframework.context.support.ClassPathXmlApplicationCc
信息: Refreshing org.springframework.context.support.ClassPathXmlApplicationContext@7
一月 18, 2018 5:51:23 上午 org.springframework.beans.factory.xml.XmlBeanDefinitionReader
信息: Loading XML bean definitions from class path resource [applicationContext.xml]
instance.BeanClass@490ab905
instance.BeanClass@490ab905
```

图 3.5 singleton 作用域的运行结果

从图 3.5 所示的运行结果可以得知，在使用 id 或 name 获取 Bean 实例时，IoC 容器仅返回同一个 Bean 实例。

3.3.2 prototype 作用域

当将 bean 的 scope 设置为 prototype 时，Spring IoC 容器将为每次请求创建一个新的实例。如果将 3.3.1 中 bean 的配置修改如下：

```
<bean id="constructorInstance" class="instance.BeanClass" scope="prototype"/>
```

则 TestInstance 的运行结果如图 3.6 所示。

```
Markers  Properties  Servers  Data Source Explorer  Snippets  Console
<terminated> TestInstance [Java Application] C:\Program Files\Java\jre1.8.0_152\bin\javaw.exe (2018
一月 18, 2018 5:58:32 上午 org.springframework.context.support.ClassPathXmlApplicationCo
信息: Refreshing org.springframework.context.support.ClassPathXmlApplicationContext@7
一月 18, 2018 5:58:32 上午 org.springframework.beans.factory.xml.XmlBeanDefinitionReader
信息: Loading XML bean definitions from class path resource [applicationContext.xml]
instance.BeanClass@490ab905
instance.BeanClass@56ac3a89
```

图 3.6 prototype 作用域的运行结果

从图 3.6 所示的运行结果可以得知，在使用 id 或 name 两次获取 Bean 实例时，IoC 容器将返回两个不同的 Bean 实例。

3.4 Bean 的生命周期

一个对象的生命周期包括创建（实例化与初始化）、使用以及销毁等阶段，在 Spring 中，Bean 对象周期也遵循这一过程，但是 Spring 提供了许多对外接口，允许开发者对 3 个过程（实例化、初始化、销毁）的前后做一些操作。在 Spring Bean 中，实例化是为 Bean 对象开辟空间，初始化则是对属性的初始化。

Spring 容器可以管理 singleton 作用域 Bean 的生命周期，在此作用域下，Spring 能够精确地知道 Bean 何时被创建，何时初始化完成，以及何时被销毁。而对于 prototype 作用域的 Bean，Spring 只负责创建，当容器创建了 Bean 的实例后，Bean 实例就交给了客户端的代码管理，Spring 容器将不再跟踪其生命周期，并且不会管理那些被配置成 prototype 作用域的 Bean。Spring 中 Bean 的生命周期的执行是一个很复杂的过程，可借鉴 Servlet 的生命周期“实例化→初始化（init）→接收请求（service）→销毁（destroy）”来理解 Bean 的生命周期。

Bean 的生命周期的整个过程如下：

（1）根据 Bean 的配置情况实例化一个 Bean。

（2）根据 Spring 上下文对实例化的 Bean 进行依赖注入，即对 Bean 的属性进行初始化。

（3）如果 Bean 实现了 BeanNameAware 接口，将调用它实现的 setBeanName(String beanId)方法，此处参数传递的是 Spring 配置文件中 Bean 的 id。

（4）如果 Bean 实现了 BeanFactoryAware 接口，将调用它实现的 setBeanFactory 方法，此处参数传递的是当前 Spring 工厂实例的引用。

（5）如果 Bean 实现了 ApplicationContextAware 接口，将调用它实现的 setApplicationContext(ApplicationContext)方法，此处参数传递的是 Spring 上下文实例的引用。

（6）如果 Bean 关联了 BeanPostProcessor 接口，将调用初始化方法 postProcessBeforeInitialization(Object obj, String s)对 Bean 进行操作。

（7）如果 Bean 实现了 InitializingBean 接口，将调用 afterPropertiesSet 方法。

（8）如果 Bean 在 Spring 配置文件中配置了 init-method 属性，将自动调用其配置的初始化方法。

（9）如果 Bean 关联了 BeanPostProcessor 接口，将调用 postProcessAfterInitialization(Object obj, String s)方法，由于是在 Bean 初始化结束时调用 After 方法，也可用于内存或缓存技术。

注意：以上工作完成后就可以使用该 Bean，由于该 Bean 的作用域是 singleton，所以调用的是同一个 Bean 实例。

（10）当 Bean 不再需要时将进入销毁阶段，如果 Bean 实现了 DisposableBean 接口，则调用其实现的 destroy 方法将 Spring 中的 Bean 销毁。

（11）如果在配置文件中通过 destroy-method 属性指定了 Bean 的销毁方法，将调用其配置的销毁方法进行销毁。

在 Spring 中，通过实现特定的接口或通过<bean>元素的属性设置可以对 Bean 的生命周期过程产生影响。开发者可以随意地配置<bean>元素的属性，但不建议过多地使用 Bean 实现接口，因为这样将使代码和 Spring 聚合比较紧密。下面通过一个实例演示 Bean 的生命周期。

❶ 创建 Bean 的实现类

在 ch3 应用的 src 目录中创建 life 包，在 life 包下创建 BeanLife 类。在 BeanLife 类中有两个方法，一个演示初始化过程，一个演示销毁过程。具体代码如下：

```
package life;
public class BeanLife {
    public void initMyself() {
        System.out.println(this.getClass().getName() + "执行自定义的初始化方法");
    }
    public void destroyMyself() {
        System.out.println(this.getClass().getName() +"执行自定义的销毁方法");
    }
}
```

❷ 配置 Bean

在 Spring 配置文件中使用实现类 BeanLife 配置一个 id 为 beanLife 的 Bean，具体代码如下：

```
<!-- 配置bean，使用init-method属性指定初始化方法，使用destroy-method属性指定销毁方法 -->
<bean id="beanLife" class="life.BeanLife" init-method="initMyself" destroy-method="destroyMyself"/>
```

❸ 测试生命周期

在 ch3 应用的 test 包中创建测试类 TestLife，具体代码如下：

```
package test;
import org.springframework.context.support.ClassPathXmlApplicationContext;
import life.BeanLife;
public class TestLife {
    public static void main(String[] args) {
        //初始化Spring容器，加载配置文件
        //为了方便演示销毁方法的执行，这里使用ClassPathXmlApplicationContext
        //实现类声明容器
        ClassPathXmlApplicationContext ctx =
        new ClassPathXmlApplicationContext("applicationContext.xml");
        System.out.println("获得对象前");
        BeanLife blife = (BeanLife)ctx.getBean("beanLife");
        System.out.println("获得对象后" + blife);
```

```
        ctx.close(); //关闭容器，销毁 Bean 对象
    }
}
```

上述测试类的运行结果如图 3.7 所示。

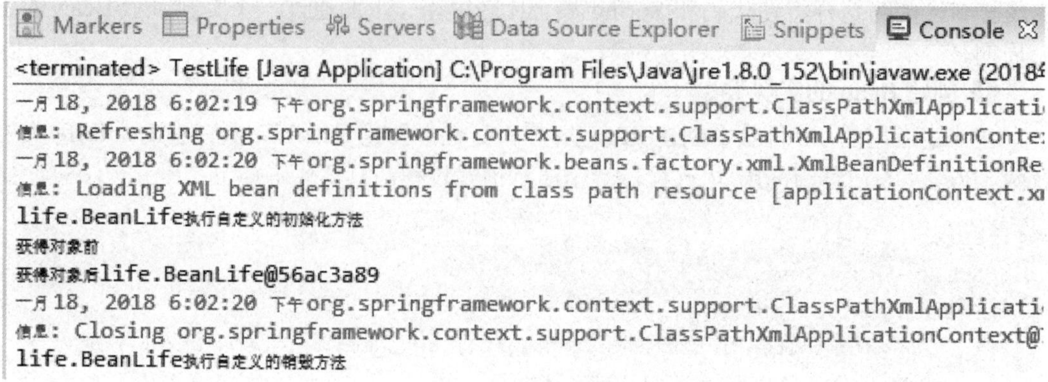

```
Markers  Properties  Servers  Data Source Explorer  Snippets  Console
<terminated> TestLife [Java Application] C:\Program Files\Java\jre1.8.0_152\bin\javaw.exe (2018
一月 18, 2018 6:02:19 下午 org.springframework.context.support.ClassPathXmlApplicati
信息: Refreshing org.springframework.context.support.ClassPathXmlApplicationConte
一月 18, 2018 6:02:20 下午 org.springframework.beans.factory.xml.XmlBeanDefinitionRe
信息: Loading XML bean definitions from class path resource [applicationContext.x
life.BeanLife执行自定义的初始化方法
获得对象前
获得对象后life.BeanLife@56ac3a89
一月 18, 2018 6:02:20 下午 org.springframework.context.support.ClassPathXmlApplicati
信息: Closing org.springframework.context.support.ClassPathXmlApplicationContext@
life.BeanLife执行自定义的销毁方法
```

图 3.7 Bean 的生命周期演示效果

从图 3.7 中可以看出，在加载配置文件时执行了 Bean 的初始化方法 initMyself；在获得对象后，关闭容器时，执行了 Bean 的销毁方法 destroyMyself。

3.5 Bean 的装配方式

视频讲解

Bean 的装配可以理解为将 Bean 依赖注入到 Spring 容器中，Bean 的装配方式即 Bean 依赖注入的方式。Spring 容器支持基于 XML 配置的装配、基于注解的装配以及自动装配等多种装配方式，其中最受青睐的装配方式是基于注解的装配（在本书后续章节中采用基于注解的装配方式装配 Bean）。本节将主要讲解基于 XML 配置的装配和基于注解的装配。

3.5.1 基于 XML 配置的装配

基于 XML 配置的装配方式已经有很久的历史了，曾经是主要的装配方式。通过 2.3 节的学习，我们知道 Spring 提供了两种基于 XML 配置的装配方式，即使用构造方法注入和使用属性的 setter 方法注入。

在使用构造方法注入方式装配 Bean 时，Bean 的实现类需要提供带参数的构造方法，并在配置文件中使用<bean>元素的子元素<constructor-arg>来定义构造方法的参数；在使用属性的 setter 方法注入方式装配 Bean 时，Bean 的实现类需要提供一个默认无参数的构造方法，并为需要注入的属性提供对应的 setter 方法，另外还需要使用<bean>元素的子元素<property>为每个属性注入值。

下面通过一个实例来演示基于 XML 配置的装配方式。

❶ 创建 Bean 的实现类

在 ch3 应用的 src 目录中创建 assemble 包，在 assemble 包下创建 ComplexUser 类。在 ComplexUser 类中分别使用构造方法注入和使用属性的 setter 方法注入。具体代码如下：

```
package assemble;
import java.util.List;
import java.util.Map;
import java.util.Set;
public class ComplexUser {
    private String uname;
    private List<String> hobbyList;
    private Map<String,String> residenceMap;
    private Set<String> aliasSet;
    private String[] array;
    /*
     * 使用构造方法注入，需要提供带参数的构造方法
     */
    public ComplexUser(String uname, List<String> hobbyList, Map<String,
    String> residenceMap, Set<String> aliasSet,String[] array) {
        super();
        this.uname=uname;
        this.hobbyList=hobbyList;
        this.residenceMap=residenceMap;
        this.aliasSet=aliasSet;
        this.array=array;
    }
    /**
     * 使用属性的 setter 方法注入，提供默认无参数的构造方法，并为注入的属性提供 setter 方法
     */
    public ComplexUser() {
        super();
    }
    /*****此处省略所有属性的 setter 方法******/
    @Override
    public String toString() {
        return "uname=" + uname + ";hobbyList=" + hobbyList + ";residenceMap="
    + residenceMap +";aliasSet=" + aliasSet + ";array=" + array;
    }
}
```

❷ 配置 Bean

在 Spring 配置文件中使用实现类 ComplexUser 配置 Bean 的两个实例，具体代码如下：

```
<!-- 使用构造方法注入方式装配 ComplexUser 实例 user1 -->
    <bean id="user1" class="assemble.ComplexUser">
```

```
<constructor-arg index="0" value="chenheng1"/>
<constructor-arg index="1">
    <list>
        <value>唱歌</value>
        <value>跳舞</value>
        <value>爬山</value>
    </list>
</constructor-arg>
<constructor-arg index="2">
    <map>
        <entry key="dalian" value="大连"/>
        <entry key="beijing" value="北京"/>
        <entry key="shanghai" value="上海"/>
    </map>
</constructor-arg>
<constructor-arg index="3">
    <set>
        <value>陈恒 100</value>
        <value>陈恒 101</value>
        <value>陈恒 102</value>
    </set>
 </constructor-arg>
 <constructor-arg index="4">
    <array>
        <value>aaaaa</value>
        <value>bbbbb</value>
    </array>
 </constructor-arg>
</bean>
<!-- 使用属性的 setter 方法注入方式装配 ComplexUser 实例 user2 -->
<bean id="user2" class="assemble.ComplexUser">
 <property name="uname" value="chenheng2"/>
 <property name="hobbyList">
    <list>
        <value>看书</value>
        <value>学习 Spring</value>
    </list>
 </property>
 <property name="residenceMap">
    <map>
        <entry key="shenzhen" value="深圳"/>
        <entry key="gaungzhou" value="广州"/>
        <entry key="tianjin" value="天津"/>
    </map>
 </property>
```

```
        <property name="aliasSet">
            <set>
                <value>陈恒 103</value>
                <value>陈恒 104</value>
                <value>陈恒 105</value>
            </set>
        </property>
        <property name="array">
            <array>
                <value>cccccc</value>
                <value>dddddd</value>
            </array>
        </property>
    </bean>
```

❸ 测试基于 XML 配置的装配方式

在 ch3 应用的 test 包中创建测试类 TestAssemble，具体代码如下：

```
package test;
import org.springframework.context.ApplicationContext;
import org.springframework.context.support.ClassPathXmlApplicationContext;
import assemble.ComplexUser;
public class TestAssemble {
    public static void main(String[] args) {
        ApplicationContext appCon = new ClassPathXmlApplicationContext
        ("applicationContext.xml");
        //使用构造方法装配测试
        ComplexUser u1=(ComplexUser)appCon.getBean("user1");
        System.out.println(u1);
        //使用 setter 方法装配测试
        ComplexUser u2 = (ComplexUser)appCon.getBean("user2");
        System.out.println(u2);
    }
}
```

上述测试代码的运行结果如图 3.8 所示。

```
Markers  Properties  Servers  Data Source Explorer  Snippets  Console
<terminated> TestAssemble [Java Application] C:\Program Files\Java\jre1.8.0_152\bin\javaw.exe (201
一月 18, 2018 7:57:20 下午 org.springframework.context.support.ClassPathXmlApplicationCo
信息: Refreshing org.springframework.context.support.ClassPathXmlApplicationContext@7
一月 18, 2018 7:57:20 下午 org.springframework.beans.factory.xml.XmlBeanDefinitionReader
信息: Loading XML bean definitions from class path resource [applicationContext.xml]
uname=chenheng1;hobbyList=[唱歌, 跳舞, 爬山];residenceMap={dalian=大连, beijing=北京, shang
uname=chenheng2;hobbyList=[看书, 学习Spring];residenceMap={shenzhen=深圳, gaungzhou=广州,
```

图 3.8　基于 XML 配置的装配方式的测试结果

3.5.2 基于注解的装配

在 Spring 框架中，尽管使用 XML 配置文件可以很简单地装配 Bean，但如果应用中有大量的 Bean 需要装配，会导致 XML 配置文件过于庞大，不方便以后的升级与维护，因此更多的时候推荐开发者使用注解（annotation）的方式去装配 Bean。

在 Spring 框架中定义了一系列的注解，下面介绍几种常用的注解。

❶ @Component

该注解是一个泛化的概念，仅仅表示一个组件对象（Bean），可以作用在任何层次上。下面通过一个实例讲解@Component。

1）创建 Bean 的实现类

在 ch3 应用的 src 目录下创建 annotation 包，在该包下创建 Bean 的实现类 AnnotationUser，代码如下：

```
package annotation;
import org.springframework.beans.factory.annotation.Value;
import org.springframework.stereotype.Component;
@Component()
/**相当于@Component("annotationUser")或@Component(value = "annotationUser"),
annotationUser 为 Bean 的 id，默认为首字母小写的类名**/
public class AnnotationUser {
    @Value("chenheng")  //只注入了简单的值，对于复杂值的注入目前使用该方式还解决不了
    private String uname;
    /**省略 setter 和 getter 方法**/
}
```

2）配置注解

现在有了 Bean 的实现类，但还不能进行测试，因为 Spring 容器并不知道去哪里扫描 Bean 对象，需要在配置文件中配置注解，方式如下：

```
<context:component-scan base-package="Bean 所在的包路径"/>
```

在 ch3 应用的 src 目录下创建配置文件 annotationContext.xml，代码如下：

```
<?xml version="1.0" encoding="UTF-8"?>
<beans xmlns="http://www.springframework.org/schema/beans"
    xmlns:xsi="http://www.w3.org/2001/XMLSchema-instance"
    xmlns:context="http://www.springframework.org/schema/context"
    xsi:schemaLocation="http://www.springframework.org/schema/beans
        http://www.springframework.org/schema/beans/spring-beans.xsd
        http://www.springframework.org/schema/context
        http://www.springframework.org/schema/context/spring-context.xsd">
    <!-- 使用 context 命名空间,通过 Spring 扫描指定包 annotation 及其子包下所有 Bean
    的实现类，进行注解解析 -->
```

```
    <context:component-scan base-package="annotation"/>
</beans>
```

3）测试 Bean 实例

在 test 包中创建测试类 TestAnnotation，测试上述 Bean，具体测试代码如下：

```
package test;
import org.springframework.context.ApplicationContext;
import org.springframework.context.support.ClassPathXmlApplicationContext;
import annotation.AnnotationUser;
public class TestAnnotation {
    public static void main(String[] args) {
        ApplicationContext appCon = new ClassPathXmlApplicationContext
       ("annotationContext.xml");
        AnnotationUser au = (AnnotationUser)appCon.getBean("annotationUser");
        System.out.println(au.getUname());
    }
}
```

注：在 Spring 4.0 以上的版本，配置注解指定包中的注解进行扫描前需要事先导入 Spring AOP 的 JAR 包 spring-aop-5.0.2.RELEASE.jar。

❷ @Repository

该注解用于将数据访问层（DAO）的类标识为 Bean，即注解数据访问层 Bean，其功能与@Component 相同。

❸ @Service

该注解用于标注一个业务逻辑组件类（Service 层），其功能与@Component 相同。

❹ @Controller

该注解用于标注一个控制器组件类（Spring MVC 的 Controller），其功能与@Component 相同。

❺ @Autowired

该注解可以对类成员变量、方法及构造方法进行标注，完成自动装配的工作。通过使用@Autowired 来消除 setter 和 getter 方法。默认按照 Bean 的类型进行装配。

❻ @Resource

该注解与@Autowired 的功能一样，区别在于该注解默认是按照名称来装配注入的，只有当找不到与名称匹配的 Bean 时才会按照类型来装配注入；而@Autowired 默认按照 Bean 的类型进行装配，如果想按照名称来装配注入，则需要和@Qualifier 注解一起使用。

@Resource 注解有两个属性——name 和 type。name 属性指定 Bean 实例名称，即按照名称来装配注入；type 属性指定 Bean 类型，即按照 Bean 的类型进行装配。

❼ @Qualifier

该注解与@Autowired 注解配合使用。当@Autowired 注解需要按照名称来装配注入时需要和该注解一起使用，Bean 的实例名称由@Qualifier 注解的参数指定。

在上面几个注解中，虽然@Repository、@Service 和@Controller 等注解的功能与

@Component 注解相同，但为了使类的标注更加清晰（层次化），在实际开发中推荐使用@Repository 标注数据访问层（DAO 层）、使用@Service 标注业务逻辑层（Service 层）、使用@Controller 标注控制器层（控制层）。

下面通过一个实例讲解如何使用这些注解。

1）创建 DAO 层

在 ch3 应用的 src 中创建 annotation.dao 包，在该包下创建 TestDao 接口和 TestDaoImpl 实现类，并将实现类 TestDaoImpl 使用@Repository 注解标注为数据访问层。

TestDao 的代码如下：

```
package annotation.dao;
public interface TestDao {
    public void save();
}
```

TestDaoImpl 的代码如下：

```
package annotation.dao;
import org.springframework.stereotype.Repository;
@Repository("testDaoImpl")
/**相当于@Repository，但如果在 service 层使用@Resource(name="testDaoImpl")，
testDaoImpl 不能省略。**/
public class TestDaoImpl implements TestDao{
    @Override
    public void save() {
        System.out.println("testDao save");
    }
}
```

2）创建 Service 层

在 ch3 应用的 src 中创建 annotation.service 包，在该包下创建 TestService 接口和 TestSeviceImpl 实现类，并将实现类 TestSeviceImpl 使用@Service 注解标注为业务逻辑层。

TestService 的代码如下：

```
package annotation.service;
public interface TestService {
    public void save();
}
```

TestSeviceImpl 的代码如下：

```
package annotation.service;
import javax.annotation.Resource;
import org.springframework.stereotype.Service;
import annotation.dao.TestDao;
@Service("testSeviceImpl")  //相当于@Service
public class TestSeviceImpl implements TestService{
```

```
    @Resource(name="testDaoImpl")
    /**相当于@Autowired，@Autowired 默认按照 Bean 类型装配**/
    private TestDao testDao;
    @Override
    public void save() {
        testDao.save();
        System.out.println("testService save");
    }
}
```

3）创建 Controller 层

在 ch3 应用的 src 中创建 annotation.controller 包，在该包下创建 TestController 类，并将 TestController 类使用@Controller 注解标注为控制器层。

TestController 的代码如下：

```
package annotation.controller;
import org.springframework.beans.factory.annotation.Autowired;
import org.springframework.stereotype.Controller;
import annotation.service.TestService;
@Controller
public class TestController {
    @Autowired
    private TestService testService;
    public void save() {
        testService.save();
        System.out.println("testController save");
    }
}
```

4）配置注解

由于 annotation.dao、annotation.service 和 annotation.controller 包都属于 annotation 包的子包，因此不需要在配置文件 annotationContext.xml 中配置注解。

5）创建测试类

在 ch3 应用的 test 包中创建测试类 TestMoreAnnotation，具体代码如下：

```
package test;
import org.springframework.context.ApplicationContext;
import org.springframework.context.support.ClassPathXmlApplicationContext;
import annotation.controller.TestController;
public class TestMoreAnnotation {
    public static void main(String[] args) {
        ApplicationContext appCon = new ClassPathXmlApplicationContext
        ("annotationContext.xml");
```

```
        TestController testcon = (TestController)appCon.getBean("testController");
        testcon.save();
    }
}
```

3.6 本章小结

本章重点介绍了如何将 Bean 装配注入到 Spring IoC 容器中，即 Bean 的装配方式。通过本章的学习，读者能够掌握 Bean 的两种常用装配方式，即基于 XML 配置的装配和基于注解的装配，其中基于注解的装配方式尤其重要，它是当前的主流装配方式。

习题 3

1．Bean 的实例化有哪几种常见的方法?

2．简述基于注解的装配方式的基本用法。

3．@Autowired 和@Resource 有什么区别?

4．Bean 的默认作用域是（　　）。

A．page　　B．request　　C．singleton　　D．prototype

5．在下面代码片段中使用@Controller 注解装配了 Bean，而 Bean 的 id 是（　　）。

```
@Controller
public class TestController {
    ...
}
```

A．TestController　　B．testController　　C．无 id　　D．任意名称

第 4 章

Spring AOP

学习目的与要求

本章主要介绍 AOP 的概念术语、动态代理、AOP 的实现以及 AspectJ 的开发等内容。通过本章的学习，读者除了需要掌握 AOP 的相关概念及实现，还需要进一步掌握动态代理。

本章主要内容

- AOP 的概念术语；
- 动态代理；
- AOP 的实现；
- AspectJ 的开发。

Spring AOP 是 Spring 框架体系结构中非常重要的功能模块之一，该模块提供了面向切面编程实现。面向切面编程在事务处理、日志记录、安全控制等操作中被广泛使用。本章将对 Spring AOP 的相关概念及实现进行详细讲解。

4.1 Spring AOP 的基本概念

4.1.1 AOP 的概念

AOP（Aspect-Oriented Programming）即面向切面编程，它与 OOP（Object-Oriented Programming，面向对象编程）相辅相成，提供了与 OOP 不同的抽象软件结构的视角。在 OOP 中，以类作为程序的基本单元，而 AOP 中的基本单元是 Aspect（切面）。Struts2 的拦截器设计就是基于 AOP 的思想，是个比较经典的应用。

在业务处理代码中通常有日志记录、性能统计、安全控制、事务处理、异常处理等操作。尽管使用 OOP 可以通过封装或继承的方式达到代码的重用，但仍然有同样的代码分散在各个方法中。因此，采用 OOP 处理日志记录等操作不仅增加了开发者的工作量，而且提高了升级维护的困难。为了解决此类问题，AOP 思想应运而生。AOP 采取横向抽取机制，即将分散在各个方法中的重复代码提取出来，然后在程序编译或运行阶段将这些抽取出来的代码应用到需要执行的地方。这种横向抽取机制采用传统的 OOP 是无法办到的，因为 OOP 实现的是父子关系的纵向重用。但是 AOP 不是 OOP 的替代品，而是 OOP 的补充，它们相辅相成。

在 AOP 中，横向抽取机制的类与切面的关系如图 4.1 所示。

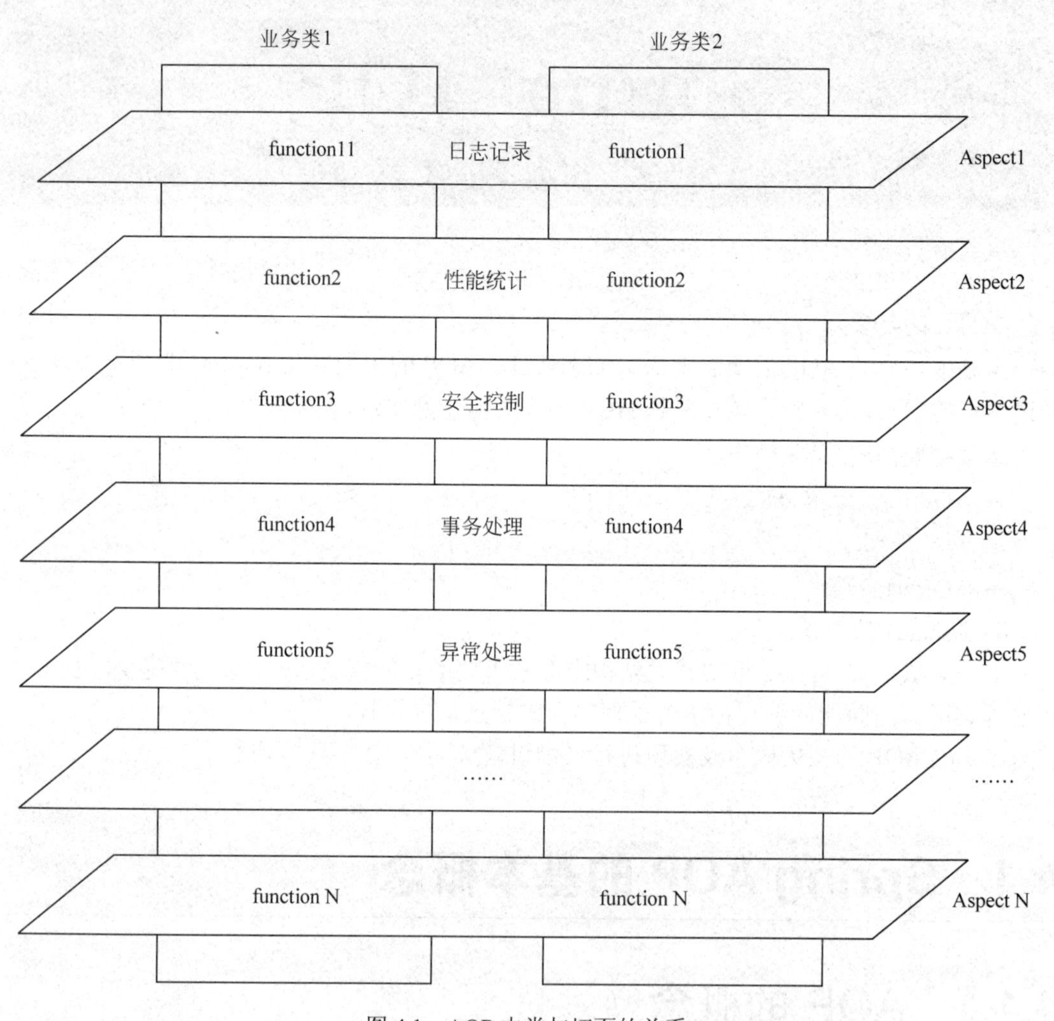

图 4.1　AOP 中类与切面的关系

从图 4.1 可以看出，通过切面 Aspect 分别在业务类 1 和业务类 2 中加入了日志记录、性能统计、安全控制、事务处理、异常处理等操作。

4.1.2　AOP 的术语

在 Spring AOP 框架中涉及以下常用术语。

❶ 切面

切面（Aspect）是指封装横切到系统功能（例如事务处理）的类。

❷ 连接点

连接点（Joinpoint）是指程序运行中的一些时间点，例如方法的调用或异常的抛出。

❸ 切入点

切入点（Pointcut）是指需要处理的连接点。在 Spring AOP 中，所有的方法执行都是连接点，而切入点是一个描述信息，它修饰的是连接点，通过切入点确定哪些连接点需要被处理。切面、连接点和切入点的关系如图 4.2 所示。

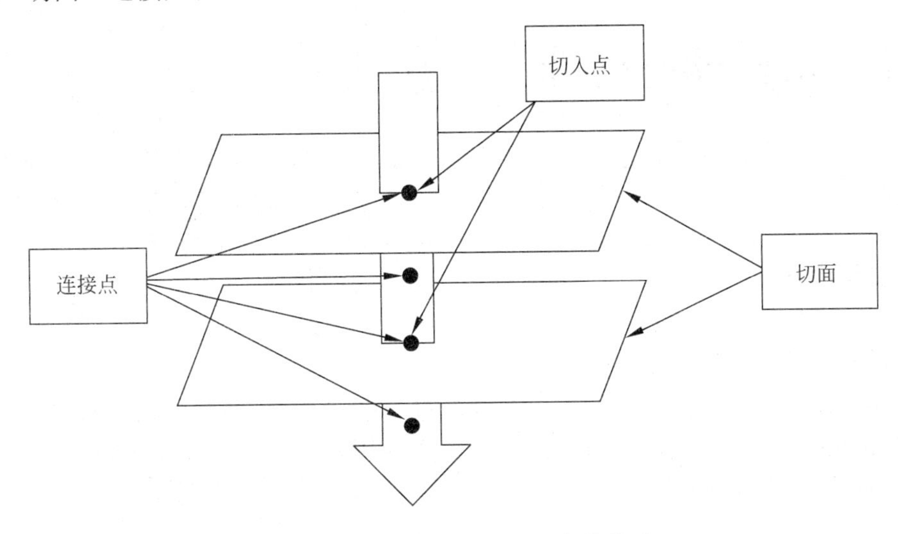

图 4.2　切面、连接点和切入点的关系

❹ 通知

通知（Advice）是由切面添加到特定的连接点（满足切入点规则）的一段代码，即在定义好的切入点处所要执行的程序代码，可以将其理解为切面开启后切面的方法，因此通知是切面的具体实现。

❺ 引入

引入（Introduction）允许在现有的实现类中添加自定义的方法和属性。

❻ 目标对象

目标对象（Target Object）是指所有被通知的对象。如果 AOP 框架使用运行时代理的方式（动态的 AOP）来实现切面，那么通知对象总是一个代理对象。

❼ 代理

代理（Proxy）是通知应用到目标对象之后被动态创建的对象。

❽ 织入

织入（Weaving）是将切面代码插入到目标对象上，从而生成代理对象的过程。根据不同的实现技术，AOP 织入有 3 种方式：编译期织入，需要有特殊的 Java 编译器；类装载期织入，需要有特殊的类装载器；动态代理织入，在运行期为目标类添加通知生成子类的方式。Spring AOP 框架默认采用动态代理织入，而 AspectJ（基于 Java 语言的 AOP 框架）

采用编译期织入和类装载期织入。

视频讲解

4.2 动态代理

在 Java 中有多种动态代理技术，例如 JDK、CGLIB、Javassist、ASM，其中最常用的动态代理技术是 JDK 和 CGLIB。目前，在 Spring AOP 中常用 JDK 和 CGLIB 两种动态代理技术。

4.2.1 JDK 动态代理

JDK 动态代理是 java.lang.reflect.*包提供的方式，它必须借助一个接口才能产生代理对象。因此，对于使用业务接口的类，Spring 默认使用 JDK 动态代理实现 AOP。下面通过一个实例演示如何使用 JDK 动态代理实现 Spring AOP，具体步骤如下：

❶ 创建应用

创建一个名为 ch4 的 Web 应用，并导入所需的 JAR 包。

❷ 创建接口及实现类

在 ch4 的 src 目录下创建一个 dynamic.jdk 包，在该包中创建接口 TestDao 和接口实现类 TestDaoImpl。该实现类作为目标类，在代理类中对其方法进行增强处理。

TestDao 的代码如下：

```
package dynamic.jdk;
public interface TestDao {
    public void save();
    public void modify();
    public void delete();
}
```

TestDaoImpl 的代码如下：

```
package dynamic.jdk;
public class TestDaoImpl implements TestDao{
    @Override
    public void save() {
        System.out.println("保存");
    }
    @Override
    public void modify() {
        System.out.println("修改");
    }
    @Override
    public void delete() {
        System.out.println("删除");
```

```
    }
}
```

❸ 创建切面类

在 ch4 的 src 目录下创建一个 aspect 包，在该包中创建切面类 MyAspect，注意在该类中可以定义多个通知（增强处理的功能方法）。

MyAspect 的代码如下：

```
package aspect;
/**
 * 切面类，可以定义多个通知，即增强处理的方法
 */
public class MyAspect {
    public void check() {
        System.out.println("模拟权限控制");
    }
    public void except() {
        System.out.println("模拟异常处理");
    }
    public void log() {
        System.out.println("模拟日志记录");
    }
    public void monitor() {
        System.out.println("性能监测");
    }
}
```

❹ 创建代理类

在 dynamic.jdk 包中创建代理类 JDKDynamicProxy。在 JDK 动态代理中代理类必须实现 java.lang.reflect.InvocationHandler 接口，并编写代理方法，在代理方法中需要通过 Proxy 实现动态代理。

JDKDynamicProxy 的代码如下：

```
package dynamic.jdk;
import java.lang.reflect.InvocationHandler;
import java.lang.reflect.Method;
import java.lang.reflect.Proxy;
import aspect.MyAspect;
public class JDKDynamicProxy implements InvocationHandler {
    //声明目标类接口对象（真实对象）
    private TestDao testDao;
    /**创建代理的方法，建立代理对象和真实对象的代理关系，并返回代理对象**/
    public Object createProxy(TestDao testDao) {
        this.testDao=testDao;
        //1.类加载器
        ClassLoader cld=JDKDynamicProxy.class.getClassLoader();
```

```
        //2.被代理对象实现的所有接口
        Class[] clazz=testDao.getClass().getInterfaces();
        //3.使用代理类进行增强，返回代理后的对象
        return Proxy.newProxyInstance(cld, clazz, this);
    }
    /**
     * 代理的逻辑方法，所有动态代理类的方法调用都交给该方法处理
     * proxy 是被代理对象
     * method 是将要被执行的方法
     * args 是执行方法时需要的参数
     * return 指返回代理结果
     */
    @Override
    public Object invoke(Object proxy, Method method, Object[] args) throws
    Throwable {
        //创建一个切面
        MyAspect myAspect=new MyAspect();
        //前增强
        myAspect.check();
        myAspect.except();
        //在目标类上调用方法并传入参数，相当于调用 testDao 中的方法
        Object obj=method.invoke(testDao, args);
        //后增强
        myAspect.log();
        myAspect.monitor();
        return obj;
    }
}
```

❺ 创建测试类

在 dynamic.jdk 包中创建测试类 JDKDynamicTest。在主方法中创建代理对象和目标对象，然后从代理对象中获取对目标对象增强后的对象，最后调用该对象的添加、修改和删除方法。

JDKDynamicTest 的代码如下：

```
package dynamic.jdk;
public class JDKDynamicTest {
    public static void main(String[] args) {
        //创建代理对象
        JDKDynamicProxy jdkProxy=new JDKDynamicProxy();
        //创建目标对象
        TestDao testDao=new TestDaoImpl();
        /**从代理对象中获取增强后的目标对象，该对象是一个被代理的对象，它会进入代理的
        逻辑方法 invoke 中**/
        TestDao testDaoAdvice=(TestDao)jdkProxy.createProxy(testDao);
```

```
        //执行方法
        testDaoAdvice.save();
        System.out.println("==============");
        testDaoAdvice.modify();
        System.out.println("==============");
        testDaoAdvice.delete();
    }
}
```

上述测试类的运行结果如图 4.3 所示。

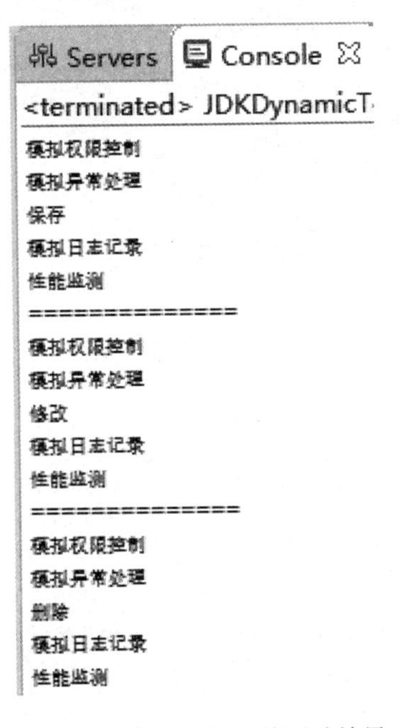

图 4.3 JDK 动态代理的测试结果

从图 4.3 可以看出，testDao 实例中的增加、修改和删除方法已经被成功调用，并且在调用前后分别增加了“模拟权限控制”“模拟异常处理”“模拟日志记录”和“性能监测”的功能。

4.2.2 CGLIB 动态代理

从 4.2.1 节可知，JDK 动态代理必须提供接口才能使用，对于没有提供接口的类，只能采用 CGLIB 动态代理。

CGLIB（Code Generation Library）是一个高性能开源的代码生成包，采用非常底层的字节码技术，对指定的目标类生成一个子类，并对子类进行增强。在 Spring Core 包中已经集成了 CGLIB 所需要的 JAR 包，不需要另外导入 JAR 包。下面通过一个实例演示 CGLIB 动态代理的实现过程，具体步骤如下：

❶ 创建目标类

在 ch4 的 src 目录下创建一个 dynamic.cglib 包，在该包中创建目标类 TestDao，注意该类不需要实现任何接口。

TestDao 的代码如下：

```
package dynamic.cglib;
public class TestDao {
    public void save() {
        System.out.println("保存");
    }
    public void modify() {
        System.out.println("修改");
    }
    public void delete() {
        System.out.println("删除");
    }
}
```

❷ 创建代理类

在 dynamic.cglib 包中创建代理类 CglibDynamicProxy，该类实现 MethodInterceptor 接口。

CglibDynamicProxy 的代码如下：

```
package dynamic.cglib;
import java.lang.reflect.Method;
import org.springframework.cglib.proxy.Enhancer;
import org.springframework.cglib.proxy.MethodInterceptor;
import org.springframework.cglib.proxy.MethodProxy;
import aspect.MyAspect;
public class CglibDynamicProxy implements MethodInterceptor{
    /**
     * 创建代理的方法，生成 CGLIB 代理对象
     * target 是目标对象，需要增强的对象
     * 返回目标对象的 CGLIB 代理对象
     */
    public Object createProxy(Object target) {
        //创建一个动态类对象，即增强类对象
        Enhancer enhancer=new Enhancer();
        //确定需要增强的类，设置其父类
        enhancer.setSuperclass(target.getClass());
        //确定代理逻辑对象为当前对象，要求当前对象实现 MethodInterceptor 的方法
        enhancer.setCallback(this);
        //返回创建的代理对象
        return enhancer.create();
    }
    /**
```

```
     * intercept 方法会在程序执行目标方法时被调用
     * proxy 是 CGLIB 根据指定父类生成的代理对象
     * method 是拦截方法
     * args 是拦截方法的参数数组
     * methodProxy 是方法的代理对象，用于执行父类的方法
     * 返回代理结果
     */
    @Override
    public Object intercept(Object proxy, Method method, Object[] args,
    MethodProxy methodProxy) throws Throwable {
        //创建一个切面
        MyAspect myAspect=new MyAspect();
        //前增强
        myAspect.check();
        myAspect.except();
        //目标方法执行，返回代理结果
        Object obj=methodProxy.invokeSuper(proxy, args);
        //后增强
        myAspect.log();
        myAspect.monitor();
        return obj;
    }
}
```

❸ 创建测试类

在 dynamic.cglib 包中创建测试类 CglibDynamicTest。在主方法中创建代理对象和目标对象，然后从代理对象中获取对目标对象增强后的对象，最后调用该对象的添加、修改和删除方法。

CglibDynamicTest 的代码如下：

```
package dynamic.cglib;
public class CglibDynamicTest {
    public static void main(String[] args) {
        //创建代理对象
        CglibDynamicProxy cdp=new CglibDynamicProxy();
        //创建目标对象
        TestDao testDao=new TestDao();
        //获取增强后的目标对象
        TestDao testDaoAdvice=(TestDao)cdp.createProxy(testDao);
        //执行方法
        testDaoAdvice.save();
        System.out.println("==============");
        testDaoAdvice.modify();
        System.out.println("==============");
```

```
        testDaoAdvice.delete();
    }
}
```

上述测试类的运行结果与图 4.3 一样，这里不再赘述。

视频讲解

4.3 基于代理类的 AOP 实现

从 4.2 节可知，在 Spring 中默认使用 JDK 动态代理实现 AOP 编程。使用 org.springframework.aop.framework.ProxyFactoryBean 创建代理是 Spring AOP 实现的最基本方式。

❶ 通知类型

在讲解 ProxyFactoryBean 之前先了解一下 Spring 的通知类型。根据 Spring 中通知在目标类方法中的连接点位置，通知可以分为 6 种类型。

1）环绕通知

环绕通知（org.aopalliance.intercept.MethodInterceptor）是在目标方法执行前和执行后实施增强，可应用于日志记录、事务处理等功能。

2）前置通知

前置通知（org.springframework.aop.MethodBeforeAdvice）是在目标方法执行前实施增强，可应用于权限管理等功能。

3）后置返回通知

后置返回通知（org.springframework.aop.AfterReturningAdvice）是在目标方法成功执行后实施增强，可应用于关闭流、删除临时文件等功能。

4）后置（最终）通知

后置通知（org.springframework.aop.AfterAdvice）是在目标方法执行后实施增强，与后置返回通知不同的是，不管是否发生异常都要执行该类通知，该类通知可应用于释放资源。

5）异常通知

异常通知（org.springframework.aop.ThrowsAdvice）是在方法抛出异常后实施增强，可应用于处理异常、记录日志等功能。

6）引入通知

引入通知（org.springframework.aop.IntroductionInterceptor）是在目标类中添加一些新的方法和属性，可应用于修改目标类（增强类）。

❷ ProxyFactoryBean

ProxyFactoryBean 是 org.springframework.beans.factory.FactoryBean 接口的实现类，FactoryBean 负责实例化一个 Bean 实例，ProxyFactoryBean 负责为其他 Bean 实例创建代理实例。ProxyFactoryBean 类的常用属性如表 4.1 所示。

表 4.1 ProxyFactoryBean 类的常用属性

属　性	描　述
target	代理的目标对象
proxyInterfaces	代理需要实现的接口列表，如果是多个接口，可以使用以下格式赋值： <list> <value></value> … <list>
interceptorNames	需要织入目标的Advice
proxyTargetClass	是否对类代理而不是接口，默认为false，使用JDK动态代理；当为true时，使用CGLIB动态代理
singleton	返回的代理实例是否为单例，默认为true
optimize	当设置为true时强制使用CGLIB动态代理

下面通过一个实现环绕通知的实例演示 Spring 使用 ProxyFactoryBean 创建 AOP 代理的过程。

1）导入相关 JAR 包

在核心 JAR 包的基础上需要向 ch4 应用的/WEB-INF/lib 目录下导入 JAR 包 spring-aop-5.0.2.RELEASE.jar 和 aopalliance-1.0.jar。

aopalliance-1.0.jar 是 AOP 联盟提供的规范包，可以通过地址"http://mvnrepository.com/artifact/aopalliance/aopalliance/1.0"下载。

2）创建切面类

由于该实例实现环绕通知，所以切面类需要实现 org.aopalliance.intercept.MethodInterceptor 接口。在 src 目录下创建一个 spring.proxyfactorybean 包，并在该包中创建切面类 MyAspect。

MyAspect 的代码如下：

```
package spring.proxyfactorybean;
import org.aopalliance.intercept.MethodInterceptor;
import org.aopalliance.intercept.MethodInvocation;
/**
 * 切面类
 */
public class MyAspect implements MethodInterceptor{
    @Override
    public Object invoke(MethodInvocation arg0) throws Throwable {
        //增强方法
        check();
        except();
        //执行目标方法
        Object obj=arg0.proceed();
        //增强方法
        log();
        monitor();
```

```
        return obj;
    }
    public void check() {
        System.out.println("模拟权限控制");
    }
    public void except() {
        System.out.println("模拟异常处理");
    }
    public void log() {
        System.out.println("模拟日志记录");
    }
    public void monitor() {
        System.out.println("性能监测");
    }
}
```

3）配置切面并指定代理

切面类需要配置为 Bean 实例，这样 Spring 容器才能识别为切面对象。在 spring.proxyfactorybean 包中创建配置文件 applicationContext.xml，并在文件中配置切面和指定代理对象。

applicationContext.xml 的代码如下：

```
<?xml version="1.0" encoding="UTF-8"?>
<beans xmlns="http://www.springframework.org/schema/beans"
    xmlns:xsi="http://www.w3.org/2001/XMLSchema-instance"
    xsi:schemaLocation="http://www.springframework.org/schema/beans
        http://www.springframework.org/schema/beans/spring-beans.xsd">
    <!--定义目标对象（使用 4.2.1 节中的实现类）-->
    <bean id="testDao" class="dynamic.jdk.TestDaoImpl"/>
    <!-- 创建一个切面 -->
    <bean id="myAspect" class="spring.proxyfactorybean.MyAspect"/>
    <!--使用 Spring 代理工厂定义一个名为 testDaoProxy 的代理对象-->
    <bean id="testDaoProxy" class="org.springframework.aop.framework.
    ProxyFactoryBean">
        <!-- 指定代理实现的接口 -->
        <property name="proxyInterfaces" value="dynamic.jdk.TestDao"/>
        <!-- 指定目标对象 -->
        <property name="target" ref="testDao"/>
        <!-- 指定切面，织入环绕通知 -->
        <property name="interceptorNames" value="myAspect"/>
        <!-- 指定代理方式，true 指定 CGLIB 动态代理；默认为 false，指定 JDK 动态代理 -->
        <property name="proxyTargetClass" value="true"/>
    </bean>
</beans>
```

在上述配置文件中首先通过<bean>元素定义了目标对象和切面，然后使用

ProxyFactoryBean 类定义了代理对象。

4）创建测试类

在 spring.proxyfactorybean 包中创建测试类 ProxyFactoryBeanTest，在主方法中使用 Spring 容器获取代理对象，并执行目标方法。

ProxyFactoryBeanTest 的代码如下：

```
package spring.proxyfactorybean;
import org.springframework.context.ApplicationContext;
import org.springframework.context.support.ClassPathXmlApplicationContext;
import dynamic.jdk.TestDao;
public class ProxyFactoryBeanTest {
    public static void main(String[] args) {
        ApplicationContext appCon=new ClassPathXmlApplicationContext
        ("/spring/proxyfactorybean/applicationContext.xml");
        //从容器中获取增强后的目标对象
        TestDao testDaoAdvice=(TestDao)appCon.getBean("testDaoProxy");
        //执行方法
        testDaoAdvice.save();
        System.out.println("================");
        testDaoAdvice.modify();
        System.out.println("================");
        testDaoAdvice.delete();
    }
}
```

上述测试类的运行结果与图 4.3 一样，这里不再赘述。

视频讲解

4.4　基于 XML 配置开发 AspectJ

AspectJ 是一个基于 Java 语言的 AOP 框架。从 Spring 2.0 以后引入了 AspectJ 的支持。对于目前的 Spring 框架，建议开发者使用 AspectJ 实现 Spring AOP。使用 AspectJ 实现 Spring AOP 的方式有两种，一是基于 XML 配置开发 AspectJ，二是基于注解开发 AspectJ。本节讲解基于 XML 配置开发 AspectJ 的相关知识，而基于注解开发 AspectJ 的相关知识将在 4.5 节讲解。

基于 XML 配置开发 AspectJ 是指通过 XML 配置文件定义切面、切入点及通知，所有这些定义都必须在<aop:config>元素内。<aop:config>元素及其子元素如表 4.2 所示。

表 4.2　<aop:config>元素及其子元素

元 素 名 称	用　　途
<aop:config>	开发AspectJ的顶层配置元素，在配置文件的<beans>下可以包含多个该元素
<aop:aspect>	配置（定义）一个切面，<aop:config>元素的子元素，属性ref指定切面的定义
<aop:pointcut>	配置切入点，<aop:aspect>元素的子元素，属性expression指定通知增强哪些方法

续表

元素名称	用途
<aop:before>	配置前置通知，<aop:aspect>元素的子元素，属性method指定前置通知方法，属性pointcut-ref指定关联的切入点
<aop:after-returning>	配置后置返回通知，<aop:aspect>元素的子元素，属性method指定后置返回通知方法，属性pointcut-ref指定关联的切入点
<aop:around>	配置环绕通知，<aop:aspect>元素的子元素，属性method指定环绕通知方法，属性pointcut-ref指定关联的切入点
<aop:after-throwing>	配置异常通知，<aop:aspect>元素的子元素，属性method指定异常通知方法，属性pointcut-ref指定关联的切入点，没有异常发生时将不会执行
<aop:after>	配置后置（最终）通知，<aop:aspect>元素的子元素，属性method指定后置（最终）通知方法，属性pointcut-ref指定关联的切入点
<aop:declare-parents>	给通知引入新的额外接口，增强功能，不要求掌握该类型的通知

下面通过一个实例演示基于 XML 配置开发 AspectJ 的过程。

❶ 导入 AspectJ 框架相关的 JAR 包

需要向 ch4 应用的/WEB-INF/lib 目录下导入 JAR 包 spring-aspects-5.0.2.RELEASE.jar 和 aspectjweaver-1.8.13.jar。

spring-aspects-5.0.2.RELEASE.jar 是 Spring 为 AspectJ 提供的实现，在 Spring 的包中已提供。

aspectjweaver-1.8.13.jar 是 AspectJ 框架所提供的规范包，可以通过地址“http://mvnrepository.com/artifact/org.aspectj/aspectjweaver/1.8.13”下载。

❷ 创建切面类

在 ch4 应用的 src 目录下创建 aspectj.xml 包，在该包中创建切面类 MyAspect，并在该类中编写各种类型的通知。

MyAspect 的代码如下：

```
package aspectj.xml;
import org.aspectj.lang.JoinPoint;
import org.aspectj.lang.ProceedingJoinPoint;
/**
 * 切面类，在此类中编写各种类型的通知
 */
public class MyAspect {
    /**
     * 前置通知，使用 JoinPoint 接口作为参数获得目标对象信息
     */
    public void before(JoinPoint jp) {
        System.out.print("前置通知：模拟权限控制");
        System.out.println("，目标类对象：" + jp.getTarget()
        + "，被增强处理的方法：" + jp.getSignature().getName());
    }
    /**
     * 后置返回通知
```

```
     */
    public void afterReturning(JoinPoint jp) {
        System.out.print("后置返回通知：" + "模拟删除临时文件");
        System.out.println("，被增强处理的方法：" + jp.getSignature().getName());
    }
    /**
     * 环绕通知
     * ProceedingJoinPoint 是 JoinPoint 的子接口，代表可以执行的目标方法
     * 返回值的类型必须是 Object
     * 必须一个参数是 ProceedingJoinPoint 类型
     * 必须 throws Throwable
     */
    public Object around(ProceedingJoinPoint pjp) throws Throwable{
        //开始
        System.out.println("环绕开始：执行目标方法前，模拟开启事务");
        //执行当前目标方法
        Object obj=pjp.proceed();
        //结束
        System.out.println("环绕结束：执行目标方法后，模拟关闭事务");
        return obj;
    }
    /**
     * 异常通知
     */
    public void except(Throwable e) {
        System.out.println("异常通知：" + "程序执行异常" + e.getMessage());
    }

    /**
     * 后置（最终）通知
     */
    public void after() {
        System.out.println("最终通知：模拟释放资源");
    }
}
```

❸ 创建配置文件，并编写相关配置

在 aspectj.xml 包中创建配置文件 applicationContext.xml，并为<aop:config>元素及其子元素编写相关配置。

applicationContext.xml 的代码如下：

```
<?xml version="1.0" encoding="UTF-8"?>
<beans xmlns="http://www.springframework.org/schema/beans"
    xmlns:xsi="http://www.w3.org/2001/XMLSchema-instance"
    xmlns:aop="http://www.springframework.org/schema/aop"
    xsi:schemaLocation="http://www.springframework.org/schema/beans
```

```
    http://www.springframework.org/schema/beans/spring-beans.xsd
    http://www.springframework.org/schema/aop
    http://www.springframework.org/schema/aop/spring-aop.xsd">
    <!-- 定义目标对象，使用 4.2.1 节的实现类 -->
    <bean id="testDao" class="dynamic.jdk.TestDaoImpl"/>
    <!-- 定义切面 -->
    <bean id="myAspect" class="aspectj.xml.MyAspect"/>
    <!-- AOP 配置 -->
    <aop:config>
    <!-- 配置切面 -->
    <aop:aspect ref="myAspect">
        <!-- 配置切入点，通知增强哪些方法 -->
        <aop:pointcut expression="execution(* dynamic.jdk.*.*(..))" id="
        myPointCut"/>
        <!-- 将通知与切入点关联 -->
        <!-- 关联前置通知 -->
        <aop:before method="before" pointcut-ref="myPointCut"/>
        <!-- 关联后置返回通知，在目标方法成功执行后执行 -->
        <aop:after-returning method="afterReturning" pointcut-ref=
        "myPointCut"/>
        <!-- 关联环绕通知 -->
        <aop:around method="around" pointcut-ref="myPointCut"/>
        <!-- 关联异常通知，没有异常发生时将不会执行增强，throwing 属性设置通知的第二
        个参数名称 -->
        <aop:after-throwing method="except" pointcut-ref="myPointCut"
        throwing="e"/>
        <!-- 关联后置（最终）通知，不管目标方法是否成功都要执行 -->
        <aop:after method="after" pointcut-ref="myPointCut"/>
    </aop:aspect>
    </aop:config>
</beans>
```

在上述配置文件中，expression="execution(* dynamic.jdk.*.*(..))" 是定义切入点表达式，该切入点表达式的意思是匹配 dynamic.jdk 包中任意类的任意方法的执行。其中，execution(* dynamic.jdk.*.*(..))是表达式的主体，第一个*表示的是返回类型，使用*代表所有类型；dynamic.jdk 表示的是需要匹配的包名，后面第二个*表示的是类名，使用*代表匹配包中所有的类；第三个*表示的是方法名，使用*表示所有方法；后面的(..)表示方法的参数，其中的“..”表示任意参数。另外，注意第一个*和包名之间有一个空格。读者如果想了解更多关于切入点表达式的配置信息，可参考 Spring 官方文档的切入点声明部分。

❹ 创建测试类

在 aspectj.xml 包中创建测试类 XMLAspectJTest，在主方法中使用 Spring 容器获取代理对象，并执行目标方法。

XMLAspectJTest 的代码如下：

```
package aspectj.xml;
import org.springframework.context.ApplicationContext;
import org.springframework.context.support.ClassPathXmlApplicationContext;
import dynamic.jdk.TestDao;
public class XMLAspectJTest {
    public static void main(String[] args) {
        ApplicationContext appCon =
         new ClassPathXmlApplicationContext("/aspectj/xml/applicationContext.xml");
        //从容器中获取增强后的目标对象
        TestDao testDaoAdvice = (TestDao)appCon.getBean("testDao");
        //执行方法
        testDaoAdvice.save();
    }
}
```

上述测试类的运行结果如图 4.4 所示。

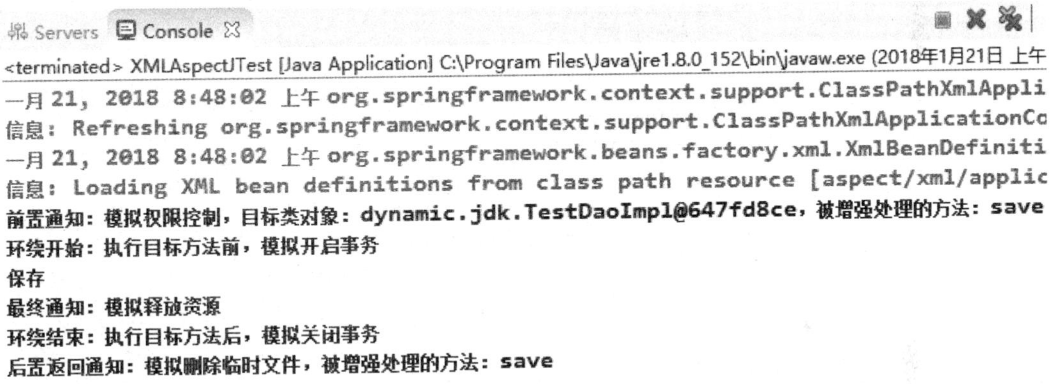

```
Servers  Console
<terminated> XMLAspectJTest [Java Application] C:\Program Files\Java\jre1.8.0_152\bin\javaw.exe (2018年1月21日 上午
一月 21, 2018 8:48:02 上午 org.springframework.context.support.ClassPathXmlAppli
信息: Refreshing org.springframework.context.support.ClassPathXmlApplicationCo
一月 21, 2018 8:48:02 上午 org.springframework.beans.factory.xml.XmlBeanDefiniti
信息: Loading XML bean definitions from class path resource [aspect/xml/applic
前置通知: 模拟权限控制，目标类对象: dynamic.jdk.TestDaoImpl@647fd8ce，被增强处理的方法: save
环绕开始: 执行目标方法前，模拟开启事务
保存
最终通知: 模拟释放资源
环绕结束: 执行目标方法后，模拟关闭事务
后置返回通知: 模拟删除临时文件，被增强处理的方法: save
```

图 4.4　基于 XML 配置开发 AspectJ 的运行结果

异常通知得到执行，需要在 dynamic.jdk.TestDaoImpl 类的 save 方法中添加异常代码，例如"int n = 100/0;"，然后重新运行测试类，运行结果如图 4.5 所示。

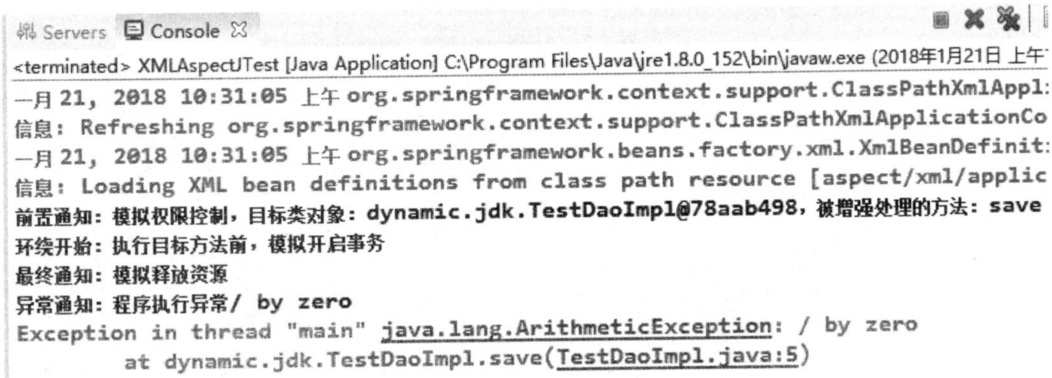

```
Servers  Console
<terminated> XMLAspectJTest [Java Application] C:\Program Files\Java\jre1.8.0_152\bin\javaw.exe (2018年1月21日 上午
一月 21, 2018 10:31:05 上午 org.springframework.context.support.ClassPathXmlAppl
信息: Refreshing org.springframework.context.support.ClassPathXmlApplicationCo
一月 21, 2018 10:31:05 上午 org.springframework.beans.factory.xml.XmlBeanDefinit
信息: Loading XML bean definitions from class path resource [aspect/xml/applic
前置通知: 模拟权限控制，目标类对象: dynamic.jdk.TestDaoImpl@78aab498，被增强处理的方法: save
环绕开始: 执行目标方法前，模拟开启事务
最终通知: 模拟释放资源
异常通知: 程序执行异常/ by zero
Exception in thread "main" java.lang.ArithmeticException: / by zero
        at dynamic.jdk.TestDaoImpl.save(TestDaoImpl.java:5)
```

图 4.5　异常通知执行结果

从图 4.4 和图 4.5 可以看出各类型通知与目标方法的执行过程，具体过程如图 4.6 所示。

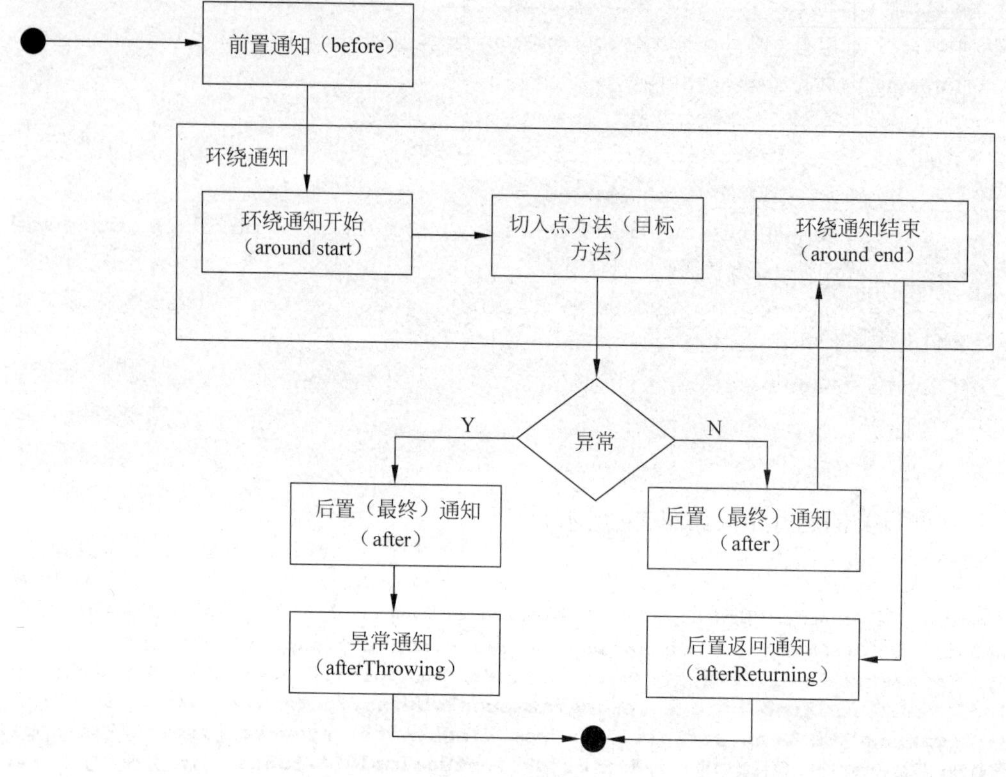

图 4.6　各类型通知的执行过程

4.5　基于注解开发 AspectJ

视频讲解

基于注解开发 AspectJ 要比基于 XML 配置开发 AspectJ 便捷许多，所以在实际开发中推荐使用注解方式。对于注解的相关内容，读者在 3.5.2 节已经接触，本节介绍 AspectJ 注解，如表 4.3 所示。

表 4.3　AspectJ 注解

注 解 名 称	描　述
@Aspect	用于定义一个切面，注解在切面类上
@Pointcut	用于定义切入点表达式。在使用时需要定义一个切入点方法，该方法是一个返回值void且方法体为空的普通方法
@Before	用于定义前置通知。在使用时通常为其指定value属性值，该值可以是已有的切入点，也可以直接定义切入点表达式
@AfterReturning	用于定义后置返回通知。在使用时通常为其指定value属性值，该值可以是已有的切入点，也可以直接定义切入点表达式
@Around	用于定义环绕通知。在使用时通常为其指定value属性值，该值可以是已有的切入点，也可以直接定义切入点表达式

续表

注 解 名 称	描 述
@AfterThrowing	用于定义异常通知。在使用时通常为其指定value属性值，该值可以是已有的切入点，也可以直接定义切入点表达式。另外，还有一个throwing属性用于访问目标方法抛出的异常，该属性值与异常通知方法中同名的形参一致
@After	用于定义后置（最终）通知。在使用时通常为其指定value属性值，该值可以是已有的切入点，也可以直接定义切入点表达式

下面通过一个实例讲解基于注解开发 AspectJ 的过程，具体步骤如下：

❶ 创建切面类，并进行注解

在 ch4 应用的 src 目录下创建 aspectj.annotation 包，在该包中创建切面类 MyAspect。在该类中首先使用@Aspect 注解定义一个切面类，由于该类在 Spring 中是作为组件使用的，所以还需要使用@Component 注解；然后使用@Pointcut 注解切入点表达式，并通过定义方法来表示切入点名称；最后在每个通知方法上添加相应的注解，并将切入点名称作为参数传递给需要执行增强的通知方法。

MyAspect 的代码如下：

```
package aspectj.annotation;
import org.aspectj.lang.JoinPoint;
import org.aspectj.lang.ProceedingJoinPoint;
import org.aspectj.lang.annotation.After;
import org.aspectj.lang.annotation.AfterReturning;
import org.aspectj.lang.annotation.AfterThrowing;
import org.aspectj.lang.annotation.Around;
import org.aspectj.lang.annotation.Aspect;
import org.aspectj.lang.annotation.Before;
import org.aspectj.lang.annotation.Pointcut;
import org.springframework.stereotype.Component;
/**
 * 切面类，在此类中编写各种类型的通知
 */
@Aspect  //对应<aop:aspect ref="myAspect">
@Component  //对应<bean id="myAspect" class="aspectj.xml.MyAspect"/>
public class MyAspect {
    /**
     * 定义切入点
     */
    @Pointcut("execution(* dynamic.jdk.*.*(..))")
    private void myPointCut() {
        //对应<aop:pointcut expression="execution(* dynamic.jdk.*.*(..))"
        id="myPointCut"/>
    }
    /**
```

```
 * 前置通知，使用 Joinpoint 接口作为参数获得目标对象信息
 */
@Before("myPointCut()")  //对应<aop:before method="before" pointcut-
ref="myPointCut"/>
public void before(JoinPoint jp) {
    System.out.print("前置通知：模拟权限控制");
    System.out.println("，目标类对象：" + jp.getTarget()
    + "，被增强处理的方法：" + jp.getSignature().getName());
}
/**
 * 后置返回通知
 */
@AfterReturning("myPointCut()")
public void afterReturning(JoinPoint jp) {
    System.out.print("后置返回通知：" + "模拟删除临时文件");
    System.out.println("，被增强处理的方法：" + jp.getSignature().getName());
}
/**
 * 环绕通知
 * ProceedingJoinPoint 是 JoinPoint 的子接口，代表可以执行的目标方法
 * 返回值的类型必须是 Object
 * 必须一个参数是 ProceedingJoinPoint 类型
 * 必须 throws Throwable
 */
@Around("myPointCut()")
public Object around(ProceedingJoinPoint pjp) throws Throwable{
    //开始
    System.out.println("环绕开始：执行目标方法前，模拟开启事务");
    //执行当前目标方法
    Object obj=pjp.proceed();
    //结束
    System.out.println("环绕结束：执行目标方法后，模拟关闭事务");
    return obj;
}
/**
 * 异常通知
 */
@AfterThrowing(value="myPointCut()",throwing="e")
public void except(Throwable e) {
    System.out.println("异常通知：" + "程序执行异常" + e.getMessage());
}
/**
 * 后置（最终）通知
```

```
    */
    @After("myPointCut()")
    public void after() {
        System.out.println("最终通知：模拟释放资源");
    }
}
```

❷ **注解目标类**

使用注解@Repository 将目标类 dynamic.jdk.TestDaoImpl 注解为目标对象，注解代码如下：

```
@Repository("testDao")
```

❸ **创建配置文件**

在 aspectj.annotation 包中创建配置文件 applicationContext.xml，并在配置文件中指定需要扫描的包，使注解生效，同时需要启动基于注解的 AspectJ 支持。

applicationContext.xml 的代码如下：

```
<?xml version="1.0" encoding="UTF-8"?>
<beans xmlns="http://www.springframework.org/schema/beans"
    xmlns:xsi="http://www.w3.org/2001/XMLSchema-instance"
    xmlns:aop="http://www.springframework.org/schema/aop"
    xmlns:context="http://www.springframework.org/schema/context"
    xsi:schemaLocation="http://www.springframework.org/schema/beans
        http://www.springframework.org/schema/beans/spring-beans.xsd
        http://www.springframework.org/schema/aop
        http://www.springframework.org/schema/aop/spring-aop.xsd
        http://www.springframework.org/schema/context
        http://www.springframework.org/schema/context/spring-context.xsd">
    <!-- 指定需要扫描的包，使注解生效 -->
    <context:component-scan base-package="aspectj.annotation"/>
    <context:component-scan base-package="dynamic.jdk"/>
    <!-- 启动基于注解的 AspectJ 支持 -->
    <aop:aspectj-autoproxy/>
</beans>
```

❹ **创建测试类**

测试类的运行结果与图 4.4 相同，这里不再赘述。

4.6　本章小结

本章主要讲解了 Spring AOP 框架的相关知识，包括 AOP 概念、AOP 术语、动态代理、基于代理类的 AOP 实现以及 AspectJ 框架的 AOP 开发方式等知识。

习题 4

1．什么是 AOP？AOP 有哪些术语？为什么要学习 AOP 编程？

2．在 Java 中有哪些常用的动态代理技术？

3．AspectJ 框架的 AOP 开发方式有哪几种？

第5章 Spring 的事务管理

学习目的与要求

本章主要介绍Spring框架所支持的事务管理，包括编程式事务管理和声明式事务管理。通过本章的学习，读者能够掌握声明式事务管理，了解编程式事务管理。

本章主要内容

- Spring 的数据库编程；
- 编程式事务管理；
- 声明式事务管理。

在数据库操作中事务管理是一个重要的概念，例如银行转账。当从 A 账户向 B 账户转 1000 元后银行系统会从 A 账户上扣除 1000 元，而在 B 账户上增加 1000 元，这是正确处理的结果。

一旦银行系统出错了怎么办，这里假设发生两种情况：

（1）A 账户少了 1000 元，但 B 账户却没有多出 1000 元。

（2）B 账户多了 1000 元钱，但 A 账户却没有被扣钱。

客户和银行都不愿意看到上面两种情况。那么有没有措施保证转账顺利进行？这种措施就是数据库事务管理机制。

Spring 的事务管理简化了传统的数据库事务管理流程，提高了开发效率，但在学习事务管理前需要了解 Spring 的数据库编程。

视频讲解

5.1 Spring 的数据库编程

数据库编程是互联网编程的基础，Spring 框架为开发者提供了 JDBC 模板模式，即 jdbcTemplate，它可以简化许多代码，但在实际应用中 jdbcTemplate 并不常用，在工作中更多的时候是使用 Hibernate 框架和 MyBatis 框架进行数据库编程。

本节仅简要介绍 Spring jdbcTemplate 的使用方法，对于 MyBatis 框架的相关内容将在本书后续章节中详细介绍。对于 Hibernate 框架，本书不再涉及，需要的读者可以查阅 Hibernate 框架的相关知识。

5.1.1 Spring JDBC 的配置

本节 Spring 数据库编程主要使用 Spring JDBC 模块的 core 和 dataSource 包。core 包是 JDBC 的核心功能包，包括常用的 JdbcTemplate 类；dataSource 包是访问数据源的工具类包。如果要使用 Spring JDBC 操作数据库，需要对其进行配置，配置文件的示例代码如下：

```
<!-- 配置数据源 -->
<bean id="dataSource" class="org.springframework.jdbc.datasource.DriverManager-
DataSource">
        <!-- MySQL 数据库驱动 -->
        <property name="driverClassName" value="com.mysql.jdbc.Driver"/>
        <!-- 连接数据库的 URL -->
        <property  name="url"  value="jdbc:mysql://localhost:3306/springtest?
        characterEncoding=utf8"/>
        <!-- 连接数据库的用户名 -->
        <property name="username" value="root"/>
        <!-- 连接数据库的密码 -->
        <property name="password" value="root"/>
  </bean>
  <!-- 配置 JDBC 模板 -->
  <bean id="jdbcTemplate" class="org.springframework.jdbc.core.JdbcTemplate">
        <property name="dataSource" ref="dataSource"/>
  </bean>
```

在上述示例代码中，配置 JDBC 模板时需要将 dataSource 注入到 jdbcTemplate，而在数据访问层（Dao 类）需要使用 jdbcTemplate 时也需要将 jdbcTemplate 注入到对应的 Bean 中。示例代码如下：

```
…
@Repository("testDao")
public class TestDaoImpl implements TestDao{
    @Autowired
    //使用配置文件中的 JDBC 模板
    private JdbcTemplate jdbcTemplate;
    …
}
```

5.1.2 Spring JdbcTemplate 的常用方法

在 5.1.1 节中获取了 JDBC 模板，那么如何使用它？这是本节将要讲述的内容。首先需要了解 JdbcTemplate 类的常用方法，该类的常用方法是 update 和 query。

- **public int update(String sql,Object args[])**：该方法可以对数据表进行增加、修改、删除等操作。使用 args[]设置 SQL 语句中的参数，并返回更新的行数。示例代码如下：

```
String insertSql="insert into user values(null,?,?)";
Object param1[]={"chenheng1", "男"};
jdbcTemplate.update(sql, param1);
```

- **public List<T> query (String sql, RowMapper<T> rowMapper, Object args[])**：该方法可以对数据表进行查询操作。rowMapper 将结果集映射到用户自定义的类中（前提是自定义类中的属性要与数据表的字段对应）。示例代码如下：

```
String selectSql="select*from user";
RowMapper<MyUser> rowMapper=new BeanPropertyRowMapper<MyUser>(MyUser.class);
List<MyUser> list=jdbcTemplate.query(sql, rowMapper, null);
```

下面通过一个实例演示 Spring JDBC 的使用过程，具体步骤如下：

❶ 创建应用并导入 JAR 包

创建一个名为 ch5 的 Web 应用，将 Spring 框架的 5 个基础 JAR 包、MySQL 数据库的驱动 JAR 包、Spring JDBC 的 JAR 包以及 Spring 事务处理的 JAR 包复制到应用的/WEB-INF/lib 目录下。ch5 应用所添加的 JAR 包如图 5.1 所示。

```
lib
  commons-logging-1.2.jar
  mysql-connector-java-5.1.45-bin.jar
  spring-aop-5.0.2.RELEASE.jar
  spring-beans-5.0.2.RELEASE.jar
  spring-context-5.0.2.RELEASE.jar
  spring-core-5.0.2.RELEASE.jar
  spring-expression-5.0.2.RELEASE.jar
  spring-jdbc-5.0.2.RELEASE.jar
  spring-tx-5.0.2.RELEASE.jar
```

图 5.1　ch5 应用所添加的 JAR 包

❷ 创建并编辑配置文件

在 src 目录下创建配置文件 applicationContext.xml，在该文件中配置数据源和 JDBC 模板，具体代码如下：

```
<?xml version="1.0" encoding="UTF-8"?>
<beans xmlns="http://www.springframework.org/schema/beans"
    xmlns:xsi="http://www.w3.org/2001/XMLSchema-instance"
    xmlns:context="http://www.springframework.org/schema/context"
    xsi:schemaLocation="http://www.springframework.org/schema/beans
        http://www.springframework.org/schema/beans/spring-beans.xsd
```

```
        http://www.springframework.org/schema/context
        http://www.springframework.org/schema/context/spring-context.xsd">
    <!-- 指定需要扫描的包（包括子包），使注解生效 -->
    <context:component-scan base-package="com.ch5"/>
    <!-- 配置数据源 -->
    <bean id="dataSource" class="org.springframework.jdbc.datasource.
    DriverManagerDataSource">
         <!-- MySQL 数据库驱动 -->
         <property name="driverClassName" value="com.mysql.jdbc.Driver"/>
         <!-- 连接数据库的 URL -->
         <property name="url" value="jdbc:mysql://localhost:3306/springtest?
         characterEncoding=utf8"/>
         <!-- 连接数据库的用户名 -->
         <property name="username" value="root"/>
         <!-- 连接数据库的密码 -->
         <property name="password" value="root"/>
    </bean>
    <!-- 配置 JDBC 模板 -->
    <bean id="jdbcTemplate" class="org.springframework.jdbc.core.JdbcTemplate">
         <property name="dataSource" ref="dataSource"/>
    </bean>
</beans>
```

❸ 创建实体类

在 src 目录下创建 com.ch5 包，在该包中创建实体类 MyUser。注意，该类的属性与数据表 user 的字段一致。数据表 user 的结构如图 5.2 所示。

名	类型	长度	小数点	允许空值 (	
uid	int	10	0	☐	🔑1
uname	varchar	20	0	☑	
usex	varchar	10	0	☑	

图 5.2　user 表的结构

实体类 MyUser 的代码如下：

```
package com.ch5;
public class MyUser {
    private Integer uid;
    private String uname;
    private String usex;
    //此处省略 setter 和 getter 方法
```

```
    public String toString() {
        return "myUser [uid=" + uid +", uname=" + uname + ", usex=" + usex + "]";
    }
}
```

❹ 创建数据访问层 Dao

在 com.ch5 包中创建 TestDao 接口和 TestDaoImpl 实现类。在实现类 TestDaoImpl 中使用 JDBC 模块 JdbcTemplate 访问数据库，并将该类注解为@Repository("testDao")。使注解生效，需要在配置文件中扫描（见第 2 个步骤）。

TestDao 接口的代码如下：

```
package com.ch5;
import java.util.List;
public interface TestDao {
    public int update(String sql, Object[] param);
    public List<MyUser> query(String sql, Object[] param);
}
```

TestDaoImpl 实现类的代码如下：

```
package com.ch5;
import java.util.List;
import org.springframework.beans.factory.annotation.Autowired;
import org.springframework.jdbc.core.BeanPropertyRowMapper;
import org.springframework.jdbc.core.JdbcTemplate;
import org.springframework.jdbc.core.RowMapper;
import org.springframework.stereotype.Repository;
@Repository("testDao")
public class TestDaoImpl implements TestDao{
     @Autowired
    //使用配置文件中的 JDBC 模板
    private JdbcTemplate jdbcTemplate;
    /**
     * 更新方法，包括添加、修改、删除
     * param 为 sql 中的参数，例如通配符？
     */
    @Override
    public int update(String sql, Object[] param) {
        return jdbcTemplate.update(sql, param);
    }
    /**
     * 查询方法
     * param 为 sql 中的参数，例如通配符？
     */
```

```
    @Override
    public List<MyUser> query(String sql, Object[] param) {
        RowMapper<MyUser> rowMapper=new BeanPropertyRowMapper<MyUser>
        (MyUser.class);
        return jdbcTemplate.query(sql, rowMapper, param);
    }
}
```

❺ 创建测试类

在 com.ch5 包中创建测试类 TestSpringJDBC。在主方法中调用数据访问层 Dao 中的方法，对数据表 user 进行操作。具体代码如下：

```
package com.ch5;
import java.util.List;
import org.springframework.context.ApplicationContext;
import org.springframework.context.support.ClassPathXmlApplicationContext;
public class TestSpringJDBC {
    public static void main(String[] args) {
        ApplicationContext appCon=new ClassPathXmlApplicationContext
        ("applicationContext.xml");
        //从容器中获取增强后的目标对象
        TestDao td=(TestDao)appCon.getBean("testDao");
        String insertSql="insert into user values(null,?,?)";
        //数组 param 的值与 insertSql 语句中的?一一对应
        Object param1[]={"chenheng1", "男"};
        Object param2[]={"chenheng2", "女"};
        Object param3[]={"chenheng3", "男"};
        Object param4[]={"chenheng4", "女"};
        //添加用户
        td.update(insertSql, param1);
        td.update(insertSql, param2);
        td.update(insertSql, param3);
        td.update(insertSql, param4);
        //查询用户
        String selectSql="select*from user";
        List<MyUser> list=td.query(selectSql, null);
        for(MyUser mu:list) {
            System.out.println(mu);
        }
    }
}
```

运行上述测试类，运行结果如图 5.3 所示。

```
Servers  Console
<terminated> TestSpringJDBC [Java Application] C:\Program Files\Java\j
一月 21, 2018 4:25:29 下午 org.springframework.con
信息: Refreshing org.springframework.context.sup
一月 21, 2018 4:25:29 下午 org.springframework.bea
信息: Loading XML bean definitions from class pa
一月 21, 2018 4:25:30 下午 org.springframework.jdb
信息: Loaded JDBC driver: com.mysql.jdbc.Driver
myUser [uid=1, uname=chenheng1, usex=男]
myUser [uid=2, uname=chenheng2, usex=女]
myUser [uid=3, uname=chenheng3, usex=男]
myUser [uid=4, uname=chenheng4, usex=女]
```

图 5.3 Spring 数据库编程的运行结果

5.2 编程式事务管理

视频讲解

在代码中显式调用 beginTransaction、commit、rollback 等与事务处理相关的方法，这就是编程式事务管理。当只有少数事务操作时，编程式事务管理才比较合适。

5.2.1 基于底层 API 的编程式事务管理

基于底层 API 的编程式事务管理就是根据 PlatformTransactionManager、TransactionDefinition 和 TransactionStatus 几个核心接口，通过编程的方式来进行事务处理。下面通过一个实例讲解基于底层 API 的编程式事务管理，具体步骤如下：

❶ 给数据源配置事务管理器

在 5.1.2 节配置文件 applicationContext.xml 的基础上使用 PlatformTransactionManager 接口的实现类 org.springframework.jdbc.datasource.DataSourceTransactionManager 为数据源添加事务管理器，具体代码如下：

```
<!--为数据源添加事务管理器 -->
<bean id="txManager"
      class="org.springframework.jdbc.datasource.DataSourceTransactionManager">
      <property name="dataSource" ref="dataSource" />
</bean>
```

❷ 创建数据访问类

在 com.ch5 包中创建数据访问类 CodeTransaction，并注解为 @Repository("codeTransaction")。在该类中使用编程的方式进行数据库事务管理。

CodeTransaction 类的代码如下：

```
package com.ch5;
import org.springframework.beans.factory.annotation.Autowired;
```

```
import org.springframework.jdbc.core.JdbcTemplate;
import org.springframework.jdbc.datasource.DataSourceTransactionManager;
import org.springframework.stereotype.Repository;
import org.springframework.transaction.TransactionDefinition;
import org.springframework.transaction.TransactionStatus;
import org.springframework.transaction.support.DefaultTransactionDefinition;
@Repository("codeTransaction")
public class CodeTransaction {
    @Autowired
    //使用配置文件中的 JDBC 模板
    private JdbcTemplate jdbcTemplate;
    //DataSourceTransactionManager 是 PlatformTransactionManager 接口的实现类
    @Autowired
    private DataSourceTransactionManager txManager;
    public String test() {
        //默认事务定义，例如隔离级别、传播行为等
        TransactionDefinition tf=new DefaultTransactionDefinition();
        //开启事务 ts
        TransactionStatus ts=txManager.getTransaction(tf);
        String message="执行成功，没有事务回滚！";
        try {
            //删除表中数据
            String sql=" delete from user  ";
            //添加数据
            String sql1=" insert into user values(?,?,?) ";
            Object param[]={ 1, "陈恒", "男" };
            //先删除数据
            jdbcTemplate.update(sql);
            //添加一条数据
            jdbcTemplate.update(sql1, param);
            //添加相同的一条数据，使主键重复
            jdbcTemplate.update(sql1, param);
            //提交事务
            txManager.commit(ts);
        } catch (Exception e) {
            //出现异常，事务回滚
            txManager.rollback(ts);
            message="主键重复，事务回滚！";
            e.printStackTrace();
        }
        return message;
    }
}
```

❸ 创建测试类

在 com.ch5 包中创建测试类 TestCodeTransaction，具体代码如下：

```
package com.ch5;
import org.springframework.context.ApplicationContext;
import org.springframework.context.support.ClassPathXmlApplicationContext;
public class TestCodeTransaction {
    public static void main(String[] args) {
        ApplicationContext appCon=new ClassPathXmlApplicationContext
        ("applicationContext.xml");
        CodeTransaction ct=(CodeTransaction)appCon.getBean("codeTransaction");
        String result=ct.test();
        System.out.println(result);
    }
}
```

上述测试类的运行结果如图 5.4 所示。

```
Servers  Console
<terminated> TestCodeTransaction [Java Application] C:\Program Files\Java\jre1.8.0_152\bin\javaw
        at com.mysql.jdbc.SQLError.createSQLException(SQLError.j
        at com.mysql.jdbc.MysqlIO.checkErrorPacket(MysqlIO.java:
        at com.mysql.jdbc.MysqlIO.checkErrorPacket(MysqlIO.java:
        at com.mysql.jdbc.MysqlIO.sendCommand(MysqlIO.java:2527)
        at com.mysql.jdbc.MysqlIO.sqlQueryDirect(MysqlIO.java:26
        at com.mysql.jdbc.ConnectionImpl.execSQL(ConnectionImpl.
        at com.mysql.jdbc.PreparedStatement.executeInternal(Prep
        at com.mysql.jdbc.PreparedStatement.executeUpdateInterna
        at com.mysql.jdbc.PreparedStatement.executeUpdateInterna
        at com.mysql.jdbc.PreparedStatement.executeLargeUpdate(P
        at com.mysql.jdbc.PreparedStatement.executeUpdate(Prepar
        at org.springframework.jdbc.core.JdbcTemplate.lambda$upd
        at org.springframework.jdbc.core.JdbcTemplate.execute(Jd
        ... 5 more
主键重复，事务回滚！
```

图 5.4　基于底层 API 的编程式事务管理测试结果

从图 5.4 所示的结果可以看出取消了主键重复前执行的删除和插入操作。

5.2.2 基于 TransactionTemplate 的编程式事务管理

事务处理的代码散落在业务逻辑代码中，破坏了原有代码的条理性，并且每一个业务方法都包含了类似的启动事务、提交以及回滚事务的样板代码。

TransactionTemplate 的 execute 方法有一个 TransactionCallback 接口类型的参数，该接口中定义了一个 doInTransaction 方法，通常以匿名内部类的方式实现 TransactionCallback 接口，并在其 doInTransaction 方法中书写业务逻辑代码。在这里可以使用默认的事务提交和回滚规则，在业务代码中不需要显式调用任何事务处理的 API。doInTransaction 方法有一个 TransactionStatus 类型的参数，可以在方法的任何位置调用该参数的 setRollbackOnly 方法将事务标识为回滚，以执行事务回滚。

根据默认规则，如果在执行回调方法的过程中抛出了未检查异常，或者显式调用了

setRollbackOnly 方法，则回滚事务；如果事务执行完成或者抛出了 checked 类型的异常，则提交事务。

基于 TransactionTemplate 的编程式事务管理的步骤如下：

❶ 为事务管理器添加事务模板

在 5.2.1 节配置文件 applicationContext.xml 的基础上使用 org.springframework.transaction.support.TransactionTemplate 类为事务管理器添加事务模板，具体代码如下：

```
<!-- 为事务管理器 txManager 创建 transactionTemplate -->
<bean id="transactionTemplate" class="org.springframework.transaction.support.
TransactionTemplate">
    <property name="transactionManager" ref="txManager"/>
</bean>
```

❷ 创建数据访问类

在 com.ch5 包中创建数据访问类 TransactionTemplateDao，并注解为@Repository("transactionTemplateDao")。在该类中使用编程的方式进行数据库事务管理。

数据访问类 TransactionTemplateDao 的代码如下：

```
package com.ch5;
import org.springframework.beans.factory.annotation.Autowired;
import org.springframework.jdbc.core.JdbcTemplate;
import org.springframework.stereotype.Repository;
import org.springframework.transaction.TransactionStatus;
import org.springframework.transaction.support.TransactionCallback;
import org.springframework.transaction.support.TransactionTemplate;
@Repository("transactionTemplateDao")
public class TransactionTemplateDao {
    @Autowired
    //使用配置文件中的 JDBC 模板
    private JdbcTemplate jdbcTemplate;
    @Autowired
    private TransactionTemplate transactionTemplate;
    String message="";
    public String test() {
        //以匿名内部类的方式实现 TransactionCallback 接口，使用默认的事务提交和回滚
          规则，在业务代码中不需要显式调用任何事务处理的 API
        transactionTemplate.execute(new TransactionCallback<Object>(){
            @Override
            public Object doInTransaction(TransactionStatus arg0) {
                //删除表中数据
                String sql=" delete from user  ";
                //添加数据
                String sql1=" insert into user values(?,?,?) ";
                Object param[]={
                        1,
```

```
                    "陈恒",
                    "男"
            };
            try{
                //先删除数据
                jdbcTemplate.update(sql);
                //添加一条数据
                jdbcTemplate.update(sql1, param);
                //添加相同的一条数据，使主键重复
                jdbcTemplate.update(sql1, param);
                message = "执行成功，没有事务回滚！";
            }catch(Exception e){
                message = "主键重复，事务回滚！";
                arg0.setRollbackOnly();
                e.printStackTrace();
            }
            return message;
        }
    });
    return message;
  }
}
```

❸ 创建测试类

在 com.ch5 包中创建测试类 TransactionTemplateTest，该类的代码及运行结果与 5.2.1 节中的测试类一样，这里不再赘述。

5.3　声明式事务管理

视频讲解

Spring 的声明式事务管理是通过 AOP 技术实现的事务管理，其本质是对方法前后进行拦截，然后在目标方法开始之前创建或者加入一个事务，在执行完目标方法之后根据执行情况提交或者回滚事务。

声明式事务管理最大的优点是不需要通过编程的方式管理事务，因而不需要在业务逻辑代码中掺杂事务处理的代码，只需相关的事务规则声明便可以将事务规则应用到业务逻辑中。通常情况下，在开发中使用声明式事务处理不仅因为其简单，更主要的是因为这样使得纯业务代码不被污染，极大地方便了后期的代码维护。

与编程式事务管理相比，声明式事务管理唯一不足的地方是最细粒度只能作用到方法级别，无法做到像编程式事务管理那样可以作用到代码块级别。但即便有这样的需求，也可以通过变通的方法进行解决，例如可以将需要进行事务处理的代码块独立为方法等。

Spring 的声明式事务管理可以通过两种方式来实现，一是基于 XML 的方式，二是基于@Transactional 注解的方式。

5.3.1 基于 XML 方式的声明式事务管理

基于 XML 方式的声明式事务管理是通过在配置文件中配置事务规则的相关声明来实现的。Spring 框架提供了 tx 命名空间来配置事务，提供了<tx:advice>元素来配置事务的通知。在配置<tx:advice>元素时一般需要指定 id 和 transaction-manager 属性，其中 id 属性是配置文件中的唯一标识，transaction-manager 属性指定事务管理器。另外还需要<tx:attributes>子元素，该子元素可配置多个<tx:method>子元素指定执行事务的细节。

在<tx:advice>元素配置了事务的增强处理后就可以通过编写 AOP 配置让 Spring 自动对目标对象生成代理。下面通过一个实例演示如何通过 XML 方式来实现 Spring 的声明式事务管理。为体现事务管理的流程，本实例创建了 Dao、Service 和 Controller 3 层，具体实现步骤如下：

❶ 导入相关的 JAR 包

在 ch5 应用的基础上导入 AOP 所需要的 JAR，导入后的 lib 目录如图 5.5 所示。

lib
- aopalliance-1.0.jar
- aspectjweaver-1.8.13.jar
- commons-logging-1.2.jar
- mysql-connector-java-5.1.45-bin.jar
- spring-aop-5.0.2.RELEASE.jar
- spring-aspects-5.0.2.RELEASE.jar
- spring-beans-5.0.2.RELEASE.jar
- spring-context-5.0.2.RELEASE.jar
- spring-core-5.0.2.RELEASE.jar
- spring-expression-5.0.2.RELEASE.jar
- spring-jdbc-5.0.2.RELEASE.jar
- spring-tx-5.0.2.RELEASE.jar

图 5.5 ch5 应用所需要的 JAR 包

❷ 创建 Dao 层

在 ch5 的 src 目录下创建 com.statement.dao 包，并在该包中创建 TestDao 接口和 TestDaoImpl 实现类。数据访问层有两个数据操作方法，即 save 和 delete 方法。

TestDao 接口的代码如下：

```
package com.statement.dao;
public interface TestDao {
    public int save(String sql, Object param[]);
    public int delete(String sql, Object param[]);
}
```

TestDaoImpl 实现类的代码如下：

```
package com.statement.dao;
import org.springframework.beans.factory.annotation.Autowired;
import org.springframework.jdbc.core.JdbcTemplate;
import org.springframework.stereotype.Repository;
@Repository("testDao")
public class TestDaoImpl implements TestDao{
    @Autowired
    private JdbcTemplate jdbcTemplate;
    @Override
    public int save(String sql, Object[] param) {
        return jdbcTemplate.update(sql,param);
    }
    @Override
    public int delete(String sql, Object[] param) {
        return jdbcTemplate.update(sql,param);
    }
}
```

❸ 创建 Service 层

在 ch5 的 src 目录下创建 com.statement.service 包，并在该包中创建 TestService 接口和 TestServiceImpl 实现类。在 Service 层依赖注入数据访问层。

TestService 接口的代码如下：

```
package com.statement.service;
public interface TestService {
    public void test();
}
```

TestServiceImpl 实现类的代码如下：

```
package com.statement.service;
import org.springframework.beans.factory.annotation.Autowired;
import org.springframework.stereotype.Service;
import com.statement.dao.TestDao;
@Service("testService")
public class TestServiceImpl implements TestService{
    @Autowired
    private TestDao testDao;
    @Override
    public void test() {
        String deleteSql ="delete from user";
        String saveSql = "insert into user values(?,?,?)";
        Object param[] = {1,"chenheng","男"};
        testDao.delete(deleteSql, null);
        testDao.save(saveSql, param);
        //插入两条主键重复的数据
        testDao.save(saveSql, param);
    }
}
```

❹ 创建 Controller 层

在 ch5 的 src 目录下创建 com.statement.controller 包，并在该包中创建 StatementController 控制器类。在控制层依赖注入 Service 层。

StatementController 类的代码如下：

```
package com.statement.controller;
import org.springframework.beans.factory.annotation.Autowired;
import org.springframework.stereotype.Controller;
import com.statement.service.TestService;
@Controller
public class StatementController {
    @Autowired
    private TestService testService;
    public void test() {
        testService.test();
    }
}
```

❺ 创建配置文件

在 ch5 的 src 目录下创建 com.statement.xml 包，并在该包中创建配置文件 XMLstatementapplicationContext.xml。在配置文件中使用<tx:advice>编写通知声明事务，使

用<aop:config>编写 AOP 让 Spring 自动对目标对象生成代理。

XMLstatementapplicationContext.xml 文件的代码如下：

```
<?xml version="1.0" encoding="UTF-8"?>
<beans xmlns="http://www.springframework.org/schema/beans"
   xmlns:xsi="http://www.w3.org/2001/XMLSchema-instance"
   xmlns:aop="http://www.springframework.org/schema/aop"
   xmlns:tx="http://www.springframework.org/schema/tx"
   xmlns:context="http://www.springframework.org/schema/context"
   xsi:schemaLocation="http://www.springframework.org/schema/beans
      http://www.springframework.org/schema/beans/spring-beans.xsd
      http://www.springframework.org/schema/context
      http://www.springframework.org/schema/context/spring-context.xsd
      http://www.springframework.org/schema/aop
      http://www.springframework.org/schema/aop/spring-aop.xsd
      http://www.springframework.org/schema/tx
      http://www.springframework.org/schema/tx/spring-tx.xsd">
  <!-- 指定需要扫描的包（包括子包），使注解生效 -->
  <context:component-scan base-package="com.statement"/>
  <!-- 配置数据源 -->
  <bean id="dataSource" class="org.springframework.jdbc.datasource.
  DriverManagerDataSource">
       <!-- MySQL 数据库驱动 -->
       <property name="driverClassName" value="com.mysql.jdbc.Driver"/>
       <!-- 连接数据库的 URL -->
       <property name="url" value="jdbc:mysql://localhost:3306/springtest?
       characterEncoding=utf8"/>
       <!-- 连接数据库的用户名 -->
       <property name="username" value="root"/>
       <!-- 连接数据库的密码 -->
       <property name="password" value="root"/>
  </bean>
  <!-- 配置 JDBC 模板 -->
  <bean id="jdbcTemplate" class="org.springframework.jdbc.core.JdbcTemplate">
       <property name="dataSource" ref="dataSource"/>
  </bean>
  <!-- 为数据源添加事务管理器 -->
  <bean id="txManager"
        class="org.springframework.jdbc.datasource.DataSourceTransaction
        Manager">
        <property name="dataSource" ref="dataSource" />
  </bean>
  <!-- 编写通知声明事务 -->
  <tx:advice id="myAdvice" transaction-manager="txManager">
       <tx:attributes>
           <!-- *表示任意方法 -->
           <tx:method name="*"/>
       </tx:attributes>
  </tx:advice>
  <!-- 编写 AOP，让 Spring 自动对目标对象生成代理，需要使用 AspectJ 的表达式 -->
  <aop:config>
       <!-- 定义切入点 -->
       <aop:pointcut expression="execution(* com.statement.service.*.*())"
       id= "txPointCut"/>
```

```
        <!-- 切面：将切入点与通知关联 -->
        <aop:advisor advice-ref="myAdvice" pointcut-ref="txPointCut"/>
    </aop:config>
</beans>
```

❻ 创建测试类

在 ch5 的 src 目录下创建 com.statement.test 包，并在该包中创建测试类 XMLTest，在测试类中通过访问 Controller 测试基于 XML 方式的声明式事务管理。

测试类 XMLTest 的代码如下：

```
package com.statement.test;
import org.springframework.context.ApplicationContext;
import org.springframework.context.support.ClassPathXmlApplicationContext;
import com.statement.controller.StatementController;
public class XMLTest {
    public static void main(String[] args) {
        ApplicationContext appCon=
    new ClassPathXmlApplicationContext("/com/statement/xml/
    XMLstatementapplicationContext.xml");
        StatementController ct=(StatementController)appCon.getBean
        ("statementController");
        ct.test();
    }
}
```

测试类 XMLTest 的运行结果与图 5.4 一样，这里不再赘述。

5.3.2　基于@Transactional 注解的声明式事务管理

@Transactional 注解可以作用于接口、接口方法、类以及类的方法上。当作用于类上时，该类的所有 public 方法都将具有该类型的事务属性，同时也可以在方法级别使用该注解来覆盖类级别的定义。虽然@Transactional 注解可以作用于接口、接口方法、类以及类的方法上，但是 Spring 小组建议不要在接口或者接口方法上使用该注解，因为它只有在使用基于接口的代理时才会生效。

下面通过实例演示使用@Transactional 注解进行事务管理的过程，该实例的 Dao、Service 和 Controller 层与 5.3.1 节中的相同，具体步骤如下：

❶ 创建配置文件

在 com.statement.xml 包中创建配置文件 annotationstatementapplicationContext.xml，在配置文件中使用<tx:annotation-driven>元素为事务管理器注册注解驱动器。

annotationstatementapplicationContext.xml 文件的代码如下：

```
<?xml version="1.0" encoding="UTF-8"?>
<beans xmlns="http://www.springframework.org/schema/beans"
   xmlns:xsi="http://www.w3.org/2001/XMLSchema-instance"
   xmlns:tx="http://www.springframework.org/schema/tx"
   xmlns:context="http://www.springframework.org/schema/context"
   xsi:schemaLocation="http://www.springframework.org/schema/beans
      http://www.springframework.org/schema/beans/spring-beans.xsd
      http://www.springframework.org/schema/context
      http://www.springframework.org/schema/context/spring-context.xsd
```

```
        http://www.springframework.org/schema/tx
        http://www.springframework.org/schema/tx/spring-tx.xsd">
    <!-- 指定需要扫描的包（包括子包），使注解生效 -->
    <context:component-scan base-package="com.statement"/>
    <!-- 配置数据源 -->
    <bean id="dataSource" class="org.springframework.jdbc.datasource.
    DriverManagerDataSource">
         <!-- MySQL 数据库驱动 -->
         <property name="driverClassName" value="com.mysql.jdbc.Driver"/>
         <!-- 连接数据库的 URL -->
         <property  name="url"  value="jdbc:mysql://localhost:3306/springtest?
         characterEncoding=utf8"/>
         <!-- 连接数据库的用户名 -->
          <property name="username" value="root"/>
         <!-- 连接数据库的密码 -->
         <property name="password" value="root"/>
    </bean>
    <!-- 配置 JDBC 模板 -->
    <bean id="jdbcTemplate" class="org.springframework.jdbc.core.JdbcTemplate">
         <property name="dataSource" ref="dataSource"/>
    </bean>
    <!-- 为数据源添加事务管理器 -->
    <bean id="txManager"
         class="org.springframework.jdbc.datasource.DataSourceTransaction
         Manager">
         <property name="dataSource" ref="dataSource" />
    </bean>
    <!-- 为事务管理器注册注解驱动器 -->
    <tx:annotation-driven transaction-manager="txManager" />
</beans>
```

❷ 为 Service 层添加@Transactional 注解

在 Spring MVC（后续章节讲解）中通常通过 Service 层进行事务管理，因此需要为 Service 层添加@Transactional 注解。

添加@Transactional 注解后的 TestServiceImpl 类的代码如下：

```
package com.statement.service;
import org.springframework.beans.factory.annotation.Autowired;
import org.springframework.stereotype.Service;
import org.springframework.transaction.annotation.Transactional;
import com.statement.dao.TestDao;
@Service("testService")
@Transactional
//加上注解@Transactional,就可以指定这个类需要受 Spring 的事务管理
//注意@Transactional 只能针对 public 属性范围内的方法添加
public class TestServiceImpl implements TestService{
    @Autowired
    private TestDao testDao;
    @Override
    public void test() {
        String deleteSql="delete from user";
        String saveSql="insert into user values(?,?,?)";
        Object param[]={1,"chenheng","男"};
        testDao.delete(deleteSql, null);
```

```
        testDao.save(saveSql, param);
        //插入两条主键重复的数据
        testDao.save(saveSql, param);
    }
}
```

测试类的运行结果与 5.3.1 节一样，这里不再赘述。

5.3.3 如何在事务处理中捕获异常

声明式事务处理的流程是：

（1）Spring 根据配置完成事务定义，设置事务属性。

（2）执行开发者的代码逻辑。

（3）如果开发者的代码产生异常（如主键重复）并且满足事务回滚的配置条件，则事务回滚；否则，事务提交。

（4）事务资源释放。

现在的问题是，如果开发者在代码逻辑中加入了 try...catch...语句，Spring 还能不能在声明式事务处理中正常得到事务回滚的异常信息？答案是不能。例如，我们将 5.3.1 节和 5.3.2 节中 TestServiceImpl 实现类的 test 方法的代码修改如下：

```
@Override
public void test() {
    String deleteSql ="delete from user";
    String saveSql = "insert into user values(?,?,?)";
    Object param[] = {1,"chenheng","男"};
    try {
        testDao.delete(deleteSql, null);
        testDao.save(saveSql, param);
        //插入两条主键重复的数据
        testDao.save(saveSql, param);
    } catch (Exception e) {
        System.out.println("主键重复，事务回滚。");
    }
}
```

这时，我们再运行测试类，发现主键重复但事务并没有回滚。这是因为默认情况下，Spring 只在发生未被捕获的 RuntimeException 时才回滚事务。现在，如何在事务处理中捕获异常呢？下面从声明式事务管理的两种实现方式来说明。

❶ 在基于 XML 方式的声明式事务管理中捕获异常

在基于 XML 方式的声明式事务管理中捕获异常，需要补充两个步骤。

（1）修改声明事务的配置。

针对 5.3.1 节，我们需要将 XMLstatementapplicationContext.xml 文件中的代码“<tx:method name="*"/>”修改为：

```
<tx:method name="*" rollback-for="java.lang.Exception"/>
<!-- rollback-for 属性指定回滚生效的异常类，多个异常类之间用逗号分隔；no-rollback-for
属性指定回滚失效的异常类-->
```

（2）在 catch 语句中添加“throw new RuntimeException();”语句，代码如下：

```
@Override
public void test() {
    String deleteSql ="delete from user";
    String saveSql = "insert into user values(?,?,?)";
    Object param[] = {1,"chenheng","男"};
    try {
        testDao.delete(deleteSql, null);
        testDao.save(saveSql, param);
        //插入两条主键重复的数据
        testDao.save(saveSql, param);
    } catch (Exception e) {
        System.out.println("主键重复，事务回滚。");
        throw new RuntimeException();
    }
}
```

❷ 在基于@Transaction 注解的声明式事务管理中捕获异常

在基于@Transaction 注解的声明式事务管理中，也同样需要补充两个步骤。

（1）修改@Transactional 注解。

针对 5.3.2 节，我们需要将 TestServiceImpl 类中的@Transactional 注解修改为：

```
@Transactional(rollbackFor= {Exception.class})
//rollbackFor 指定回滚生效的异常类，多个异常类之间用逗号分隔
//noRollbackFor 指定回滚失效的异常类
```

（2）也需要在 catch 语句中添加“throw new RuntimeException();”语句。

注意：在实际工程应用中，经常在 catch 语句中添加“TransactionAspectSupport.currentTransactionStatus().setRollbackOnly();”语句。也就是说，不需要在 XML 配置文件 tx:method 元素中添加 rollback-for 属性或在@Transaction 注解中添加 rollbackFor 属性。

5.4 本章小结

基于 TransactionDefinition、PlatformTransactionManager、TransactionStatus 的编程式事务管理是 Spring 提供的最原始的方式，通常在实际工程中不推荐使用，但了解这种方式对理解 Spring 事务处理的本质有很大帮助。

基于 TransactionTemplate 的编程式事务管理是对上一种方式的封装，使得编码更简单、清晰。基于 XML 和@Transactional 的方式将事务管理简化到了极致，极大地提高了编程开发效率。

习题 5

1．什么是编程式事务管理？在 Spring 中有哪几种编程式事务管理？
2．简述声明式事务管理的处理方式。

第 2 部分

MyBatis

第 6 章

MyBatis 开发入门

学习目的与要求

本章讲解 MyBatis 环境的构建、MyBatis 的工作原理以及与 Spring 框架的整合开发。通过本章的学习，读者能够了解 MyBatis 的工作原理，掌握 MyBatis 环境的构建以及与 Spring 框架的整合开发。

本章主要内容

- MyBatis 环境的构建;
- MyBatis 的工作原理;
- 与 Spring 框架的整合开发;
- MyBatis 的入门程序。

MyBatis 是主流的 Java 持久层框架之一，它与 Hibernate 一样，也是一种 ORM（Object/Relational Mapping，即对象关系映射）框架。其因性能优异，且具有高度的灵活性、可优化性、易于维护以及简单易学等特点，受到了广大互联网企业和编程爱好者的青睐。

6.1 MyBatis 简介

MyBatis 本是 Apache 的一个开源项目——iBatis，2010 年这个项目由 Apache Software Foundation 迁移到了 Google Code，并且改名为 MyBatis。

MyBatis 是一个基于 Java 的持久层框架。MyBatis 提供的持久层框架包括 SQL Maps 和 Data Access Objects（DAO），它消除了几乎所有的 JDBC 代码和参数的手工设置以及结果集的检索。MyBatis 使用简单的 XML 或注解用于配置和原始映射，将接口和 Java 的 POJOs（Plain Old Java Objects，普通的 Java 对象）映射成数据库中的记录。

目前，Java 的持久层框架产品有许多，常见的有 Hibernate 和 MyBatis。MyBatis 是一个半自动映射的框架，因为 MyBatis 需要手动匹配 POJO、SQL 和映射关系；而 Hibernate 是一个全表映射的框架，只需提供 POJO 和映射关系即可。MyBatis 是一个小巧、方便、高效、简单、直接、半自动化的持久层框架；Hibernate 是一个强大、方便、高效、复杂、间接、全自动化的持久层框架。两个持久层框架各有优缺点，开发者应根据实际应用选择它们。

视频讲解

6.2 MyBatis 环境的构建

在编写本书时 MyBatis 的最新版本是 3.4.5，因此编者选择这个版本作为本书的实践环境，也希望读者下载该版本，以便于学习。

MyBatis 的 3.4.5 版本可以通过“https://github.com/mybatis/mybatis-3/releases”网址下载。在下载时只需选择 mybatis-3.4.5.zip 即可，解压后得到如图 6.1 所示的目录。

图 6.1 MyBatis 的目录

图 6.1 中的 mybatis-3.4.5.jar 是 MyBatis 的核心包，mybatis-3.4.5.pdf 是 MyBatis 的使用手册，lib 文件夹下的 JAR 是 MyBatis 的依赖包。

在使用 MyBatis 框架时需要将它的核心包和依赖包引入到应用程序中。如果是 Web 应用，只需将核心包和依赖包复制到/WEB-INF/lib 目录中。

6.3 MyBatis 的工作原理

在学习 MyBatis 程序之前，读者需要了解一下 MyBatis 的工作原理，以便于理解程序。MyBatis 的工作原理如图 6.2 所示。

下面对图 6.2 中的每步流程进行说明。

（1）读取 MyBatis 配置文件：mybatis-config.xml 为 MyBatis 的全局配置文件，配置了 MyBatis 的运行环境等信息，例如数据库连接信息。

（2）加载映射文件。映射文件即 SQL 映射文件，该文件中配置了操作数据库的 SQL 语句，需要在 MyBatis 配置文件 mybatis-config.xml 中加载。mybatis-config.xml 文件可以加载多个映射文件，每个文件对应数据库中的一张表。

（3）构造会话工厂：通过 MyBatis 的环境等配置信息构建会话工厂 SqlSessionFactory。

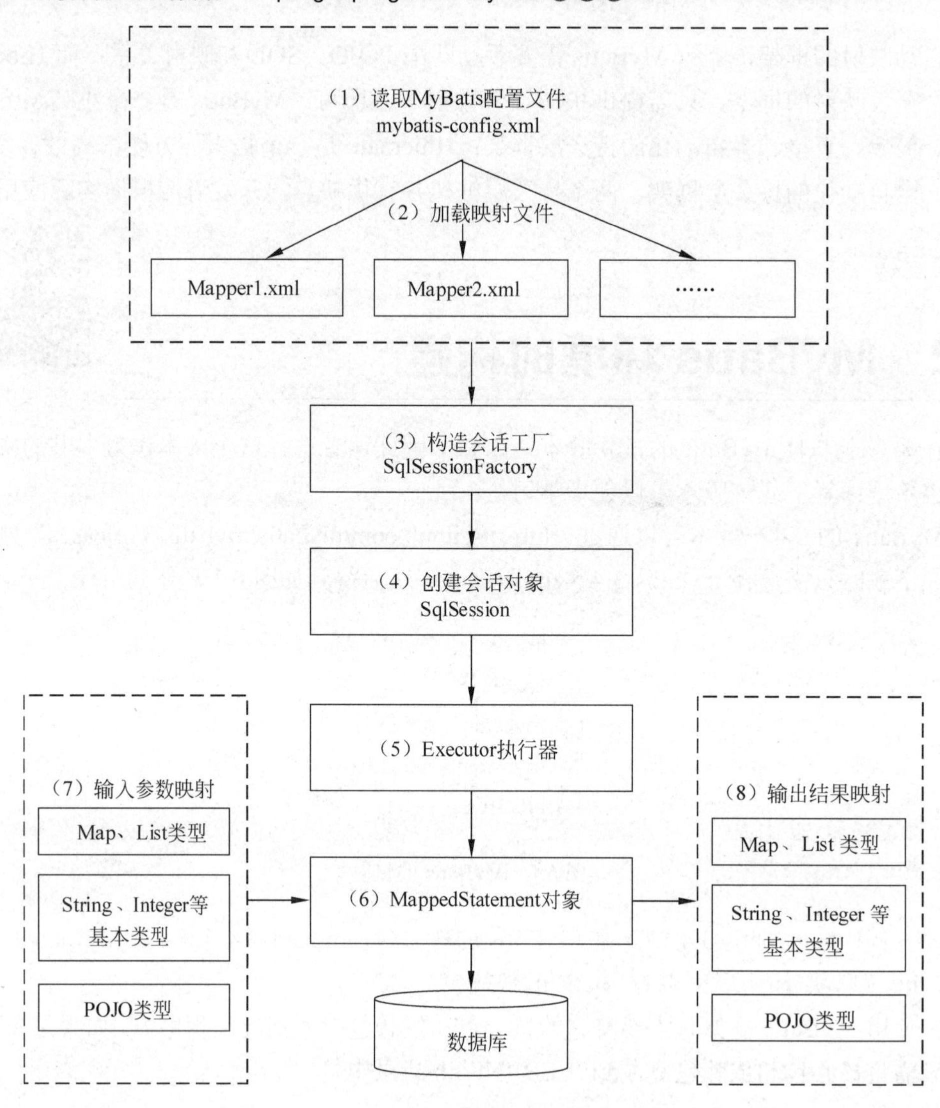

图 6.2　MyBatis 框架的执行流程图

（4）创建会话对象：由会话工厂创建 SqlSession 对象，该对象中包含了执行 SQL 语句的所有方法。

（5）Executor 执行器：MyBatis 底层定义了一个 Executor 接口来操作数据库，它将根据 SqlSession 传递的参数动态地生成需要执行的 SQL 语句，同时负责查询缓存的维护。

（6）MappedStatement 对象：在 Executor 接口的执行方法中有一个 MappedStatement 类型的参数，该参数是对映射信息的封装，用于存储要映射的 SQL 语句的 id、参数等信息。

（7）输入参数映射：输入参数类型可以是 Map、List 等集合类型，也可以是基本数据类型和 POJO 类型。输入参数映射过程类似于 JDBC 对 preparedStatement 对象设置参数的过程。

（8）输出结果映射：输出结果类型可以是 Map、List 等集合类型，也可以是基本数据类型和 POJO 类型。输出结果映射过程类似于 JDBC 对结果集的解析过程。

通过上面的讲解，读者对 MyBatis 框架应该有了一个初步的了解，在后续的学习中将慢慢加深理解。

6.4　使用 Eclipse 开发 MyBatis 入门程序

本节使用第 5 章中 MySQL 数据库 springtest 的 user 数据表进行讲解，下面通过一个实例讲解如何使用 Eclipse 开发 MyBatis 入门程序。

视频讲解

❶ 创建 Web 应用，并添加相关 JAR 包

在 Eclipse 中创建一个名为 ch6 的 Web 应用，将 MyBatis 的核心 JAR 包、依赖 JAR 包以及 MySQL 数据库的驱动 JAR 包一起复制到/WEB-INF/lib 目录下。添加后的 lib 目录如图 6.3 所示。

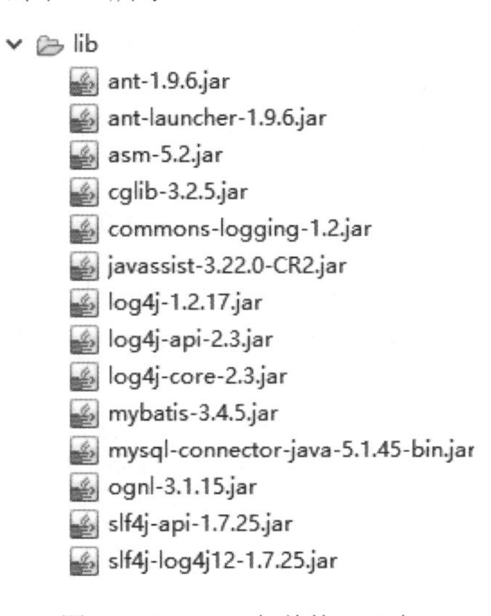

图 6.3　MyBatis 相关的 JAR 包

❷ 创建日志文件

MyBatis 默认使用 log4j 输出日志信息，如果开发者需要查看控制台输出的 SQL 语句，那么需要在 classpath 路径下配置其日志文件。在 ch6 应用的 src 目录下创建 log4j.properties 文件，其内容如下：

```
# Global logging configuration
log4j.rootLogger=ERROR, stdout
# MyBatis logging configuration...
log4j.logger.com.mybatis=DEBUG
# Console output...
log4j.appender.stdout=org.apache.log4j.ConsoleAppender
log4j.appender.stdout.layout=org.apache.log4j.PatternLayout
log4j.appender.stdout.layout.ConversionPattern=%5p [%t] - %m%n
```

在日志文件中配置了全局的日志配置、MyBatis 的日志配置和控制台输出，其中

MyBatis 的日志配置用于将 com.mybatis 包下所有类的日志记录级别设置为 DEBUG。该配置文件内容不需要开发者全部手写，可以从 MyBatis 使用手册中的 Logging 小节复制，然后进行简单修改。

❸ 创建持久化类

在 src 目录下创建一个名为 com.mybatis.po 的包，在该包中创建持久化类 MyUser，注意在类中声明的属性与数据表 user（创建表的代码参见源代码中的 ch7.sql）的字段一致。

MyUser 的代码如下：

```
package com.mybatis.po;
/**
 *springtest 数据库中 user 表的持久化类
 */
public class MyUser {
    private Integer uid;  //主键
    private String uname;
    private String usex;
    //此处省略 setter 和 getter 方法
    @Override
    public String toString() {  //为了方便查看结果，重写了 toString 方法
        return "User [uid=" + uid +",uname=" + uname + ",usex=" + usex +"]";
    }
}
```

❹ 创建映射文件

在 src 目录下创建一个名为 com.mybatis.mapper 的包，在该包中创建映射文件 UserMapper.xml。

UserMapper.xml 文件的内容如下：

```
<?xml version="1.0" encoding="UTF-8" ?>
<!DOCTYPE mapper
PUBLIC "-//mybatis.org//DTD Mapper 3.0//EN"
"http://mybatis.org/dtd/mybatis-3-mapper.dtd">
<mapper namespace="com.mybatis.mapper.UserMapper">
    <!-- 根据 uid 查询一个用户信息 -->
    <select id="selectUserById" parameterType="Integer"
        resultType="com.mybatis.po.MyUser">
        select * from user where uid = #{uid}
    </select>
    <!-- 查询所有用户信息 -->
    <select id="selectAllUser"  resultType="com.mybatis.po.MyUser">
        select * from user
    </select>
    <!-- 添加一个用户，#{uname}为 com.mybatis.po.MyUser 的属性值 -->
    <insert id="addUser" parameterType="com.mybatis.po.MyUser">
        insert into user (uname,usex) values(#{uname},#{usex})
```

```
    </insert>
    <!-- 修改一个用户 -->
    <update id="updateUser" parameterType="com.mybatis.po.MyUser">
      update user set uname = #{uname},usex = #{usex} where uid = #{uid}
    </update>
    <!-- 删除一个用户 -->
    <delete id="deleteUser" parameterType="Integer">
      delete from user where uid = #{uid}
    </delete>
</mapper>
```

在上述映射文件中，<mapper>元素是配置文件的根元素，它包含了一个 namespace 属性，该属性值通常设置为"包名+SQL 映射文件名"，指定了唯一的命名空间；子元素<select>、<insert>、<update>以及<delete>中的信息是用于执行查询、添加、修改以及删除操作的配置。在定义的 SQL 语句中，"#{}"表示一个占位符，相当于"?"，而"#{uid}"表示该占位符待接收参数的名称为 uid。

❺ 创建 MyBatis 的配置文件

在 src 目录下创建 MyBatis 的核心配置文件 mybatis-config.xml，在该文件中配置了数据库环境和映射文件的位置，具体内容如下：

```
<?xml version="1.0" encoding="UTF-8" ?>
<!DOCTYPE configuration
PUBLIC "-//mybatis.org//DTD Config 3.0//EN"
"http://mybatis.org/dtd/mybatis-3-config.dtd">
<configuration>
   <!-- 配置环境 -->
   <environments default="development">
      <environment id="development">
         <!-- 使用 JDBC 的事务管理 -->
         <transactionManager type="JDBC"/>
         <dataSource type="POOLED">
            <!-- MySQL 数据库驱动 -->
            <property name="driver" value="com.mysql.jdbc.Driver"/>
            <!-- 连接数据库的 URL -->
               <property name="url" value="jdbc:mysql://localhost:3306/
               springtest? characterEncoding=utf8"/>
            <property name="username" value="root"/>
            <property name="password" value="root"/>
         </dataSource>
      </environment>
   </environments>
   <mappers>
   <!-- 映射文件的位置 -->
   <mapper resource="com/mybatis/mapper/UserMapper.xml"/>
   </mappers>
</configuration>
```

上述映射文件和配置文件都不需要读者完全手动编写，都可以从 MyBatis 使用手册中复制，然后做简单修改。

❻ 创建测试类

在 src 目录下创建一个名为 com.mybatis.test 的包，在该包中创建 MyBatisTest 测试类。在测试类中首先使用输入流读取配置文件，然后根据配置信息构建 SqlSessionFactory 对象。接下来通过 SqlSessionFactory 对象创建 SqlSession 对象，并使用 SqlSession 对象的方法执行数据库操作。

MyBatisTest 测试类的代码如下：

```
package com.mybatis.test;
import java.io.IOException;
import java.io.InputStream;
import java.util.List;
import org.apache.ibatis.io.Resources;
import org.apache.ibatis.session.SqlSession;
import org.apache.ibatis.session.SqlSessionFactory;
import org.apache.ibatis.session.SqlSessionFactoryBuilder;
import com.mybatis.po.MyUser;
public class MyBatisTest {
    public static void main(String[] args) {
        try {
            //读取配置文件 mybatis-config.xml
            InputStream config=Resources.getResourceAsStream("mybatis-
            config.xml");
            //根据配置文件构建 SqlSessionFactory
            SqlSessionFactory ssf=new SqlSessionFactoryBuilder().build
            (config);
            //通过 SqlSessionFactory 创建 SqlSession
            SqlSession ss=ssf.openSession();
            //SqlSession 执行映射文件中定义的 SQL，并返回映射结果
            /*com.mybatis.mapper.UserMapper.selectUserById 为
              UserMapper.xml 中的命名空间+select 的 id*/
            //查询一个用户
            MyUser mu = ss.selectOne("com.mybatis.mapper.UserMapper.selectUser
            ById", 1);
            System.out.println(mu);
            //添加一个用户
            MyUser addmu=new MyUser();
            addmu.setUname("陈恒");
            addmu.setUsex("男");
            ss.insert("com.mybatis.mapper.UserMapper.addUser",addmu);
            //修改一个用户
            MyUser updatemu=new MyUser();
            updatemu.setUid(1);
```

```
            updatemu.setUname("张三");
            updatemu.setUsex("女");
            ss.update("com.mybatis.mapper.UserMapper.updateUser", updatemu);
            //删除一个用户
            ss.delete("com.mybatis.mapper.UserMapper.deleteUser", 3);
            //查询所有用户
            List<MyUser> listMu = ss.selectList("com.mybatis.mapper.UserMapper.
            selectAllUser");
            for (MyUser myUser : listMu) {
                System.out.println(myUser);
            }
            //提交事务
            ss.commit();
            //关闭 SqlSession
            ss.close();
        } catch (IOException e) {
            // TODO Auto-generated catch block
            e.printStackTrace();
        }
    }
}
```

上述测试类的运行结果如图 6.4 所示。

```
Servers  Console
<terminated> MyBatisTest [Java Application] C:\Program Files\Java\jre1.8.0_152\bin\javaw.exe (2018年1月22日 上午9:02:25)
DEBUG [main] - ==>  Preparing: insert into user (uname,usex) values(?,?)
DEBUG [main] - ==> Parameters: 陈恒(String), 男(String)
DEBUG [main] - <==    Updates: 1
DEBUG [main] - ==>  Preparing: update user set uname = ?,usex = ? where uid = ?
DEBUG [main] - ==> Parameters: 张三(String), 女(String), 1(Integer)
DEBUG [main] - <==    Updates: 1
DEBUG [main] - ==>  Preparing: delete from user where uid = ?
DEBUG [main] - ==> Parameters: 3(Integer)
DEBUG [main] - <==    Updates: 0
DEBUG [main] - ==>  Preparing: select * from user
DEBUG [main] - ==> Parameters: 
DEBUG [main] - <==      Total: 2
User [uid=1,uname=张三,usex=女]
User [uid=9,uname=陈恒,usex=男]
```

图 6.4 MyBatis 入门程序的运行结果

6.5 MyBatis 与 Spring 的整合

视频讲解

从 6.4 节测试类的代码中可以看出直接使用 MyBatis 框架的 SqlSession

访问数据库并不简便。MyBatis 框架的重点是 SQL 映射文件，为方便后续学习，本节讲解 MyBatis 与 Spring 的整合。在本书 MyBatis 的后续讲解中将使用整合后的框架进行演示。

6.5.1 导入相关 JAR 包

实现 MyBatis 与 Spring 的整合需要导入相关 JAR 包，包括 MyBatis、Spring 以及其他 JAR 包。

❶ MyBatis 框架所需的 JAR 包

MyBatis 框架所需的 JAR 包包括它的核心包和依赖包，包的详情见 6.2 节。

❷ Spring 框架所需的 JAR 包

Spring 框架所需的 JAR 包包括它的核心模块 JAR、AOP 开发使用的 JAR、JDBC 和事务的 JAR 包（其中依赖包不需要再导入，因为 MyBatis 已提供），具体如下：

```
aopalliance-1.0.jar
aspectjweaver-1.8.13.jar
spring-aop-5.0.2.RELEASE.jar
spring-aspects-5.0.2.RELEASE.jar
spring-beans-5.0.2.RELEASE.jar
spring-context-5.0.2.RELEASE.jar
spring-core-5.0.2.RELEASE.jar
spring-expression-5.0.2.RELEASE.jar
spring-jdbc-5.0.2.RELEASE.jar
spring-tx-5.0.2.RELEASE.jar
```

❸ MyBatis 与 Spring 整合的中间 JAR 包

在编写本书时该中间 JAR 包的最新版本为 mybatis-spring-1.3.1.jar，此版本可以从网址“http://mvnrepository.com/artifact/org.mybatis/mybatis-spring/1.3.1”下载。

❹ 数据库驱动 JAR 包

本书所使用的 MySQL 数据库驱动包为 mysql-connector-java-5.1.45-bin.jar。

❺ 数据源所需的 JAR 包

在整合时使用的是 DBCP 数据源，需要准备 DBCP 和连接池的 JAR 包。在编写本书时最新版本的 DBCP 的 JAR 包为 commons-dbcp2-2.2.0.jar，可以从网址“http://commons.apache.org/proper/commons-dbcp/download_dbcp.cgi”下载；最新版本的连接池的 JAR 包为 commons-pool2-2.5.0.jar，可以从网址“http://commons.apache.org/proper/commons-pool/download_pool.cgi”下载。

6.5.2 在 Spring 中配置 MyBatis 工厂

通过与 Spring 的整合，MyBatis 的 SessionFactory 交由 Spring 来构建，在构建时需要在 Spring 的配置文件中添加如下代码：

```
<!-- 配置数据源 -->
<bean id="dataSource" class="org.apache.commons.dbcp2.BasicDataSource">
    <property name="driverClassName" value="com.mysql.jdbc.Driver" />
    <property name="url" value="jdbc:mysql://localhost:3306/springtest?
    characterEncoding=utf8" />
    <property name="username" value="root" />
    <property name="password" value="root" />
    <!-- 最大连接数 -->
    <property name="maxTotal" value="30"/>
    <!-- 最大空闲连接数 -->
    <property name="maxIdle" value="10"/>
    <!-- 初始化连接数 -->
    <property name="initialSize" value="5"/>
</bean>
<!-- 配置 MyBatis 工厂，同时指定数据源，并与 MyBatis 完美整合 -->
<bean id="sqlSessionFactory" class="org.mybatis.spring.SqlSessionFactoryBean">
    <property name="dataSource" ref="dataSource" />
    <!-- configLocation 的属性值为 MyBatis 的核心配置文件 -->
    <property name="configLocation" value="classpath:com/mybatis/mybatis-
    config.xml"/>
</bean>
```

6.5.3　使用 Spring 管理 MyBatis 的数据操作接口

使用 Spring 管理 MyBatis 数据操作接口的方式有多种，其中最常用、最简洁的一种是基于 MapperScannerConfigurer 的整合。该方式需要在 Spring 的配置文件中加入以下内容：

```
<!-- Mapper 代理开发，使用 Spring 自动扫描 MyBatis 的接口并装配
（Spring 将指定包中所有接口自动装配为 MyBatis 的 Mapper 接口的实现类） -->
<bean class="org.mybatis.spring.mapper.MapperScannerConfigurer">
    <!-- mybatis-spring 组件的扫描器，com.dao 只需要接口（接口方法与 SQL 映射文件
    中的相同）-->
    <property name="basePackage" value="com.dao"/>
    <property name="sqlSessionFactoryBeanName" value="sqlSessionFactory"/>
</bean>
```

6.5.4　框架整合示例

下面通过一个实例实现 MyBatis 与 Spring 的整合，具体实现过程如下：

❶ 创建应用并导入相关 JAR 包

创建一个名为 ch6SS 的 Web 应用，并将 6.5.1 节的 JAR 导入/WEB-INF/lib 目录下。

❷ 创建持久化类

在 src 目录下创建一个名为 com.po 的包，将 6.4 节的持久化类复制到包中。

❸ 创建 SQL 映射文件和 MyBatis 核心配置文件

在 src 目录下创建一个名为 com.mybatis 的包，在该包中创建 MyBatis 核心配置文件 mybatis-config.xml 和 SQL 映射文件 UserMapper.xml。

UserMapper.xml 的代码如下：

```
<?xml version="1.0" encoding="UTF-8" ?>
<!DOCTYPE mapper
PUBLIC "-//mybatis.org//DTD Mapper 3.0//EN"
"http://mybatis.org/dtd/mybatis-3-mapper.dtd">
<mapper namespace="com.dao.UserDao">
    <!-- 根据 uid 查询一个用户信息 -->
    <select id="selectUserById" parameterType="Integer"
        resultType="com.po.MyUser">
        select * from user where uid = #{uid}
    </select>
    <!-- 查询所有用户信息 -->
    <select id="selectAllUser"  resultType="com.po.MyUser">
        select * from user
    </select>
    <!-- 添加一个用户，#{uname}为 com.po.MyUser 的属性值-->
    <insert id="addUser" parameterType="com.po.MyUser">
        insert into user (uname,usex) values(#{uname},#{usex})
    </insert>
    <!-- 修改一个用户 -->
    <update id="updateUser" parameterType="com.po.MyUser">
        update user set uname = #{uname},usex = #{usex} where uid = #{uid}
    </update>
    <!-- 删除一个用户 -->
    <delete id="deleteUser" parameterType="Integer">
        delete from user where uid = #{uid}
    </delete>
</mapper>
```

mybatis-config.xml 的代码如下：

```
<?xml version="1.0" encoding="UTF-8" ?>
<!DOCTYPE configuration
PUBLIC "-//mybatis.org//DTD Config 3.0//EN"
"http://mybatis.org/dtd/mybatis-3-config.dtd">
<configuration>
    <!-- 告诉 MyBatis 到哪里去找映射文件 -->
     <mappers>
        <mapper resource="com/mybatis/UserMapper.xml"/>
    </mappers>
</configuration>
```

❹ 创建数据访问接口

在 src 目录下，创建一个名为 com.dao 的包，在该包中创建 UserDao 接口，接口中的方法与 SQL 映射文件一致。

UserDao 接口的代码如下：

```
package com.dao;
import java.util.List;
import org.springframework.stereotype.Repository;
import com.po.MyUser;
@Repository("userDao")
//可有可无，但有时提示依赖注入找不到（不影响运行），加上后可以消去该提示。
public interface UserDao {
    /**
     * 接口方法对应 SQL 映射文件 UserMapper.xml 中的 id
     */
    public MyUser selectUserById(Integer uid);
    public List<MyUser> selectAllUser();
    public int addUser(MyUser user);
    public int updateUser(MyUser user);
    public int deleteUser(Integer uid);
}
```

❺ 创建日志文件

在 src 目录下创建日志文件 log4j.properties，文件内容如下：

```
# Global logging configuration
log4j.rootLogger=ERROR, stdout
# MyBatis logging configuration...
log4j.logger.com.dao=DEBUG
# Console output...
log4j.appender.stdout=org.apache.log4j.ConsoleAppender
log4j.appender.stdout.layout=org.apache.log4j.PatternLayout
log4j.appender.stdout.layout.ConversionPattern=%5p [%t] - %m%n
```

❻ 创建控制层

在 src 目录下创建一个名为 com.controller 的包，在包中创建 UserController 类，在该类中调用数据访问接口中的方法。

UserController 类的代码如下：

```
package com.controller;
import java.util.List;
import org.springframework.beans.factory.annotation.Autowired;
import org.springframework.stereotype.Controller;
```

```
import com.dao.UserDao;
import com.po.MyUser;
@Controller("userController")
public class UserController {
    @Autowired
    private UserDao userDao;
    public void test() {
        //查询一个用户
        MyUser auser=userDao.selectUserById(1);
        System.out.println(auser);
        System.out.println("================");
        //添加一个用户
        MyUser addmu=new MyUser();
        addmu.setUname("陈恒");
        addmu.setUsex("男");
        int add=userDao.addUser(addmu);
        System.out.println("添加了" + add + "条记录");
        System.out.println("================");
        //修改一个用户
        MyUser updatemu=new MyUser();
        updatemu.setUid(1);
        updatemu.setUname("张三");
        updatemu.setUsex("女");
        int up=userDao.updateUser(updatemu);
        System.out.println("修改了" + up + "条记录");
        System.out.println( "================");
        //删除一个用户
        int dl=userDao.deleteUser(9);
        System.out.println("删除了" + dl + "条记录");
        System.out.println("================");
        //查询所有用户
        List<MyUser> list=userDao.selectAllUser();
        for (MyUser myUser : list) {
            System.out.println(myUser);
        }
    }
}
```

❼ 创建 Spring 的配置文件

在 src 目录下创建配置文件 applicationContext.xml，在配置文件中配置数据源、MyBatis 工厂以及 Mapper 代理开发等信息。

applicationContext.xml 的代码如下：

```
<?xml version="1.0" encoding="UTF-8"?>
<beans xmlns="http://www.springframework.org/schema/beans"
    xmlns:xsi="http://www.w3.org/2001/XMLSchema-instance"
```

```
xmlns:context="http://www.springframework.org/schema/context"
xmlns:tx="http://www.springframework.org/schema/tx"
xsi:schemaLocation="http://www.springframework.org/schema/beans
   http://www.springframework.org/schema/beans/spring-beans.xsd
   http://www.springframework.org/schema/context
   http://www.springframework.org/schema/context/spring-context.xsd
   http://www.springframework.org/schema/tx
   http://www.springframework.org/schema/tx/spring-tx.xsd">
<!-- 指定需要扫描的包（包括子包），使注解生效 -->
<context:component-scan base-package="com.dao"/>
<context:component-scan base-package="com.controller"/>
<!-- 配置数据源 -->
<bean id="dataSource" class="org.apache.commons.dbcp2.BasicDataSource">
      <property name="driverClassName" value="com.mysql.jdbc.Driver" />
      <property name="url" value="jdbc:mysql://localhost:3306/springtest?
      characterEncoding=utf8" />
      <property name="username" value="root" />
      <property name="password" value="root" />
      <!-- 最大连接数 -->
      <property name="maxTotal" value="30"/>
      <!-- 最大空闲连接数 -->
      <property name="maxIdle" value="10"/>
      <!-- 初始化连接数 -->
      <property name="initialSize" value="5"/>
</bean>
<!-- 添加事务支持 -->
<bean id="txManager"
   class="org.springframework.jdbc.datasource.DataSourceTransaction
   Manager">
   <property name="dataSource" ref="dataSource" />
</bean>
<!-- 开启事务注解-->
<tx:annotation-driven transaction-manager="txManager" />
 <!-- 配置 MyBatis 工厂，同时指定数据源，并与 MyBatis 完美整合 -->
 <bean  id="sqlSessionFactory"  class="org.mybatis.spring.SqlSession
FactoryBean">
   <property name="dataSource" ref="dataSource" />
   <!-- configLocation 的属性值为 MyBatis 的核心配置文件 -->
   <property name="configLocation" value="classpath:com/mybatis/mybatis-
   config.xml"/>
 </bean>
 <!-- Mapper 代理开发，使用 Spring 自动扫描 MyBatis 的接口并装配
  （Spring 将指定包中所有被@Mapper 注解标注的接口自动装配为 MyBatis 的映射接口）-->
 <bean class="org.mybatis.spring.mapper.MapperScannerConfigurer">
    <!-- mybatis-spring 组件的扫描器 -->
```

```
        <property name="basePackage" value="com.dao"/>
        <property name="sqlSessionFactoryBeanName" value="sqlSessionFactory"/>
    </bean>
</beans>
```

❽ 创建测试类

在 com.controller 包中创建测试类 TestController，代码如下：

```
package com.controller;
import org.springframework.context.ApplicationContext;
import org.springframework.context.support.ClassPathXmlApplicationContext;
public class TestController {
    public static void main(String[] args) {
        ApplicationContext appCon=new ClassPathXmlApplicationContext
       ("applicationContext.xml");
        UserController uc=(UserController)appCon.getBean("userController");
        uc.test();
    }
}
```

上述测试类的运行结果如图 6.5 所示。

```
Servers  Console
<terminated> TestController [Java Application] C:\P
User [uid=1,uname=张三,usex=女]
=================
添加了1条记录
=================
修改了1条记录
=================
删除了0条记录
=================
User [uid=1,uname=张三,usex=女]
User [uid=10,uname=陈恒,usex=男]
```

图 6.5　框架整合测试结果

从第 6 步中的 UserController 类可以看出，开发者只需要进行业务处理，不需要再写 SqlSession 对象的创建、数据库事务的处理等烦琐代码。因此，MyBatis 整合 Spring 后方便了数据库访问操作，提高了开发效率。

6.6 使用 MyBatis Generator 插件自动生成映射文件

使用 MyBatis Generator 插件自动生成 MyBatis 所需要的 DAO 接口、实体模型类、

Mapping 映射文件，这样省去了很多工夫，将生成的代码复制到项目工程中即可，把更多精力放在业务逻辑上。

MyBatis Generator 有 3 种常用方法自动生成代码，即命令行、Eclipse 插件和 Maven 插件。本节使用比较简单的方法（命令行）自动生成相关代码，具体步骤如下：

❶ 准备相关 JAR 包

需要准备的 JAR 包是 mysql-connector-java-5.1.45-bin.jar 和 mybatis-generator-core-1.3.6.jar（http://mvnrepository.com/artifact/org.mybatis.generator/mybatis-generator-core/1.3.6）。

❷ 创建文件目录

在某磁盘根目录下新建一个文件目录，例如 D:\generator，并将 mysql-connector-java-5.1.45-bin.jar 和 mybatis-generator-core-1.3.6.jar 文件复制到 generator 目录下。另外，在 generator 目录下创建 src 子目录存放生成的相关代码文件。

❸ 创建配置文件

在第 2 步创建的文件目录（D:\generator）下创建配置文件，例如 D:\generator\generator.xml，如图 6.6 所示。

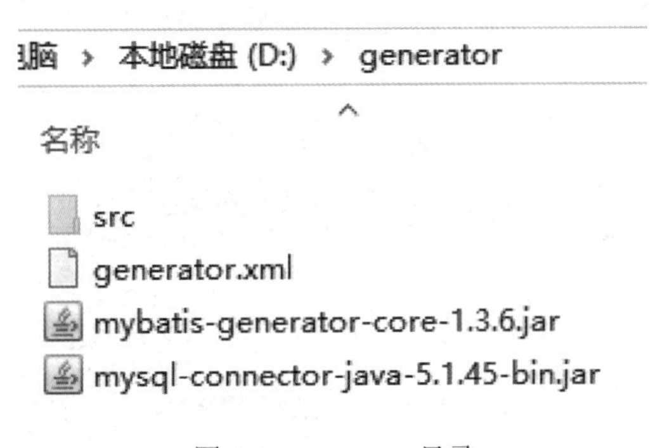

图 6.6　generator 目录

generator.xml 配置文件的内容如下（具体含义见注释）：

```
<?xml version="1.0" encoding="UTF-8"?>
<!DOCTYPE generatorConfiguration PUBLIC "-//mybatis.org//DTD MyBatis
Generator Configuration 1.0//EN" "http://mybatis.org/dtd/mybatis-
generator-config_1_0.dtd">
<generatorConfiguration>
   <!-- 数据库驱动包位置 -->
   <classPathEntry location="D:\generator\mysql-connector-java-5.1.45-
   bin.jar" />
   <context id="mysqlTables" targetRuntime="MyBatis3">
      <commentGenerator>
         <property name="suppressAllComments" value="true" />
      </commentGenerator>
      <!-- 数据库链接 URL、用户名、密码（前提是数据库 springtest 存在） -->
       <jdbcConnection
            driverClass="com.mysql.jdbc.Driver"
```

```
            connectionURL="jdbc:mysql://localhost:3306/springtest?character
            Encoding=utf8"
            userId="root" password="root">
        </jdbcConnection>
        <javaTypeResolver>
            <property name="forceBigDecimals" value="false" />
        </javaTypeResolver>
        <!-- 生成模型（MyBatis 里面用到实体类）的包名和位置 -->
        <javaModelGenerator targetPackage="com.po" targetProject="D:\
        generator\src">
            <property name="enableSubPackages" value="true" />
            <property name="trimStrings" value="true" />
        </javaModelGenerator>
        <!-- 生成的映射文件（MyBatis 的 SQL 语句 xml 文件）包名和位置-->
        <sqlMapGenerator targetPackage="mybatis" targetProject="D:\
        generator\src">
            <property name="enableSubPackages" value="true" />
        </sqlMapGenerator>
        <!-- 生成 DAO 的包名和位置 -->
        <javaClientGenerator type="XMLMAPPER" targetPackage="com.dao"
        targetProject="D:\generator\src">
            <property name="enableSubPackages" value="true" />
        </javaClientGenerator>
        <!-- 生成表(更改 tableName 和 domainObjectName 就可以，前提是数据库 springtest
        中的 user 表已创建) -->
        <table tableName="user" domainObjectName="User" enableCountByExample=
        "false" enableUpdateByExample="false" enableDeleteByExample="false"
        enableSelectByExample="false" selectByExampleQueryId="false" />
    </context>
</generatorConfiguration>
```

❹ 使用命令生成代码

打开命令提示符，进入 D:\generator，输入命令 java -jar mybatis-generator-core-1.3.6.jar -configfile generator.xml-overwrite，如图 6.7 所示。

```
D:\>cd generator

D:\generator>java -jar mybatis-generator-core-1.3.6.jar -configfile generator.xml -overwrite
MyBatis Generator finished successfully.
```

图 6.7 使用命令行生成映射文件

6.7 本章小结

本章首先简单介绍了 MyBatis 环境的构建与工作原理；其次详细讲解了在 Eclipse 中如

何开发 MyBatis 入门程序；为了方便学习 MyBatis 的相关知识，最后以 ch6SS 应用为例介绍了 MyBatis 与 Spring 的整合开发。对于整合后的框架，开发者不再需要编写 SqlSession 对象的创建、数据库事务的处理等烦琐代码，提高了开发效率。

习题 6

1．简述 MyBatis 的工作原理。
2．简述 MyBatis 与 Spring 的整合过程。
3．除了 MyBatis 持久化框架以外，您还知道哪些持久化框架？

第 7 章

映射器

学习目的与要求

本章重点讲解 MyBatis 的 SQL 映射文件。通过本章的学习，读者能够了解 MyBatis 的核心配置文件的配置信息，熟练掌握 MyBatis 的 SQL 映射文件的编写，熟悉级联查询的 MyBatis 实现。

本章主要内容

- 核心配置文件；
- SQL 映射文件；
- 级联查询。

MyBatis 框架的强大之处体现在 SQL 映射文件的编写上，因此本章将重点讲解 SQL 映射文件的编写。

视频讲解

7.1 MyBatis 配置文件概述

MyBatis 的核心配置文件配置了很多影响 MyBatis 行为的信息，这些信息通常只会配置在一个文件中，并且不会轻易改动。另外，与 Spring 框架整合后，MyBatis 的核心配置文件信息将配置到 Spring 的配置文件中。因此，在实际开发中需要编写或修改 MyBatis 的核心配置文件的情况不多。本节只是了解一下 MyBatis 的核心配置文件中的主要元素。

MyBatis 的核心配置文件的模板代码如下：

```
<?xml version="1.0" encoding="UTF-8" ?>
<!DOCTYPE configuration
PUBLIC "-//mybatis.org//DTD Config 3.0//EN"
"http://mybatis.org/dtd/mybatis-3-config.dtd">
<configuration>
```

```
        <properties/><!-- 属性 -->
        <settings><!-- 设置 -->
           <setting name="" value=""/>
        </settings>
        <typeAliases/><!-- 类型的命名（别名） -->
        <typeHandlers/><!-- 类型处理器 -->
        <objectFactory type=""/><!-- 对象工厂 -->
        <plugins><!-- 插件 -->
           <plugin interceptor=""></plugin>
        </plugins>
        <environments default=""><!-- 配置环境 -->
           <environment id=""><!-- 环境变量 -->
              <transactionManager type=""/><!-- 事务管理器 -->
              <dataSource type=""/><!-- 数据源 -->
           </environment>
        </environments>
        <databaseIdProvider type=""/><!-- 数据库厂商标识 -->
         <mappers><!-- 映射器，告诉 MyBatis 到哪里去找映射文件 -->
           <mapper resource="com/mybatis/UserMapper.xml"/>
         </mappers>
    </configuration>
```

MyBatis 的核心配置文件中的元素配置顺序不能颠倒，一旦颠倒，在 MyBatis 启动阶段将发生异常。

7.2 映射器概述

映射器是 MyBatis 最复杂且最重要的组件，由一个接口加上 XML 文件（SQL 映射文件）组成（见 6.5.4 节）。MyBatis 的映射器也可以使用注解完成，但在实际应用中使用不广泛，原因主要来自以下几个方面：其一，面对复杂的 SQL 会显得无力；其二，注解的可读性较差；其三，注解丢失了 XML 上下文相互引用的功能。因此，推荐使用 XML 文件开发映射器。

SQL 映射文件的常用配置元素如表 7.1 所示。

表 7.1　SQL 映射文件的常用配置元素

元素名称	描　述	备　注
select	查询语句，最常用、最复杂的元素之一	可以自定义参数，返回结果集等
insert	插入语句	执行后返回一个整数，代表插入的行数
update	更新语句	执行后返回一个整数，代表更新的行数
delete	删除语句	执行后返回一个整数，代表删除的行数
sql	定义一部分SQL，在多个位置被引用	例如一张表，列名一次定义，可以在多个SQL语句中使用
resultMap	用来描述从数据库结果集中来加载对象，是最复杂、最强大的元素	提供映射规则

7.3 <select>元素

在 SQL 映射文件中<select>元素用于映射 SQL 的 select 语句，其示例代码如下：

```
<!-- 根据 uid 查询一个用户信息 -->
<select id="selectUserById" parameterType="Integer"
    resultType="com.po.MyUser">
    select * from user where uid = #{uid}
</select>
```

在上述示例代码中，id 的值是唯一标识符，它接收一个 Integer 类型的参数，返回一个 MyUser 类型的对象，结果集自动映射到 MyUser 属性。

<select>元素除了有上述示例代码中的几个属性以外，还有一些常用的属性，如表 7.2 所示。

表 7.2　<select>元素的常用属性

属性名称	描　述
id	它和Mapper的命名空间组合起来使用，是唯一标识符，供MyBatis调用
parameterType	表示传入SQL语句的参数类型的全限定名或别名。它是一个可选属性，MyBatis能推断出具体传入语句的参数
resultType	SQL语句执行后返回的类型（全限定名或者别名）。如果是集合类型，返回的是集合元素的类型，返回时可以使用resultType或resultMap之一
resultMap	它是映射集的引用，与<resultMap>元素一起使用，返回时可以使用resultType或resultMap之一
flushCache	用于设置在调用SQL语句后是否要求MyBatis清空之前查询的本地缓存和二级缓存，默认值为false，如果设置为true，则任何时候只要SQL语句被调用都将清空本地缓存和二级缓存
useCache	启动二级缓存的开关，默认值为true，表示将查询结果存入二级缓存中
timeout	用于设置超时参数，单位是秒（s），超时将抛出异常
fetchSize	获取记录的总条数设定
statementType	告诉MyBatis使用哪个JDBC的Statement工作，取值为STATEMENT（Statement）、PREPARED（PreparedStatement）、CALLABLE（CallableStatement）
resultSetType	这是针对JDBC的ResultSet接口而言，其值可设置为FORWARD_ONLY（只允许向前访问）、SCROLL_SENSITIVE（双向滚动，但不及时更新）、SCROLL_INSENSITIVE（双向滚动，及时更新）

7.3.1 使用 Map 接口传递多个参数

在实际开发中，查询 SQL 语句经常需要多个参数，例如多条件查询。当传递多个参数时，<select>元素的 parameterType 属性值的类型是什么呢？在 MyBatis 中允许 Map 接口通过键值对传递多个参数。

假设数据操作接口中有个实现查询陈姓男性用户信息功能的方法：

```
public List<MyUser> selectAllUser(Map<String, Object> param);
```

此时，传递给映射器的是一个 Map 对象，使用它在 SQL 文件中设置对应的参数，对应 SQL 文件的代码如下：

```
<!-- 查询陈姓男性用户信息 -->
<select id="selectAllUser" resultType="com.po.MyUser" parameterType="map">
    select * from user
    where uname like concat('%',#{u_name},'%')
    and usex=#{u_sex}
</select>
```

在上述 SQL 文件中，参数名 u_name 和 u_sex 是 Map 的 key。

为了测试该示例，首先创建一个 Web 应用 ch7，将 6.5.4 节中 ch6SS 应用的所有 JAR 包复制到/WEB-INF/lib 下，同时将 ch6SS 应用的 src 目录下的所有包和文件复制到 ch7 应用的 src 目录下，然后将 com.mybatis 包中的 SQL 映射文件 UserMapper.xml 中的“查询所有用户信息”的代码片段修改为上述“查询陈姓男性用户信息”的代码片段，最后将 com.controller 包中 UserController 的代码简单修改即可运行测试类了。

com.controller 包中 UserController 的代码片段如下：

```
...
@Controller("userController")
public class UserController {
    @Autowired
    private UserDao userDao;
    public void test() {
        ...
        //查询多个用户
        Map<String, Object> map=new HashMap<>();
        map.put("u_name", "陈");
        map.put("u_sex", "男");
        List<MyUser> list=userDao.selectAllUser(map);
        for (MyUser myUser : list) {
            System.out.println(myUser);
        }
        ...
    }
```

Map 是一个键值对应的集合，使用者要通过阅读它的键才能了解其作用。另外，使用 Map 不能限定其传递的数据类型，所以业务性不强，可读性较差。如果 SQL 语句很复杂，参数很多，使用 Map 将很不方便。幸运的是，MyBatis 还提供了使用 Java Bean 传递多个参数的形式。

7.3.2　使用 Java Bean 传递多个参数

首先在 ch7 应用的 src 目录下创建一个名为 com.pojo 的包，在包中创建一个 POJO 类

SeletUserParam，代码如下：

```
package com.pojo;
public class SeletUserParam {
    private String u_name;
    private String u_sex;
    //此处省略setter和getter方法
}
```

接着将 Dao 接口中的 selectAllUser 方法修改为如下：

```
public List<MyUser> selectAllUser(SeletUserParam param);
```

然后将 com.mybatis 包中的 SQL 映射文件 UserMapper.xml 中的“查询陈姓男性用户信息”的代码修改为如下：

```
<select id="selectAllUser"  resultType="com.po.MyUser" parameterType="com.
pojo.SeletUserParam">
    select * from user
    where uname like concat('%',#{u_name},'%')
    and usex=#{u_sex}
</select>
```

最后将 com.controller 包中 UserController 的“查询多个用户”的代码片段做如下修改：

```
SeletUserParam su=new SeletUserParam();
su.setU_name("陈");
su.setU_sex("男");
List<MyUser> list=userDao.selectAllUser(su);
for (MyUser myUser : list) {
    System.out.println(myUser);
}
```

在实际应用中是选择 Map 还是选择 Java Bean 传递多个参数应根据实际情况而定，如果参数较少，建议选择 Map；如果参数较多，建议选择 Java Bean。

7.4 <insert>元素

视频讲解

<insert>元素用于映射插入语句，MyBatis 执行完一条插入语句后将返回一个整数表示其影响的行数。它的属性与<select>元素的属性大部分相同，在本节讲解它的几个特有属性。

- keyProperty：该属性的作用是将执行插入操作时的主键值赋值给 PO 类的某个属性，通常会设置为主键对应的属性。如果是联合主键，可以将多个值用逗号隔开。
- keyColumn：该属性用于设置第几列是主键，当主键列不是表中的第 1 列时需要设置。如果是联合主键，可以将多个值用逗号隔开。

- useGeneratedKeys：该属性将使 MyBatis 使用 JDBC 的 getGeneratedKeys()方法获取由数据库内部产生的主键，例如 MySQL、SQL Server 等自动递增的字段，其默认值为 false。

7.4.1 主键（自动递增）回填

MySQL、SQL Server 等数据库的表格可以采用自动递增的字段作为主键，有时可能需要使用这个刚刚产生的主键，用于关联其他业务。因为本书采用的数据库是 MySQL 数据库，所以可以直接使用 ch7 应用讲解自动递增主键的回填使用方法。

首先为 com.mybatis 包中的 SQL 映射文件 UserMapper.xml 中 id 为 addUser 的<insert>元素添加 keyProperty 和 useGeneratedKeys 属性，具体代码如下：

```
<!-- 添加一个用户，成功后将主键值回填给 uid（po 类的属性）-->
<insert id="addUser" parameterType="com.po.MyUser"
keyProperty="uid" useGeneratedKeys="true">
    insert into user (uname,usex) values(#{uname},#{usex})
</insert>
```

然后在 com.controller 包的 UserController 类中进行调用，具体代码如下：

```
//添加一个用户
MyUser addmu=new MyUser();
addmu.setUname("陈恒");
addmu.setUsex("男");
int add=userDao.addUser(addmu);
System.out.println("添加了" + add + "条记录");
System.out.println("添加记录的主键是" + addmu.getUid());
```

7.4.2 自定义主键

如果在实际工程中使用的数据库不支持主键自动递增（例如 Oracle），或者取消了主键自动递增的规则，可以使用 MyBatis 的<selectKey>元素来自定义生成主键。具体配置示例代码如下：

```
<insert id="insertUser" parameterType="com.po.MyUser">
    <!-- 先使用 selectKey 元素定义主键，然后再定义 SQL 语句 -->
    <selectKey keyProperty="uid" resultType="Integer" order="BEFORE">
        select decode(max(uid), null, 1 , max(uid)+1) as newUid from user
    </selectKey>
    insert into user (uid,uname,usex) values(#{uid},#{uname},#{usex})
</insert>
```

在执行上述示例代码时，<selectKey>元素首先被执行，该元素通过自定义的语句设置数据表的主键，然后执行插入语句。

<selectKey>元素的 keyProperty 属性指定了新生主键值返回给 PO 类（com.po.MyUser）的哪个属性。order 属性可以设置为 BEFORE 或 AFTER，BEFORE 表示先执行<selectKey>元素然后执行插入语句；AFTER 表示先执行插入语句再执行<selectKey>元素。

7.5 <update>与<delete>元素

视频讲解

<update>和<delete>元素比较简单，它们的属性和<insert>元素、<select>元素的属性差不多，执行后也返回一个整数，表示影响了数据库的记录行数。配置示例代码如下：

```
<!-- 修改一个用户 -->
<update id="updateUser" parameterType="com.po.MyUser">
    update user set uname=#{uname},usex=#{usex} where uid = #{uid}
</update>
<!-- 删除一个用户 -->
<delete id="deleteUser" parameterType="Integer">
    delete from user where uid = #{uid}
</delete>
```

7.6 <sql>元素

视频讲解

<sql>元素的作用在于可以定义 SQL 语句的一部分（代码片段），以方便后面的 SQL 语句引用它，例如反复使用的列名。在 MyBatis 中只需使用<sql>元素编写一次便能在其他元素中引用它。配置示例代码如下：

```
<sql id="comColumns">id,uname,usex</sql>
<select id="selectUser" resultType="com.po.MyUser">
    select <include refid="comColumns"/>  from user
</select>
```

在上述代码中使用<include>元素的 refid 属性引用了自定义的代码片段。

7.7 <resultMap>元素

视频讲解

<resultMap>元素表示结果映射集，是 MyBatis 中最重要也是最强大的元素，主要用来定义映射规则、级联的更新以及定义类型转化器等。

7.7.1 <resultMap>元素的结构

<resultMap>元素包含了一些子元素，结构如下：

```
<resultMap type="" id="">
    <constructor><!-- 类在实例化时用来注入结果到构造方法 -->
        <idArg/><!-- ID 参数，结果为 ID -->
        <arg/><!-- 注入到构造方法的一个普通结果 -->
    </constructor>
    <id/><!-- 用于表示哪个列是主键 -->
    <result/><!-- 注入到字段或 JavaBean 属性的普通结果 -->
    <association property=""/><!-- 用于一对一关联 -->
    <collection property=""/><!-- 用于一对多、多对多关联 -->
    <discriminator javaType=""><!-- 使用结果值来决定使用哪个结果映射 -->
        <case value=""/>    <!-- 基于某些值的结果映射 -->
    </discriminator>
</resultMap>
```

<resultMap>元素的 type 属性表示需要的 POJO，id 属性是 resultMap 的唯一标识。子元素<constructor>用于配置构造方法（当 POJO 未定义无参数的构造方法时使用）。子元素<id>用于表示哪个列是主键。子元素<result>用于表示 POJO 和数据表普通列的映射关系。子元素<association>、<collection>和<discriminator>用在级联的情况下。关于级联的问题比较复杂，将在 7.8 节学习。

一条查询 SQL 语句执行后将返回结果，而结果可以使用 Map 存储，也可以使用 POJO 存储。

7.7.2　使用 Map 存储结果集

任何 select 语句都可以使用 Map 存储结果，示例代码如下：

```
<!-- 查询所有用户信息存到 Map 中 -->
<select id="selectAllUserMap"  resultType="map">
    select * from user
</select>
```

测试上述 SQL 配置文件的过程如下：

首先在 com.dao.UserDao 接口中添加以下接口方法。

```
public List<Map<String, Object>> selectAllUserMap();
```

然后在 com.controller 包的 UserController 类中调用接口方法，具体代码如下。

```
//查询所有用户信息存到 Map 中
List<Map<String, Object>> lmp=userDao.selectAllUserMap();
for (Map<String, Object> map : lmp) {
    System.out.println(map);
}
```

上述 Map 的 key 是 select 语句查询的字段名（必须完全一样），而 Map 的 value 是查询返回结果中字段对应的值，一条记录映射到一个 Map 对象中。Map 用起来很方便，但可读性稍差，有的开发者不太喜欢使用 Map，更多时候喜欢使用 POJO 的方式。

7.7.3 使用 POJO 存储结果集

有的开发者喜欢使用 POJO 的方式存储结果集，一方面可以使用自动映射，例如使用 resultType 属性，但有时候需要更为复杂的映射或级联，这时候就需要使用<select>元素的 resultMap 属性配置映射集合。具体步骤如下：

❶ 创建 POJO 类

在 ch7 应用的 com.pojo 包中创建 POJO 类 MapUser。MapUser 类的代码如下：

```
package com.pojo;
public class MapUser {
    private Integer m_uid;
    private String m_uname;
    private String m_usex;
    //此处省略 setter 和 getter 方法
    @Override
    public String toString() {
        return "User [uid=" + m_uid +",uname=" + m_uname + ",usex=" + m_usex +"]";
    }
}
```

❷ 配置<resultMap>元素

在 SQL 映射文件 UserMapper.xml 中配置<resultMap>元素，其属性 type 引用 POJO 类。具体配置如下：

```
<!-- 使用自定义结果集类型 -->
<resultMap type="com.pojo.MapUser" id="myResult">
    <!-- property 是 com.pojo.MapUser 类中的属性-->
    <!-- column 是查询结果的列名，可以来自不同的表 -->
    <id property="m_uid" column="uid"/>
    <result property="m_uname" column="uname"/>
    <result property="m_usex" column="usex"/>
</resultMap>
```

❸ 配置<select>元素

在 SQL 映射文件 UserMapper.xml 中配置<select>元素，其属性 resultMap 引用了<resultMap>元素的 id。具体配置如下：

```
<!-- 使用自定义结果集类型查询所有用户 -->
<select id="selectResultMap" resultMap="myResult">
    select * from user
</select>
```

❹ 添加接口方法

在 com.dao.UserDao 接口中添加以下接口方法：

```
public List<MapUser> selectResultMap();
```

❺ 调用接口方法

在 com.controller 包的 UserController 类中调用接口方法，具体代码如下：

```
//使用 resultMap 映射结果集
List<MapUser> listResultMap = userDao.selectResultMap();
for (MapUser myUser : listResultMap) {
    System.out.println(myUser);
}
```

7.8　级联查询

级联关系是一个数据库实体的概念，有 3 种级联关系，分别是一对一级联、一对多级联以及多对多级联。级联的优点是获取关联数据十分方便，但是级联过多会增加数据库系统的复杂度，同时降低系统的性能。在实际开发中要根据实际情况判断是否需要使用级联。更新和删除的级联关系很简单，由数据库内在机制即可完成。本节只讲述级联查询的相关实现。

如果表 A 中有一个外键引用了表 B 的主键，A 表就是子表，B 表就是父表。当查询表 A 的数据时，通过表 A 的外键将表 B 的相关记录返回，这就是级联查询。例如，当查询一个人的信息时，同时根据外键（身份证号）将他的身份证信息返回。

7.8.1　一对一级联查询

视频讲解

一对一级联关系在现实生活中是十分常见的，例如一个大学生只有一张一卡通，一张一卡通只属于一个学生。再如人与身份证的关系也是一对一的级联关系。

MyBatis 如何处理一对一级联查询呢？在 MyBatis 中，通过<resultMap>元素的子元素<association>处理这种一对一级联关系。在<association>元素中通常使用以下属性。

- property：指定映射到实体类的对象属性。
- column：指定表中对应的字段（即查询返回的列名）。
- javaType：指定映射到实体对象属性的类型。
- select：指定引入嵌套查询的子 SQL 语句，该属性用于关联映射中的嵌套查询。

下面以个人与身份证之间的关系为例讲解一对一级联查询的处理过程，读者只需参考该实例即可学会一对一级联查询的 MyBatis 实现。

❶ 创建数据表

本实例需要两张数据表，一张是身份证表 idcard，一张是个人信息表 person。这两张表具有一对一的级联关系，它们的创建代码如下：

```
CREATE TABLE 'idcard' (
  'id' int(10) NOT NULL AUTO_INCREMENT,
  'code' varchar(18) COLLATE utf8_unicode_ci DEFAULT NULL,
```

```
  PRIMARY KEY ('id')
);
CREATE TABLE 'person' (
  'id' int(10) NOT NULL,
  'name' varchar(20) COLLATE utf8_unicode_ci DEFAULT NULL,
  'age' int(11) DEFAULT NULL,
  'idcard_id' int(10) DEFAULT NULL,
  PRIMARY KEY ('id'),
  KEY 'idcard_id' ('idcard_id'),
  CONSTRAINT 'idcard_id' FOREIGN KEY ('idcard_id') REFERENCES 'idcard' ('id')
);
```

❷ 创建持久化类

在 ch7 应用的 com.po 包中创建数据表对应的持久化类 Idcard 和 Person。

Idcard 的代码如下：

```
package com.po;
/**
 * springtest 数据库中 idcard 表的持久化类
 */
public class Idcard {
    private Integer id;
    private String code;
    //省略 setter 和 getter 方法
    /**
     * 为方便测试，重写了 toString 方法
     */
    @Override
    public String toString() {
        return "Idcard [id=" + id + ",code="+ code + "]";
    }
}
```

Person 的代码如下：

```
package com.po;
/**
 * springtest 数据库中 person 表的持久化类
 */
public class Person {
    private Integer id;
    private String name;
    private Integer age;
    //个人身份证关联
    private Idcard card;
    //省略 setter 和 getter 方法
    @Override
```

```
    public String toString() {
        return "Person [id=" + id + ",name=" + name + ",age=" + age +",card="
        + card +"]" ;
    }
}
```

❸ 创建映射文件

首先，在 MyBatis 的核心配置文件 mybatis-config.xml（com.mybatis）中打开延迟加载开关，代码如下：

```
<!-- 在使用 MyBatis 嵌套查询方式进行关联查询时，使用 MyBatis 的延迟加载可以在一定程度上提高查询效率 -->
<settings>
    <!-- 打开延迟加载的开关 -->
    <setting name="lazyLoadingEnabled" value="true"/>
    <!-- 将积极加载改为按需加载 -->
    <setting name="aggressiveLazyLoading" value="false"/>
</settings>
```

然后，在 ch7 应用的 com.mybatis 中创建两张表对应的映射文件 IdCardMapper.xml 和 PersonMapper.xml。在 PersonMapper.xml 文件中以 3 种方式实现“根据 id 查询个人信息”的功能，详情请看代码备注。

IdCardMapper.xml 的代码如下：

```
<?xml version="1.0" encoding="UTF-8" ?>
<!DOCTYPE mapper
PUBLIC "-//mybatis.org//DTD Mapper 3.0//EN"
"http://mybatis.org/dtd/mybatis-3-mapper.dtd">
<mapper namespace="com.dao.IdCardDao">
    <select id="selectCodeById" parameterType="Integer" resultType=
    "com.po.Idcard">
        select * from idcard where id=#{id}
    </select>
</mapper>
```

PersonMapper.xml 的代码如下：

```
<?xml version="1.0" encoding="UTF-8" ?>
<!DOCTYPE mapper
PUBLIC "-//mybatis.org//DTD Mapper 3.0//EN"
"http://mybatis.org/dtd/mybatis-3-mapper.dtd">
<mapper namespace="com.dao.PersonDao">
    <!-- 一对一 根据 id 查询个人信息：级联查询的第一种方法（嵌套查询，执行两个 SQL 语句） -->
    <resultMap type="com.po.Person" id="cardAndPerson1">
        <id property="id" column="id"/>
        <result property="name" column="name"/>
        <result property="age" column="age"/>
```

```
        <!-- 一对一级联查询 -->
        <association property="card" column="idcard_id" javaType=
        "com.po.Idcard"
        select="com.dao.IdCardDao.selectCodeById"/>
    </resultMap>
    <select id="selectPersonById1" parameterType="Integer" resultMap=
    "cardAndPerson1">
        select * from person where id=#{id}
    </select>
    <!-- 一对一 根据 id 查询个人信息：级联查询的第二种方法（嵌套结果，执行一个 SQL 语句） -->
    <resultMap type="com.po.Person" id="cardAndPerson2">
        <id property="id" column="id"/>
        <result property="name" column="name"/>
        <result property="age" column="age"/>
        <!-- 一对一级联查询 -->
        <association property="card" javaType="com.po.Idcard">
            <id property="id" column="idcard_id"/>
            <result property="code" column="code"/>
        </association>
    </resultMap>
    <select id="selectPersonById2" parameterType="Integer" resultMap=
    "cardAndPerson2">
        select p.*,ic.code
        from person p, idcard ic
        where p.idcard_id=ic.id and p.id=#{id}
    </select>
    <!-- 一对一 根据 id 查询个人信息：连接查询（使用 POJO 存储结果） -->
    <select id="selectPersonById3" parameterType="Integer" resultType=
    "com.pojo.SelectPersonById">
        select p.*,ic.code
        from person p, idcard ic
        where p.idcard_id = ic.id and p.id=#{id}
    </select>
</mapper>
```

❹ 创建 POJO 类

在 ch7 应用的 com.pojo 包中创建在第 3 步中使用的 POJO 类 com.pojo.SelectPersonById。SelectPersonById 的代码如下：

```
package com.pojo;
public class SelectPersonById {
    private Integer id;
    private String name;
    private Integer age;
    private String code;
    //省略 setter 和 getter 方法
```

```
    @Override
    public String toString() {
        return "Person [id=" + id + ",name=" + name + ",age="
            + age + ",code=" + code + "]";
    }
}
```

❺ 创建数据操作接口

在 ch7 应用的 com.dao 包中创建第 3 步中映射文件对应的数据操作接口 IdCardDao 和 PersonDao。

IdCardDao 的代码如下：

```
package com.dao;
import org.springframework.stereotype.Repository;
import com.po.Idcard;
@Repository("idCardDao")
public interface IdCardDao {
    public Idcard selectCodeById(Integer i);
}
```

PersonDao 的代码如下：

```
package com.dao;
import org.springframework.stereotype.Repository;
import com.po.Person;
import com.pojo.SelectPersonById;
@Repository("personDao")
public interface PersonDao {
    public Person selectPersonById1(Integer id);
    public Person selectPersonById2(Integer id);
    public SelectPersonById selectPersonById3(Integer id);
}
```

❻ 调用接口方法及测试

在 ch7 应用的 com.controller 包中创建 OneToOneController 类，在该类中调用第 5 步的接口方法，同时创建测试类 TestOneToOne。

OneToOneController 的代码如下：

```
package com.controller;
import org.springframework.beans.factory.annotation.Autowired;
import org.springframework.stereotype.Controller;
import com.dao.PersonDao;
import com.po.Person;
import com.pojo.SelectPersonById;
```

```
@Controller("oneToOneController")
public class OneToOneController {
    @Autowired
    private PersonDao personDao;
    public void test() {
        Person p1=personDao.selectPersonById1(1);
        System.out.println(p1);
        System.out.println("=======================");
        Person p2=personDao.selectPersonById2(1);
        System.out.println(p2);
        System.out.println("=======================");
        SelectPersonById p3 = personDao.selectPersonById3(1);
        System.out.println(p3);
    }
}
```

TestOneToOne 的代码如下：

```
package com.controller;
import org.springframework.context.ApplicationContext;
import org.springframework.context.support.ClassPathXmlApplicationContext;
public class TestOneToOne {
    public static void main(String[] args) {
        ApplicationContext appCon=new ClassPathXmlApplicationContext
       ("applicationContext.xml");
        OneToOneController oto = (OneToOneController)appCon.getBean ("oneToOne-
        Controller");
        oto.test();
    }
}
```

上述测试类的运行结果如图 7.1 所示。

```
Servers  Console
<terminated> TestOneToOne [Java Application] C:\Program Files\Java\jre1.8.0_152\bin\javaw.exe (2018年1月23
DEBUG [main] - ==>  Preparing: select * from person where id=?
DEBUG [main] - ==> Parameters: 1(Integer)
DEBUG [main] - <==      Total: 1
DEBUG [main] - ==>  Preparing: select * from idcard where id=?
DEBUG [main] - ==> Parameters: 1(Integer)
DEBUG [main] - <==      Total: 1
Person [id=1,name=陈恒,age=88,card=Idcard [id=1,code=123456789123456789]]
=======================
DEBUG [main] - ==>  Preparing: select p.*,ic.code from person p, idcard
DEBUG [main] - ==> Parameters: 1(Integer)
DEBUG [main] - <==      Total: 1
Person [id=1,name=陈恒,age=88,card=Idcard [id=1,code=123456789123456789]]
=======================
DEBUG [main] - ==>  Preparing: select p.*,ic.code from person p, idcard
DEBUG [main] - ==> Parameters: 1(Integer)
DEBUG [main] - <==      Total: 1
Person [id=1,name=陈恒,age=88,code=123456789123456789]
```

图 7.1　一对一级联查询结果

7.8.2 一对多级联查询

视频讲解

在7.8.1节学习了MyBatis如何处理一对一级联查询，那么MyBatis又是如何处理一对多级联查询的呢？在实际生活中一对多级联关系有许多，例如一个用户可以有多个订单，而一个订单只属于一个用户。

下面以用户和订单之间的关系为例讲解一对多级联查询（实现“根据uid查询用户及其关联的订单信息”的功能）的处理过程，读者只需参考该实例即可学会一对多级联查询的MyBatis实现。

❶ 创建数据表

本实例需要两张数据表，一张是用户表user，一张是订单表orders，这两张表具有一对多的级联关系。user表在前面已创建，orders表的创建代码如下：

```
CREATE TABLE 'orders' (
  'id' int(10) NOT NULL AUTO_INCREMENT,
  'ordersn' varchar(10) COLLATE utf8_unicode_ci DEFAULT NULL,
  'user_id' int(10) DEFAULT NULL,
  PRIMARY KEY ('id'),
  KEY 'user_id' ('user_id'),
  CONSTRAINT 'user_id' FOREIGN KEY ('user_id') REFERENCES 'user' ('uid')
);
```

❷ 创建持久化类

在ch7应用的com.po包中创建数据表orders对应的持久化类Orders，user表对应的持久化类MyUser在前面已创建，但需要为MyUser添加如下属性：

```
//一对多级联查询，用户关联的订单
private List<Orders> ordersList;
```

同时，需要为该属性添加setter和getter方法。

Orders类的代码如下：

```
package com.po;
import java.util.List;

/**
 *springtest 数据库中 orders 表的持久化类
 */
public class Orders {
    private Integer id;
    private String ordersn;
    //省略 setter 和 getter 方法
    @Override
    public String toString() {
```

```
        return "Orders [id=" + id + ",ordersn=" + ordersn + "]";
    }
}
```

❸ 创建映射文件

在 ch7 应用的 com.mybatis 中创建两张表对应的映射文件 UserMapper.xml 和 OrdersMapper.xml。映射文件 UserMapper.xml 在前面已创建，但需要添加以下配置才能实现一对多级联查询（根据 uid 查询用户及其关联的订单信息）：

```
<!-- 一对多 根据uid查询用户及其关联的订单信息：级联查询的第一种方法（嵌套查询） -->
<resultMap type="com.po.MyUser" id="userAndOrders1">
    <id property="uid" column="uid"/>
    <result property="uname" column="uname"/>
    <result property="usex" column="usex"/>
    <!-- 一对多级联查询，ofType表示集合中的元素类型，将uid传递给selectOrdersById -->
    <collection property="ordersList" ofType="com.po.Orders" column="uid"
     select="com.dao.OrdersDao.selectOrdersById"/>
</resultMap>
<select  id="selectUserOrdersById1"  parameterType="Integer"  resultMap=
"userAndOrders1">
    select * from user where uid = #{id}
</select>
<!-- 一对多 根据uid查询用户及其关联的订单信息：级联查询的第二种方法（嵌套结果） -->
<resultMap type="com.po.MyUser" id="userAndOrders2">
    <id property="uid" column="uid"/>
    <result property="uname" column="uname"/>
    <result property="usex" column="usex"/>
    <!-- 一对多级联查询，ofType表示集合中的元素类型 -->
    <collection property="ordersList" ofType="com.po.Orders" >
        <id property="id" column="id"/>
        <result property="ordersn" column="ordersn"/>
    </collection>
</resultMap>
<select  id="selectUserOrdersById2"  parameterType="Integer"  resultMap=
"userAndOrders2">
    select u.*,o.id,o.ordersn from user u, orders o where u.uid = o.user_id
    and u.uid=#{id}
</select>
<!-- 一对多 根据uid查询用户及其关联的订单信息：连接查询（使用POJO存储结果） -->
<select id="selectUserOrdersById3" parameterType="Integer"
    esultType="com.pojo.SelectUserOrdersById">
    select u.*,o.id,o.ordersn from user u, orders o where u.uid = o.user_id
    and u.uid=#{id}
</select>
```

OrdersMapper.xml 的配置代码如下：

```
<?xml version="1.0" encoding="UTF-8" ?>
<!DOCTYPE mapper
PUBLIC "-//mybatis.org//DTD Mapper 3.0//EN"
"http://mybatis.org/dtd/mybatis-3-mapper.dtd">
<mapper namespace="com.dao.OrdersDao">
    <!-- 根据用户 uid 查询订单信息 -->
    <select id="selectOrdersById" parameterType="Integer" resultType="com.
    po.Orders">
        select * from orders where user_id=#{id}
    </select>
</mapper>
```

❹ 创建 POJO 类

在 ch7 应用的 com.pojo 包中创建在第 3 步中使用的 POJO 类 com.pojo.SelectUserOrdersById。

SelectUserOrdersById 的代码如下：

```
package com.pojo;
public class SelectUserOrdersById {
    private Integer uid;
    private String uname;
    private String usex;
    private Integer id;
    private String ordersn;
    //省略 setter 和 getter 方法
    @Override
    public String toString() {
        return "User [uid=" + uid +",uname=" + uname + ",usex=" + usex + ",oid="
        + id +",ordersn=" + ordersn + "]";
    }
}
```

❺ 创建数据操作接口

在 ch7 应用的 com.dao 包中创建第 3 步中映射文件对应的数据操作接口 OrdersDao 和 UserDao。

OrdersDao 的代码如下：

```
package com.dao;
import java.util.List;
import org.springframework.stereotype.Repository;
import com.po.Orders;
@Repository("ordersDao")
public interface OrdersDao {
    public List<Orders> selectOrdersById(Integer uid);
}
```

UserDao 接口在前面已创建，这里只需添加如下接口方法：

```
public MyUser selectUserOrdersById1(Integer uid);
public MyUser selectUserOrdersById2(Integer uid);
public List<SelectUserOrdersById> selectUserOrdersById3(Integer uid);
```

❻ 调用接口方法及测试

在 ch7 应用的 com.controller 包中创建 OneToMoreController 类，在该类中调用第 5 步的接口方法，同时创建测试类 TestOneToMore。

OneToMoreController 的代码如下：

```
package com.controller;
import java.util.List;
import org.springframework.beans.factory.annotation.Autowired;
import org.springframework.stereotype.Controller;
import com.dao.UserDao;
import com.po.MyUser;
import com.pojo.SelectUserOrdersById;
@Controller("oneToMoreController")
public class OneToMoreController {
   @Autowired
   private UserDao userDao;
   public void test() {
      //查询一个用户及订单信息
      MyUser auser1=userDao.selectUserOrdersById1(1);
      System.out.println(auser1);
      System.out.println("=====================================");
      MyUser auser2=userDao.selectUserOrdersById2(1);
      System.out.println(auser2);
      System.out.println("=====================================");
      List<SelectUserOrdersById> auser3 = userDao.selectUserOrdersById3(1);
      System.out.println(auser3);
      System.out.println("=====================================");
   }
}
```

TestOneToMore 的代码如下：

```
package com.controller;
import org.springframework.context.ApplicationContext;
import org.springframework.context.support.ClassPathXmlApplicationContext;
public class TestOneToMore {
   public static void main(String[] args) {
      ApplicationContext appCon=new ClassPathXmlApplicationContext
      ("applicationContext.xml");
      OneToMoreController otm=(OneToMoreController)appCon.getBean
      ("oneToMoreController");
```

```
        otm.test();
    }
}
```

测试类的运行结果如图 7.2 所示。

```
Servers  Console
<terminated> TestOneToMore [Java Application] C:\Program Files\Java\jre1.8.0_152\bin\javaw.exe (2018年1月24日 上午6:10:26)
DEBUG [main] - ==>  Preparing: select * from user where uid = ?
DEBUG [main] - ==> Parameters: 1(Integer)
DEBUG [main] - <==      Total: 1
DEBUG [main] - ==>  Preparing: select * from orders where user_id=?
DEBUG [main] - ==> Parameters: 1(Integer)
DEBUG [main] - <==      Total: 2
User [uid=1,uname=张三,usex=女,ordersList=[Orders [id=1,ordersn=999999], Orders [id=2,ordersn=88888]]]
===================================
DEBUG [main] - ==>  Preparing: select u.*,o.id,o.ordersn from user u, orders o where u.uid = o.user_id ar
DEBUG [main] - ==> Parameters: 1(Integer)
DEBUG [main] - <==      Total: 2
User [uid=1,uname=张三,usex=女,ordersList=[Orders [id=1,ordersn=999999], Orders [id=2,ordersn=88888]]]
===================================
DEBUG [main] - ==>  Preparing: select u.*,o.id,o.ordersn from user u, orders o where u.uid = o.user_id ar
DEBUG [main] - ==> Parameters: 1(Integer)
DEBUG [main] - <==      Total: 2
[User [uid=1,uname=张三,usex=女,oid=1,ordersn=999999], User [uid=1,uname=张三,usex=女,oid=2,ordersn=88888]]
```

图 7.2 一对多级联查询结果

7.8.3 多对多级联查询

视频讲解

其实，MyBatis 没有实现多对多级联，这是因为多对多级联可以通过两个一对多级联进行替换。例如，一个订单可以有多种商品，一种商品可以对应多个订单，订单与商品就是多对多的级联关系，使用一个中间表（订单记录表）就可以将多对多级联转换成两个一对多的关系。下面以订单和商品（实现“查询所有订单以及每个订单对应的商品信息”的功能）为例讲解多对多级联查询。

❶ 创建数据表

订单表在前面已创建，这里需要创建商品表 product 和订单记录表 orders_detail，创建代码如下：

```
CREATE TABLE 'product' (
  'id' int(10) NOT NULL,
  'name' varchar(50) COLLATE utf8_unicode_ci DEFAULT NULL,
  'price' double DEFAULT NULL,
  PRIMARY KEY ('id')
);

CREATE TABLE 'orders_detail' (
  'id' int(10) NOT NULL AUTO_INCREMENT,
  'orders_id' int(10) DEFAULT NULL,
  'product_id' int(10) DEFAULT NULL,
  PRIMARY KEY ('id'),
```

```
  KEY 'orders_id' ('orders_id'),
  KEY 'product_id' ('product_id'),
  CONSTRAINT 'orders_id' FOREIGN KEY ('orders_id') REFERENCES 'orders' ('id'),
  CONSTRAINT 'product_id' FOREIGN KEY ('product_id') REFERENCES 'product' ('id')
);
```

❷ 创建持久化类

在 ch7 应用的 com.po 包中创建数据表 product 对应的持久化类 Product，而中间表 orders_detail 不需要持久化类，但需要在订单表 orders 对应的持久化类 Orders 中添加关联属性。

Product 的代码如下：

```
package com.po;
import java.util.List;
public class Product {
    private Integer id;
    private String name;
    private Double price;
    //多对多中的一个一对多
    private List<Orders> orders;
    //省略 setter 和 getter 方法
    @Override
    public String toString() {
        return "Product [id=" + id + ",name=" + name + ",price=" + price + "]";
    }
}
```

Orders 的代码如下：

```
package com.po;
import java.util.List;

/**
 *springtest 数据库中 orders 表的持久化类
 */
public class Orders {
    private Integer id;
    private  String ordersn;
    //多对多中的另一个一对多
    private List<Product> products;
    //省略 setter 和 getter 方法
    @Override
    public String toString() {
        return "Orders [id=" + id + ",ordersn=" + ordersn + ",products=" +
        products + "]";
```

```
    }
}
```

❸ **创建映射文件**

本实例只需在 com.mybatis 的 OrdersMapper.xml 文件中追加以下配置即可实现多对多级联查询。

```
<!-- 多对多级联 查询所有订单以及每个订单对应的商品信息（嵌套结果） -->
<resultMap type="com.po.Orders" id="allOrdersAndProducts">
    <id property="id" column="id"/>
    <result property="ordersn" column="ordersn"/>
    <!-- 多对多级联 -->
    <collection property="products" ofType="com.po.Product">
        <id property="id" column="pid"/>
        <result property="name" column="name"/>
        <result property="price" column="price"/>
    </collection>
</resultMap>
<select id="selectallOrdersAndProducts" resultMap="allOrdersAndProducts">
    select o.*,p.id as pid,p.name,p.price
    from orders o,orders_detail od,product p
    where od.orders_id = o.id
    and od.product_id = p.id
</select>
```

❹ **创建 POJO 类**

该实例不需要创建 POJO 类。

❺ **添加数据操作接口方法**

在 Orders 接口中添加以下接口方法：

```
public List<Orders> selectallOrdersAndProducts();
```

❻ **调用接口方法及测试**

在 ch7 应用的 com.controller 包中创建 MoreToMoreController 类，在该类中调用第 5 步的接口方法，同时创建测试类 TestMoreToMore。

MoreToMoreController 的代码如下：

```
package com.controller;
import java.util.List;
import org.springframework.beans.factory.annotation.Autowired;
import org.springframework.stereotype.Controller;
import com.dao.OrdersDao;
import com.po.Orders;
@Controller("moreToMoreController")
public class MoreToMoreController {
    @Autowired
```

```
    private OrdersDao ordersDao;
    public void test() {
        List<Orders> os=ordersDao.selectallOrdersAndProducts();
        for (Orders orders:os) {
            System.out.println(orders);
        }
    }
}
```

TestMoreToMore 的代码如下：

```
package com.controller;
import org.springframework.context.ApplicationContext;
import org.springframework.context.support.ClassPathXmlApplicationContext;
public class TestMoreToMore {
    public static void main(String[] args) {
        ApplicationContext appCon=new ClassPathXmlApplicationContext
       ("applicationContext.xml");
     MoreToMoreController otm=(MoreToMoreController)appCon.getBean
     ("moreToMoreController");
        otm.test();
    }
}
```

上述测试类的运行结果如图 7.3 所示。

```
Servers  Console
<terminated> TestMoreToMore [Java Application] C:\Program Files\Java\jre1.8.0_152\bin\javaw.exe (2018年1月24日 上
DEBUG [main] - ==>  Preparing: select o.*,p.id as pid,p.name,p.price from or
DEBUG [main] - ==> Parameters:
DEBUG [main] - <==      Total: 5
Orders [id=1,ordersn=999999,products=[Product [id=1,name=好书,price=88.0], Pr
Orders [id=2,ordersn=88888,products=[Product [id=1,name=好书,price=88.0], Pro
Orders [id=3,ordersn=7777777,products=[Product [id=1,name=好书,price=88.0]]]
```

图 7.3　多对多级联查询结果

7.9　本章小结

本章首先简要介绍了 MyBatis 的核心配置文件，这是因为核心配置文件不会轻易改动；然后详细讲解了 SQL 映射文件的主要元素，包括<select>、<resultMap>等元素；最后以实例为主讲解了级联查询的 MyBatis 实现，读者参考这些实例即可实现自己需要的级联查询。

习题 7

1．MyBatis 实现查询时返回的结果集有几种常见的存储方式？请举例说明。
2．在 MyBatis 中针对不同的数据库软件，<insert>元素如何将主键回填？
3．在 MyBatis 中如何给 SQL 语句传递参数？

第 8 章

动态 SQL

学习目的与要求

本章重点讲解如何拼接 MyBatis 的动态 SQL 语句。通过本章的学习，读者能够掌握 MyBatis 动态 SQL 语句的拼接语法。

本章主要内容

- <if>元素；
- <choose>、<when>、<otherwise>元素；
- <trim>、<where>、<set>元素；
- <foreach>元素；
- <bind>元素。

开发人员通常根据需求手动拼接 SQL 语句，这是一个极其麻烦的工作，而 MyBatis 提供了对 SQL 语句动态组装的功能，恰能解决这一问题。MyBatis 的动态 SQL 元素与 JSTL 或 XML 文本处理器相似，常用<if>、<choose>、<when>、<otherwise>、<trim>、<where>、<set>、<foreach>和<bind>等元素。

为测试动态 SQL 元素，本章创建 ch8 应用，并将第 6 章的 ch6SS 应用的所有 JAR 包和 src 中所有 Java 程序与 XML 文件都复制到 ch8 的相应位置。

8.1 <if>元素

视频讲解

动态 SQL 通常要做的事情是有条件地包含 where 子句的一部分，所以在 MyBatis 中<if>元素是最常用的元素，它类似于 Java 中的 if 语句。在 ch8 应用中测试<if>元素，具体过程如下：

❶ 添加 SQL 映射语句

在 com.mybatis 包的 UserMapper.xml 文件中添加如下 SQL 映射语句：

```
<!-- 使用 if 元素根据条件动态查询用户信息 -->
<select  id="selectUserByIf"  resultType="com.po.MyUser"  parameterType=
"com.po.MyUser">
    select * from user where 1=1
    <if test="uname!=null and uname!=''">
        and uname like concat('%',#{uname},'%')
    </if>
    <if test="usex!=null and usex!=''">
        and usex=#{usex}
    </if>
</select>
```

❷ 添加数据操作接口方法

在 com.dao 包的 UserDao 接口中添加如下数据操作接口方法：

```
public List<MyUser> selectUserByIf(MyUser user);
```

❸ 调用数据操作接口方法

在 com.controller 包的 UserController 类中添加如下程序调用数据操作接口方法。

```
//使用 if 元素查询用户信息
MyUser ifmu=new MyUser();
ifmu.setUname("张");
ifmu.setUsex("女");
List<MyUser> listByif=userDao.selectUserByIf(ifmu);
System.out.println("if 元素=================");
for (MyUser myUser:listByif) {
    System.out.println(myUser);
}
```

❹ 测试动态 SQL 语句

运行 com.controller 包中的 TestController 主类，测试动态 SQL 语句。

8.2 <choose>、<when>、<otherwise>元素

有些时候不想用到所有的条件语句，而只想从中择取一二，针对这种情况，MyBatis 提供了<choose>元素，它有点像 Java 中的 switch 语句。在 ch8 应用中测试<choose>元素，具体过程如下：

视频讲解

❶ 添加 SQL 映射语句

在 com.mybatis 包的 UserMapper.xml 文件中添加如下 SQL 映射语句：

```
<!-- 使用 choose、when、otherwise 元素根据条件动态查询用户信息 -->
<select id="selectUserByChoose" resultType="com.po.MyUser" parameterType=
"com.po.MyUser">
    select*from user where 1=1
```

```
        <choose>
        <when test="uname!=null and uname!=''">
            and uname like concat('%',#{uname},'%')
        </when>
        <when test="usex !=null and usex!=''">
            and usex=#{usex}
        </when>
        <otherwise>
            and uid > 10
        </otherwise>
        </choose>
</select>
```

❷ **添加数据操作接口方法**

在 com.dao 包的 UserDao 接口中添加如下数据操作接口方法：

```
public List<MyUser> selectUserByChoose(MyUser user);
```

❸ **调用数据操作接口方法**

在 com.controller 包的 UserController 类中添加如下程序调用数据操作接口方法。

```
//使用 choose 元素查询用户信息
MyUser choosemu=new MyUser();
choosemu.setUname("");
choosemu.setUsex("");
List<MyUser> listByChoose=userDao.selectUserByChoose(choosemu);
System.out.println("choose 元素=================");
for (MyUser myUser:listByChoose) {
    System.out.println(myUser);
}
```

❹ **测试动态 SQL 语句**

运行 com.controller 包中的 TestController 主类，测试动态 SQL 语句。

8.3 <trim>、<where>、<set>元素

视频讲解

8.3.1 <trim>元素

<trim>元素的主要功能是可以在自己包含的内容前加上某些前缀，也可以在其后加上某些后缀，与之对应的属性是 prefix 和 suffix；可以把包含内容的首部某些内容覆盖，即忽略，也可以把尾部的某些内容覆盖，对应的属性是 prefixOverrides 和 suffixOverrides。正因为<trim>元素有这样的功能，所以也可以非常简单地利用<trim>来代替<where>元素的功能。在 ch8 应用中测试<trim>元素，具体过程如下：

❶ **添加 SQL 映射语句**

在 com.mybatis 包的 UserMapper.xml 文件中添加如下 SQL 映射语句：

```
<!-- 使用 trim 元素根据条件动态查询用户信息 -->
<select id="selectUserByTrim" resultType="com.po.MyUser" parameterType=
"com.po.MyUser">
    select * from user
    <trim prefix="where" prefixOverrides="and |or">
        <if test="uname!=null and uname!=''">
            and uname like concat('%',#{uname},'%')
        </if>
        <if test="usex!=null and usex!=''">
            and usex=#{usex}
        </if>
    </trim>
</select>
```

❷ **添加数据操作接口方法**

在 com.dao 包的 UserDao 接口中添加如下数据操作接口方法：

```
public List<MyUser> selectUserByTrim(MyUser user);
```

❸ **调用数据操作接口方法**

在 com.controller 包的 UserController 类中添加如下程序调用数据操作接口方法。

```
//使用 trim 元素查询用户信息
MyUser trimmu=new MyUser();
trimmu.setUname("张");
trimmu.setUsex("男");
List<MyUser> listByTrim=userDao.selectUserByTrim(trimmu);
System.out.println("trim 元素=================");
for (MyUser myUser:listByTrim) {
    System.out.println(myUser);
}
```

❹ **测试动态 SQL 语句**

运行 com.controller 包中的 TestController 主类，测试动态 SQL 语句。

8.3.2　<where>元素

<where>元素的作用是会在写入<where>元素的地方输出一个 where 语句，另外一个好处是不需要考虑<where>元素里面的条件输出是什么样子的，MyBatis 将智能处理。如果所有的条件都不满足，那么 MyBatis 就会查出所有的记录，如果输出后是以 and 开头的，MyBatis 会把第一个 and 忽略，当然如果是以 or 开头的，MyBatis 也会把它忽略；此外，在<where>元素中不需要考虑空格的问题，MyBatis 将智能加上。在 ch8 应用中测试<where>元素，具体过程如下：

❶ 添加 SQL 映射语句

在 com.mybatis 包的 UserMapper.xml 文件中添加如下 SQL 映射语句：

```
<!-- 使用 where 元素根据条件动态查询用户信息 -->
<select id="selectUserByWhere" resultType="com.po.MyUser" parameterType=
"com.po.MyUser">
    select * from user
    <where>
        <if test="uname!=null and uname!=''">
            and uname like concat('%',#{uname},'%')
        </if>
        <if test="usex!=null and usex!=''">
            and usex=#{usex}
        </if>
    </where>
</select>
```

❷ 添加数据操作接口方法

在 com.dao 包的 UserDao 接口中添加如下数据操作接口方法：

```
public List<MyUser> selectUserByWhere(MyUser user);
```

❸ 调用数据操作接口方法

在 com.controller 包的 UserController 类中添加如下程序调用数据操作接口方法。

```
//使用 where 元素查询用户信息
MyUser wheremu=new MyUser();
wheremu.setUname("张");
wheremu.setUsex("男");
List<MyUser> listByWhere=userDao.selectUserByWhere(wheremu);
System.out.println("where 元素=================");
for (MyUser myUser:listByWhere) {
    System.out.println(myUser);
}
```

❹ 测试动态 SQL 语句

运行 com.controller 包中的 TestController 主类，测试动态 SQL 语句。

8.3.3 <set>元素

在动态 update 语句中可以使用<set>元素动态更新列。在 ch8 应用中测试<set>元素，具体过程如下：

❶ 添加 SQL 映射语句

在 com.mybatis 包的 UserMapper.xml 文件中添加如下 SQL 映射语句：

```
<!-- 使用 set 元素动态修改一个用户 -->
<update id="updateUserBySet" parameterType="com.po.MyUser">
```

```
    update user
    <set>
        <if test="uname!=null">uname=#{uname},</if>
        <if test="usex!=null">usex=#{usex}</if>
    </set>
    where uid=#{uid}
</update>
```

❷ **添加数据操作接口方法**

在 com.dao 包的 UserDao 接口中添加如下数据操作接口方法：

```
public int updateUserBySet(MyUser user);
```

❸ **调用数据操作接口方法**

在 com.controller 包的 UserController 类中添加如下程序调用数据操作接口方法。

```
//使用 set 元素修改一个用户
MyUser setmu=new MyUser();
setmu.setUid(1);
setmu.setUname("张九");
int setup=userDao.updateUserBySet(setmu);
System.out.println("set 元素修改了" + setup + "条记录");
System.out.println( "================");
```

❹ **测试动态 SQL 语句**

运行 com.controller 包中的 TestController 主类，测试动态 SQL 语句。

8.4 <foreach>元素

视频讲解

<foreach>元素主要用在构建 in 条件中，它可以在 SQL 语句中迭代一个集合。<foreach>元素的属性主要有 item、index、collection、open、separator、close。item 表示集合中每一个元素进行迭代时的别名，index 指定一个名字，用于表示在迭代过程中每次迭代到的位置，open 表示该语句以什么开始，separator 表示在每次进行迭代之间以什么符号作为分隔符，close 表示以什么结束。在使用<foreach>元素时，最关键、最容易出错的是 collection 属性，该属性是必选的，但在不同情况下该属性的值是不一样的，主要有以下 3 种情况：

- 如果传入的是单参数且参数类型是一个 List，collection 属性值为 list。
- 如果传入的是单参数且参数类型是一个 array 数组，collection 的属性值为 array。
- 如果传入的参数是多个，需要把它们封装成一个 Map，当然单参数也可以封装成 Map。Map 的 key 是参数名，collection 属性值是传入的 List 或 array 对象在自己封装的 Map 中的 key。

在 ch8 应用中测试<foreach>元素，具体过程如下：

❶ **添加 SQL 映射语句**

在 com.mybatis 包的 UserMapper.xml 文件中添加如下 SQL 映射语句：

```
<!-- 使用 foreach 元素查询用户信息 -->
<select id="selectUserByForeach" resultType="com.po.MyUser" parameterType=
"List">
    select*from user where uid in
    <foreach item="item" index="index" collection="list"
    open="(" separator="," close=")">
        #{item}
    </foreach>
</select>
```

❷ **添加数据操作接口方法**

在 com.dao 包的 UserDao 接口中添加如下数据操作接口方法：

```
public List<MyUser> selectUserByForeach(List<Integer> listId);
```

❸ **调用数据操作接口方法**

在 com.controller 包的 UserController 类中添加如下程序调用数据操作接口方法。

```
//使用 foreach 元素查询用户信息
List<Integer> listId=new ArrayList<Integer>();
listId.add(34);
listId.add(37);
List<MyUser> listByForeach = userDao.selectUserByForeach(listId);
System.out.println("foreach 元素================");
for(MyUser myUser : listByForeach) {
    System.out.println(myUser);
}
```

❹ **测试动态 SQL 语句**

运行 com.controller 包中的 TestController 主类，测试动态 SQL 语句。

8.5 <bind>元素

在进行模糊查询时，如果使用“${}”拼接字符串，则无法防止 SQL 注入问题；如果使用字符串拼接函数或连接符号，但不同数据库的拼接函数或连接符号不同，例如 MySQL 的 concat 函数、Oracle 的连接符号“||”，这样 SQL 映射文件就需要根据不同的数据库提供不同的实现，显然比较麻烦，且不利于代码的移植。幸运的是，MyBatis 提供了<bind>元素来解决这一问题。在 ch8 应用中测试<bind>元素，具体过程如下：

视频讲解

❶ **添加 SQL 映射语句**

在 com.mybatis 包的 UserMapper.xml 文件中添加如下 SQL 映射语句：

```
<!-- 使用 bind 元素进行模糊查询 -->
<select id="selectUserByBind" resultType="com.po.MyUser" parameterType=
"com.po.MyUser">
```

```
    <!-- bind中的uname是com.po.MyUser的属性名 -->
    <bind name="paran_uname" value="'%' + uname + '%'"/>
    select*from user where uname like #{paran_uname}
</select>
```

❷ **添加数据操作接口方法**

在 com.dao 包的 UserDao 接口中添加如下数据操作接口方法：

```
public List<MyUser> selectUserByBind(MyUser user);
```

❸ **调用数据操作接口方法**

在 com.controller 包的 UserController 类中添加如下程序调用数据操作接口方法。

```
//使用bind元素查询用户信息
MyUser bindmu=new MyUser();
bindmu.setUname("张");
List<MyUser> listByBind=userDao.selectUserByBind(bindmu);
System.out.println("bind元素================");
for(MyUser myUser:listByBind) {
    System.out.println(myUser);
}
```

❹ **测试动态 SQL 语句**

运行 com.controller 包中的 TestController 主类，测试动态 SQL 语句。

8.6 本章小结

本章以 ch8 应用为例详细讲解了动态 SQL 语句的使用过程，在 MyBatis 框架中这些动态 SQL 元素是十分重要的，熟练地掌握它们可以提高开发效率。

习题 8

1．简述<bind>元素的作用。

2．在动态 SQL 元素中类似分支语句的元素有哪些？如何使用它们？

第 3 部分

Spring MVC

第 9 章

Spring MVC 入门

学习目的与要求

本章重点讲解 MVC 的设计思想以及 Spring MVC 的工作原理。通过本章的学习，读者能够了解 Spring MVC 的工作原理，掌握 Spring MVC 应用的开发步骤。

本章主要内容

- Spring MVC 的工作原理;
- 第一个 Spring MVC 应用。

MVC 思想将一个应用分成 3 个基本部分，即 Model（模型）、View（视图）和 Controller（控制器），让这 3 个部分以最低的耦合进行协同工作，从而提高应用的可扩展性及可维护性。Spring MVC 是一款优秀的基于 MVC 思想的应用框架，它是 Spring 提供的一个实现了 Web MVC 设计模式的轻量级 Web 框架。

9.1 MVC 模式与 Spring MVC 工作原理

视频讲解

9.1.1 MVC 模式

❶ MVC 的概念

MVC 是 Model、View 和 Controller 的缩写，分别代表 Web 应用程序中的 3 种职责。

- 模型：用于存储数据以及处理用户请求的业务逻辑。
- 视图：向控制器提交数据，显示模型中的数据。
- 控制器：根据视图提出的请求判断将请求和数据交给哪个模型处理，将处理后的有关结果交给哪个视图更新显示。

❷ 基于 Servlet 的 MVC 模式

基于 Servlet 的 MVC 模式的具体实现如下。

- 模型：一个或多个 JavaBean 对象，用于存储数据（实体模型，由 JavaBean 类创建）和处理业务逻辑（业务模型，由一般的 Java 类创建）。
- 视图：一个或多个 JSP 页面，向控制器提交数据和为模型提供数据显示，JSP 页面主要使用 HTML 标记和 JavaBean 标记来显示数据。
- 控制器：一个或多个 Servlet 对象，根据视图提交的请求进行控制，即将请求转发给处理业务逻辑的 JavaBean，并将处理结果存放到实体模型 JavaBean 中，输出给视图显示。

基于 Servlet 的 MVC 模式的流程如图 9.1 所示。

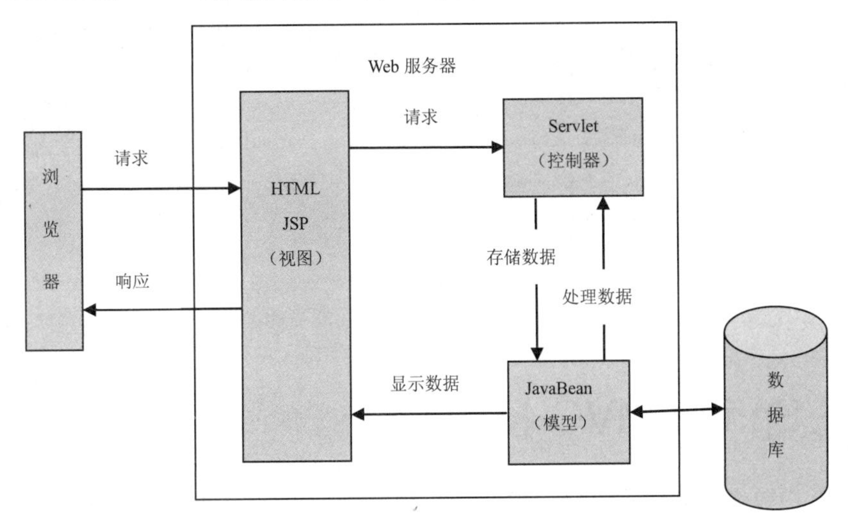

图 9.1　JSP 中的 MVC 模式

9.1.2　Spring MVC 工作原理

Spring MVC 框架是高度可配置的，包含多种视图技术，例如 JSP 技术、Velocity、Tiles、iText 和 POI。Spring MVC 框架并不关心使用的视图技术，也不会强迫开发者只使用 JSP 技术，但本书使用的视图是 JSP。

Spring MVC 框架主要由 DispatcherServlet、处理器映射、控制器、视图解析器、视图组成，其工作原理如图 9.2 所示。

从图 9.2 可总结出 Spring MVC 的工作流程如下：

（1）客户端请求提交到 DispatcherServlet；

（2）由 DispatcherServlet 控制器寻找一个或多个 HandlerMapping，找到处理请求的 Controller；

（3）DispatcherServlet 将请求提交到 Controller；

（4）Controller 调用业务逻辑处理后返回 ModelAndView；

（5）DispatcherServlet 寻找一个或多个 ViewResolver 视图解析器，找到 ModelAndView 指定的视图；

（6）视图负责将结果显示到客户端。

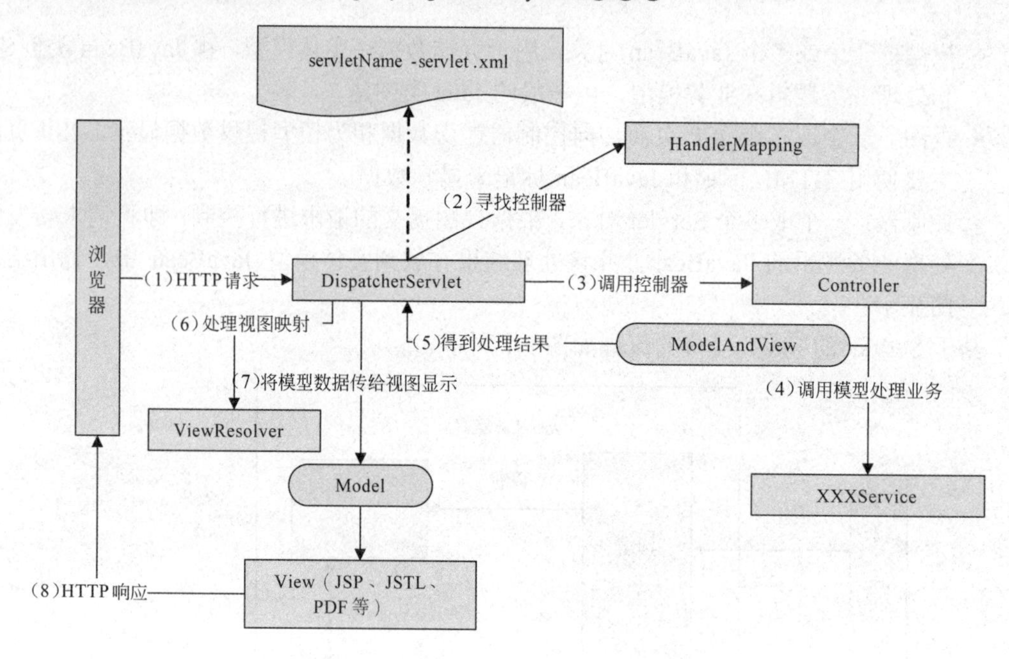

图 9.2　Spring MVC 工作原理图

9.1.3　Spring MVC 接口

在图 9.2 中包含 4 个 Spring MVC 接口，即 DispatcherServlet、HandlerMapping、Controller 和 ViewResolver。

Spring MVC 所有的请求都经过 DispatcherServlet 来统一分发，在 DispatcherServlet 将请求分发给 Controller 之前需要借助 Spring MVC 提供的 HandlerMapping 定位到具体的 Controller。

HandlerMapping 接口负责完成客户请求到 Controller 映射。

Controller 接口将处理用户请求，这和 Java Servlet 扮演的角色是一致的。一旦 Controller 处理完用户请求，将返回 ModelAndView 对象给 DispatcherServlet 前端控制器，ModelAndView 中包含了模型（Model）和视图（View）。从宏观角度考虑，DispatcherServlet 是整个 Web 应用的控制器；从微观考虑，Controller 是单个 Http 请求处理过程中的控制器，而 ModelAndView 是 Http 请求过程中返回的模型（Model）和视图（View）。

ViewResolver 接口（视图解析器）在 Web 应用中负责查找 View 对象，从而将相应结果渲染给客户。

9.2　第一个 Spring MVC 应用

视频讲解

本节通过一个简单的 Web 应用 ch9 来演示 Spring MVC 入门程序的实现过程。

9.2.1 创建 Web 应用并引入 JAR 包

在 Eclipse 中创建一个名为 ch9 的 Web 应用，在 ch9 的 lib 目录中添加 Spring MVC 程序所需要的 JAR 包，包括 Spring 的 4 个核心 JAR 包、commons-logging 的 JAR 包以及两个与 Web 相关的 JAR 包（spring-web-5.0.2.RELEASE.jar 和 spring-webmvc-5.0.2.RELEASE.jar）。

另外，在 Spring MVC 应用中使用注解时不要忘记添加 spring-aop-5.0.2.RELEASE.jar 包，添加后的 JAR 包如图 9.3 所示。

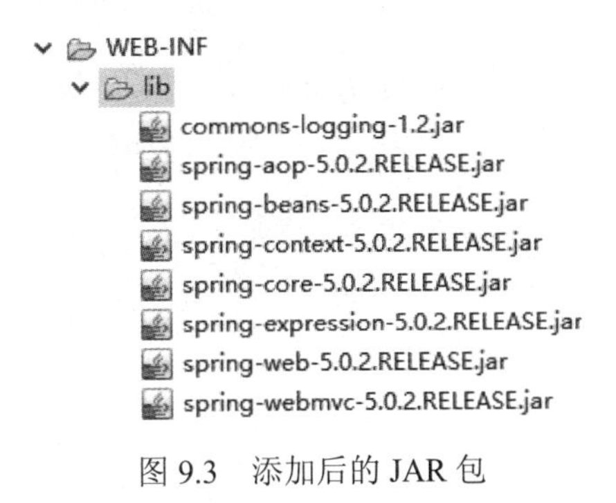

图 9.3 添加后的 JAR 包

9.2.2 在 web.xml 文件中部署 DispatcherServlet

在开发 Spring MVC 应用时需要在 web.xml 中部署 DispatcherServlet，代码如下：

```
<?xml version="1.0" encoding="UTF-8"?>
<web-app
xmlns:xsi="http://www.w3.org/2001/XMLSchema-instance"
xmlns="http://xmlns.jcp.org/xml/ns/javaee"
xsi:schemaLocation="http://xmlns.jcp.org/xml/ns/javaee
http://xmlns.jcp.org/xml/ns/javaee/web-app_3_1.xsd"
id="WebApp_ID" version="3.1">
<!-- 部署 DispatcherServlet -->
<servlet>
    <servlet-name>springmvc</servlet-name>
    <servlet-class>org.springframework.web.servlet.DispatcherServlet
    </servlet-class>
    <!-- 表示容器在启动时立即加载 servlet -->
    <load-on-startup>1</load-on-startup>
</servlet>
<servlet-mapping>
    <servlet-name>springmvc</servlet-name>
    <!-- 处理所有 URL -->
    <url-pattern>/</url-pattern>
</servlet-mapping>
</web-app>
```

上述 DispatcherServlet 的 servlet 对象 springmvc 初始化时将在应用程序的 WEB-INF 目录下查找一个配置文件（见 9.2.5 节），该配置文件的命名规则是“servletName-servlet.xml”，例如 springmvc-servlet.xml。

另外，也可以将 Spring MVC 配置文件存放在应用程序目录中的任何地方，但需要使用 servlet 的 init-param 元素加载配置文件。示例代码如下：

```
<!-- 部署 DispatcherServlet -->
<servlet>
    <servlet-name>springmvc</servlet-name>
    <servlet-class>org.springframework.web.servlet.DispatcherServlet
    </servlet-class>
    <init-param>
        <param-name>contextConfigLocation</param-name>
        <param-value>/WEN-INF/spring-config/springmvc-servlet.xml</param-value>
    </init-param>
    <load-on-startup>1</load-on-startup>
</servlet>
<servlet-mapping>
    <servlet-name>springmvc</servlet-name>
    <url-pattern>/</url-pattern>
</servlet-mapping>
```

9.2.3 创建 Web 应用首页

在 ch9 应用的 WebContent 目录下有个应用首页 index.jsp。index.jsp 的代码如下：

```
<%@ page language="java" contentType="text/html; charset=UTF-8"
   pageEncoding="UTF-8"%>
<!DOCTYPE html PUBLIC "-//W3C//DTD HTML 4.01 Transitional//EN"
"http://www.w3.org/TR/html4/loose.dtd">
<html>
<head>
<meta http-equiv="Content-Type" content="text/html; charset=UTF-8">
<title>Insert title here</title>
</head>
<body>
    未注册的用户，请<a href="${pageContext.request.contextPath }/register">
    注册</a>! <br>
    已注册的用户，去<a href="${pageContext.request.contextPath }/login">登录</a>!
</body>
</html>
```

9.2.4 创建 Controller 类

在 src 目录下创建 controller 包，并在该包中创建 RegisterController 和 LoginController 两个传统风格的控制器类（实现了 Controller 接口），分别处理首页中“注册”和“登录”

超链接的请求。

RegisterController 的具体代码如下：

```
package controller;
import javax.servlet.http.HttpServletRequest;
import javax.servlet.http.HttpServletResponse;
import org.springframework.web.servlet.ModelAndView;
import org.springframework.web.servlet.mvc.Controller;
public class RegisterController implements Controller{
    @Override
    public ModelAndView handleRequest(HttpServletRequest arg0, HttpServlet
    Response arg1) throws Exception {
        return new ModelAndView("/WEB-INF/jsp/register.jsp");
    }
}
```

LoginController 的具体代码如下：

```
package controller;
import javax.servlet.http.HttpServletRequest;
import javax.servlet.http.HttpServletResponse;
import org.springframework.web.servlet.ModelAndView;
import org.springframework.web.servlet.mvc.Controller;
public class LoginController implements Controller{
    @Override
    public ModelAndView handleRequest(HttpServletRequest arg0, HttpServlet
    Response arg1) throws Exception {
        return new ModelAndView("/WEB-INF/jsp/login.jsp");
    }
}
```

9.2.5　创建 Spring MVC 配置文件并配置 Controller 映射信息

传统风格的控制器定义之后，需要在 Spring MVC 配置文件中部署它们（学习基于注解的控制器后不再需要部署控制器）。在 WEB-INF 目录下创建名为 springmvc-servlet.xml 的配置文件（文件名的命名规则见 9.2.2 节），具体代码如下：

```
<?xml version="1.0" encoding="UTF-8"?>
<beans xmlns="http://www.springframework.org/schema/beans"
   xmlns:xsi="http://www.w3.org/2001/XMLSchema-instance"
   xsi:schemaLocation="
      http://www.springframework.org/schema/beans
      http://www.springframework.org/schema/beans/spring-beans.xsd">
```

```
    <!-- LoginController 控制器类，映射到"/login" -->
    <bean name="/login" class="controller.LoginController"/>
    <!-- RegisterController 控制器类，映射到"/register" -->
    <bean name="/register" class="controller.RegisterController"/>
</beans>
```

9.2.6 应用的其他页面

RegisterController 控制器处理成功后跳转到/WEB-INF/jsp 下的 register.jsp 视图，LoginController 控制器处理成功后跳转到/WEB-INF/jsp 下的 login.jsp 视图，因此在应用的/WEB-INF/jsp 目录下应有 register.jsp 和 login.jsp 页面，这两个 JSP 页面的代码在此省略。

9.2.7 发布并运行 Spring MVC 应用

在 Eclipse 中第一次运行 Spring MVC 应用时需要将应用发布到 Tomcat。例如在运行 ch9 应用时可以选中应用名称 ch9 并右击，然后选择 Run As→Run on Server 命令，打开如图 9.4 所示的对话框，在对话框中单击 Finish 按钮完成发布并运行。

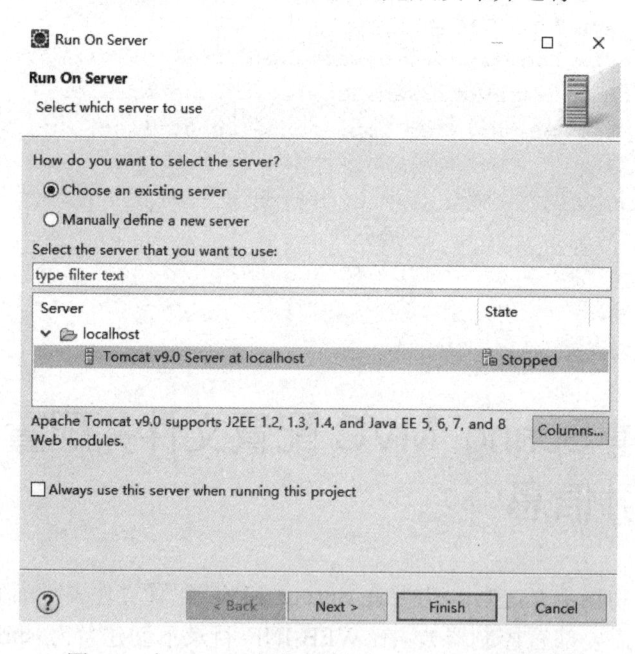

图 9.4　在 Eclipse 中发布并运行 Spring MVC 应用

通过地址"http://localhost:8080/ch9"首先访问 index.jsp 页面，如图 9.5 所示。

图 9.5　index.jsp 页面

在如图 9.5 所示的页面中，当用户单击“注册”超链接时，根据 springmvc-servlet.xml 文件中的映射将请求转发给 RegisterController 控制器处理，处理后跳转到/WEB-INF/jsp 下的 register.jsp 视图。同理，当单击“登录”超链接时，控制器处理后转到/WEB-INF/jsp 下的 login.jsp 视图。

9.3　视图解析器

用户可以在配置文件中定义 Spring MVC 的一个视图解析器（ViewResolver），示例代码如下：

```
<bean class="org.springframework.web.servlet.view.InternalResourceViewResolver"
        id="internalResourceViewResolver">
    <!-- 前缀 -->
    <property name="prefix" value="/WEB-INF/jsp/" />
    <!-- 后缀 -->
    <property name="suffix" value=".jsp" />
</bean>
```

上述视图解析器配置了前缀和后缀两个属性，因此 9.2.4 节的 RegisterController 和 LoginController 控制器类的视图路径仅需提供 register 和 login，视图解析器将会自动添加前缀和后缀。

9.4　本章小结

本章首先简单介绍了 MVC 设计模式，然后详细讲解了 Spring MVC 的工作原理，最后以 ch9 应用为例简要介绍了 Spring MVC 应用的开发步骤。

在 Spring MVC 中开发者无须编写自己的 DispatcherServlet，传统的控制器类需要实现 Controller 接口，但从 Spring 2.5 版本开始提供了基于注解的控制器。本书在 3.5.2 节简要介绍了注解的基本用法，后面所有章节的应用程序尽量使用注解的形式。

习题 9

1. 在开发 Spring MVC 应用时如何部署 DispatcherServlet？如何创建 Spring MVC 的配置文件？

2. 简述 Spring MVC 的工作流程。

第 10 章

Spring MVC 的 Controller

学习目的与要求

本章重点讲解基于注解的控制器、Controller 接收请求参数的方式以及编写请求处理方法。通过本章的学习，读者能够掌握基于注解的控制器的编写方法，掌握在 Controller 中接收请求参数以及编写请求处理方法。

本章主要内容

- 基于注解的控制器；
- 编写请求处理方法；
- Controller 接收请求参数的方式；
- 重定向和转发；
- 应用@Autowired 和@Service 进行依赖注入；
- @ModelAttribute。

在使用 Spring MVC 进行 Web 应用开发时 Controller 是 Web 应用的核心，Controller 实现类包含了对用户请求的处理逻辑，是用户请求和业务逻辑之间的“桥梁”，是 Spring MVC 框架的核心部分，负责具体的业务逻辑处理。

10.1 基于注解的控制器

在 9.2 节“第一个 Spring MVC 应用”中创建了两个传统风格的控制器，它们是实现 Controller 接口的类。传统风格的控制器不仅需要在配置文件中部署映射，而且只能编写一个处理方法，不够灵活。使用基于注解的控制器具有以下两个优点：

视频讲解

（1）在基于注解的控制器类中可以编写多个处理方法，进而可以处理多个请求（动作），这就允许将相关的操作编写在同一个控制器类中，从而减少控制器类的数量，方便以后的维护。

（2）基于注解的控制器不需要在配置文件中部署映射，仅需要使用 RequestMapping 注释类型注解一个方法进行请求处理。

在 Spring MVC 中最重要的两个注解类型是 Controller 和 RequestMapping，本章将重点介绍它们。在本章将创建一个 Spring MVC 应用 ch10 来演示相关知识，ch10 的 JAR 包、web.xml 与第 9 章 ch9 应用的 JAR 包、web.xml 完全一样。

10.1.1 Controller 注解类型

在 Spring MVC 中使用 org.springframework.stereotype.Controller 注解类型声明某类的实例是一个控制器。例如，在 ch10 应用的 src 目录下创建 controller 包，并在该包中创建 Controller 注解的控制器类 IndexController，示例代码如下：

```
package controller;
import org.springframework.stereotype.Controller;
/** “@Controller”表示 IndexController 的实例是一个控制器
 * @Controller 相当于@Controller("indexController")
 * 或@Controller(value="indexController")
 */
@Controller
public class IndexController {
    //处理请求的方法
}
```

在 Spring MVC 中使用扫描机制找到应用中所有基于注解的控制器类，所以，为了让控制器类被 Spring MVC 框架扫描到，需要在配置文件中声明 spring-context，并使用 <context:component-scan/>元素指定控制器类的基本包（请确保所有控制器类都在基本包及其子包下）。例如，在 ch10 应用的/WEB-INF/目录下创建配置文件 springmvc-servlet.xml，示例代码如下：

```
<?xml version="1.0" encoding="UTF-8"?>
<beans xmlns="http://www.springframework.org/schema/beans"
    xmlns:xsi="http://www.w3.org/2001/XMLSchema-instance"
    xmlns:p="http://www.springframework.org/schema/p"
    xmlns:context="http://www.springframework.org/schema/context"
    xsi:schemaLocation="
     http://www.springframework.org/schema/beans
    http://www.springframework.org/schema/beans/spring-beans.xsd
    http://www.springframework.org/schema/context
    http://www.springframework.org/schema/context/spring-context.xsd">
    <!-- 使用扫描机制扫描控制器类，控制器类都在 controller 包及其子包下 -->
    <context:component-scan base-package="controller"/>

   <bean class="org.springframework.web.servlet.view.InternalResourceViewResolver"
            id="internalResourceViewResolver">
      <!-- 前缀 -->
```

```
        <property name="prefix" value="/WEB-INF/jsp/" />
        <!-- 后缀 -->
        <property name="suffix" value=".jsp" />
    </bean>
</beans>
```

10.1.2 RequestMapping 注解类型

在基于注解的控制器类中可以为每个请求编写对应的处理方法。那么如何将请求与处理方法一一对应呢？需要使用 org.springframework.web.bind.annotation.RequestMapping 注解类型将请求与处理方法一一对应。

❶ 方法级别注解

方法级别注解的示例代码如下：

```
package controller;
import org.springframework.stereotype.Controller;
import org.springframework.web.bind.annotation.RequestMapping;
@Controller
public class IndexController {
    @RequestMapping(value="/index/login")
    public String login() {
        /**login 代表逻辑视图名称，需要根据 Spring MVC 配置
         * 文件中 internalResourceViewResolver 的前缀和后缀找到对应的物理视图
         */
        return "login";
    }
    @RequestMapping(value="/index/register")
    public String register() {
        return "register";
    }
}
```

上述示例中有两个 RequestMapping 注解语句，它们都作用在处理方法上。注解的 value 属性将请求 URI 映射到方法，value 属性是 RequestMapping 注解的默认属性，如果只有一个 value 属性，则可以省略该属性。用户可以使用如下 URL 访问 login 方法（请求处理方法），在访问 login 方法之前需要事先在/ WEB-INF/jsp/目录下创建 login.jsp。

```
http://localhost:8080/ch10/index/login
```

❷ 类级别注解

类级别注解的示例代码如下：

```
package controller;
import org.springframework.stereotype.Controller;
import org.springframework.web.bind.annotation.RequestMapping;
@Controller
```

```
@RequestMapping("/index")
public class IndexController {
    @RequestMapping("/login")
    public String login() {
        return "login";
    }
    @RequestMapping("/register")
    public String register() {
        return "register";
    }
}
```

在类级别注解的情况下，控制器类中的所有方法都将映射为类级别的请求。用户可以使用如下 URL 访问 login 方法。

```
http://localhost:8080/ch10/index/login
```

为了方便维护程序，建议开发者采用类级别注解，将相关处理放在同一个控制器类中。例如，对商品的增、删、改、查处理方法都可以放在 GoodsOperate 控制类中。

10.1.3　编写请求处理方法

在控制类中每个请求处理方法可以有多个不同类型的参数，以及一个多种类型的返回结果。

❶ 请求处理方法中常出现的参数类型

如果需要在请求处理方法中使用 Servlet API 类型，那么可以将这些类型作为请求处理方法的参数类型。Servlet API 参数类型的示例代码如下：

```
package controller;
import javax.servlet.http.HttpServletRequest;
import javax.servlet.http.HttpSession;
import org.springframework.stereotype.Controller;
import org.springframework.web.bind.annotation.RequestMapping;
@Controller
@RequestMapping("/index")
public class IndexController {
    @RequestMapping("/login")
    public String login(HttpSession session, HttpServletRequest request) {
        session.setAttribute("skey", "session 范围的值");
        request.setAttribute("rkey", "request 范围的值");
        return "login";
    }
}
```

除了 Servlet API 参数类型以外，还有输入输出流、表单实体类、注解类型、与 Spring 框架相关的类型等，这些类型在后续章节中使用时再详细介绍。其中特别重要的类型是

org.springframework.ui.Model 类型，该类型是一个包含 Map 的 Spring 框架类型。在每次调用请求处理方法时 Spring MVC 都将创建 org.springframework.ui.Model 对象。Model 参数类型的示例代码如下：

```
package controller;
import org.springframework.stereotype.Controller;
import org.springframework.ui.Model;
import org.springframework.web.bind.annotation.RequestMapping;
@Controller
@RequestMapping("/index")
public class IndexController {
    @RequestMapping("/register")
    public String register(Model model) {
    /*在视图中可以使用 EL 表达式${success}取出 model 中的值，对于 EL 的相关知识请读者
    参考本书的有关内容*/
        model.addAttribute("success", "注册成功");
        return "register";
    }
}
```

❷ 请求处理方法常见的返回类型

最常见的返回类型就是代表逻辑视图名称的 String 类型，例如前面章节中的请求处理方法。除了 String 类型以外，还有 ModelAndView（例如第 9 章的传统控制器）、Model、View 以及其他任意的 Java 类型。

10.2 Controller 接收请求参数的常见方式

Controller 接收请求参数的方式有很多种，有的适合 get 请求方式，有的适合 post 请求方式，有的两者都适合。下面分别介绍这些方式，读者可以根据实际情况选择合适的接收方式。

视频讲解

10.2.1 通过实体 Bean 接收请求参数

通过一个实体 Bean 来接收请求参数，适用于 get 和 post 提交请求方式。需要注意的是，Bean 的属性名称必须与请求参数名称相同。下面通过具体应用 ch10 讲解“通过实体 Bean 接收请求参数”。

❶ 创建首页面

在 ch10 应用的 WebContent 目录下创建 index.jsp 页面，代码如下：

```
<%@ page language="java" contentType="text/html; charset=UTF-8" pageEncoding=
"UTF-8"%>
<!DOCTYPE html PUBLIC "-//W3C//DTD HTML 4.01 Transitional//EN" "http://www.
```

```
    w3.org/TR/html4/loose.dtd">
<html>
  <head>
    <meta http-equiv="Content-Type" content="text/html; charset=UTF-8">
    <title>My JSP 'index.jsp' starting page</title>
  </head>
  <body>
    没注册的用户，请<a href="${pageContext.request.contextPath }/index/
    register">注册</a>! <br>
    已注册的用户，去<a href="${pageContext.request.contextPath }/index/
    login">登录</a>!
  </body>
</html>
```

❷ 完善配置文件

完善配置文件 springmvc-servlet.xml，代码如下：

```
<?xml version="1.0" encoding="UTF-8"?>
<beans xmlns="http://www.springframework.org/schema/beans"
    xmlns:xsi="http://www.w3.org/2001/XMLSchema-instance"
    xmlns:context="http://www.springframework.org/schema/context"
    xmlns:mvc="http://www.springframework.org/schema/mvc"
    xsi:schemaLocation="
    http://www.springframework.org/schema/beans
    http://www.springframework.org/schema/beans/spring-beans.xsd
    http://www.springframework.org/schema/context
    http://www.springframework.org/schema/context/spring-context.xsd
    http://www.springframework.org/schema/mvc
    http://www.springframework.org/schema/mvc/spring-mvc.xsd">
    <!-- 使用扫描机制，扫描控制器类 -->
    <context:component-scan base-package="controller"/>
    <mvc:annotation-driven />
    <!-- annotation-driven 用于简化开发的配置，
    注解 DefaultAnnotationHandlerMapping 和 AnnotationMethodHandlerAdapter -->
    <!-- 使用 resources 过滤掉不需要 dispatcherservlet 的资源(即静态资源，例如 css、
    js、html、images)。
    在使用 resources 时必须使用 annotation-driven，否则 resources 元素会阻止任意控
    制器被调用。
    -->
    <!-- 允许 css 目录下的所有文件可见 -->
    <mvc:resources location="/css/" mapping="/css/**"></mvc:resources>
    <!-- 允许 html 目录下的所有文件可见 -->
    <mvc:resources location="/html/" mapping="/html/**"></mvc:resources>
     <!-- 允许 images 目录下的所有文件可见 -->
    <mvc:resources location="/images/" mapping="/images/**"> </mvc:resources>
     <!-- 配置视图解析器 -->
```

```
    <bean class="org.springframework.web.servlet.view.InternalResource
    ViewResolver"
          id="internalResourceViewResolver">
      <!-- 前缀 -->
      <property name="prefix" value="/WEB-INF/jsp/" />
      <!-- 后缀 -->
      <property name="suffix" value=".jsp" />
  </bean>
</beans>
```

❸ 创建 POJO 实体类

在 ch10 应用的 src 目录下创建 pojo 包，并在该包中创建实体类 UserForm，代码如下：

```
package pojo;
public class UserForm {
    private String uname;  //与请求参数名称相同
    private String upass;
    private String reupass;
    //省略 getter 和 setter 方法
}
```

❹ 创建控制器类

在 ch10 应用的 controller 包中创建控制器类 IndexController 和 UserController。

IndexController 的代码如下：

```
package controller;
import org.springframework.stereotype.Controller;
import org.springframework.web.bind.annotation.RequestMapping;
@Controller
@RequestMapping("/index")
public class IndexController {
    @RequestMapping("/login")
    public String login() {
        return "login";  //跳转到/WEB-INF/jsp 下的 login.jsp
    }
    @RequestMapping("/register")
    public String register() {
        return "register";
    }
}
```

UserController 的代码如下：

```
package controller;
import javax.servlet.http.HttpSession;
import org.apache.commons.logging.Log;
import org.apache.commons.logging.LogFactory;
```

```
import org.springframework.stereotype.Controller;
import org.springframework.ui.Model;
import org.springframework.web.bind.annotation.RequestMapping;
import pojo.UserForm;
@Controller
@RequestMapping("/user")
public class UserController {
    //得到一个用来记录日志的对象，这样在打印信息的时候能够标记打印的是哪个类的信息
    private static final Log logger = LogFactory.getLog(UserController.class);
    /**
     * 处理登录
     * 使用 UserForm 对象（实体 Bean）user 接收注册页面提交的请求参数
     */
    @RequestMapping("/login")
    public String login(UserForm user, HttpSession session, Model model) {
            if("zhangsan".equals(user.getUname())
                    && "123456".equals(user.getUpass())) {
                session.setAttribute("u", user);
                logger.info("成功");
                return "main";  //登录成功，跳转到 main.jsp
            }else{
                logger.info("失败");
                model.addAttribute("messageError", "用户名或密码错误");
                return "login";
            }

    }

    /**
     *处理注册
     *使用 UserForm 对象（实体 Bean）user 接收注册页面提交的请求参数
     */
    @RequestMapping("/register")
    public String register(UserForm user, Model model) {
        if("zhangsan".equals(user.getUname())
                && "123456".equals(user.getUpass())) {
            logger.info("成功");
            return "login";  //注册成功，跳转到 login.jsp
        }else{
            logger.info("失败");
            //在 register.jsp 页面上可以使用 EL 表达式取出 model 的 uname 值
            model.addAttribute("uname", user.getUname());
            return "register";  //返回 register.jsp
        }
    }
}
```

❺ 创建页面视图

在 ch10 应用的/WEB-INF/jsp 目录下创建 register.jsp 和 login.jsp。

register.jsp 的核心代码（详细代码请参考本章源码 ch10）如下：

```
<form action="${pageContext.request.contextPath }/user/register" method=
"post" name="registForm">
    <table border=1 bgcolor="lightblue" align="center">
            <tr>
                <td>姓名：</td>
                <td>
                    <input class="textSize" type="text" name="uname" value=
                    "${ uname }"/>
                </td>
            </tr>
            <tr>
                <td>密码：</td>
                <td><input class="textSize" type="password" maxlength="20"
                name="upass"/></td>
            </tr>
            <tr>
                <td>确认密码：</td>
            <td><input class="textSize" type="password" maxlength="20"
            name="reupass"/></td>
            </tr>
            <tr>
    <td colspan="2" align="center"><input type="button" value="注册"
    onclick="allIsNull()"/></td>
            </tr>
        </table>
    </form>
```

在 register.jsp 的代码中使用了 EL 语句“${uname }”取出“model.addAttribute("uname", user.getUname())”中的值。对于 EL 和 JSTL 的相关知识，请读者参考本书的相关内容。

login.jsp 的核心代码如下：

```
<form  action="${pageContext.request.contextPath  }/user/login"  method=
"post">
<table>
        <tr>
            <td colspan="2"><img src="${pageContext.request.contextPath }
            /images/ login.gif"></td>
        </tr>
        <tr>
            <td>姓名：</td>
            <td><input type="text" name="uname" class="textSize"></td>
        </tr>
```

```
        <tr>
            <td>密码：</td>
            <td><input type="password" name="upass" class="textSize"></td>
        </tr>
        <tr>
            <td colspan="2">
<input  type="image"  src="${pageContext.request.contextPath  }/images/
ok.gif" onclick="gogo()" >
<input  type="image"  src="${pageContext.request.contextPath  }/images/
cancel.gif" onclick="cancel()" >
            </td>
        </tr>
    </table>
    ${messageError }
    </form>
```

❻ **测试应用**

运行 ch10 应用的首页面，进行程序测试。

10.2.2　通过处理方法的形参接收请求参数

通过处理方法的形参接收请求参数也就是直接把表单参数写在控制器类相应方法的形参中，即形参名称与请求参数名称完全相同。该接收参数方式适用于 get 和 post 提交请求方式。用户可以将 10.2.1 节中控制器类 UserController 中 register 方法的代码修改如下：

```
@RequestMapping("/register")
/**
 * 通过形参接收请求参数，形参名称与请求参数名称完全相同
 */
public String register(String uname, String upass, Model model) {
    if("zhangsan".equals(uname)
            && "123456".equals(upass)) {
        logger.info("成功");
        return "login";  //注册成功，跳转到 login.jsp
    }else{
        logger.info("失败");
        //在 register.jsp 页面上可以使用 EL 表达式取出 model 的 uname 值
        model.addAttribute("uname", uname);
        return "register";  //返回 register.jsp
    }
}
```

10.2.3　通过 HttpServletRequest 接收请求参数

通过 HttpServletRequest 接收请求参数适用于 get 和 post 提交请求方式，可以将 10.2.1

节中控制器类 UserController 中 register 方法的代码修改如下：

```
@RequestMapping("/register")
/*
 * 通过 HttpServletRequest 接收请求参数
 */
public String register(HttpServletRequest request, Model model) {
    String uname=request.getParameter("uname");
    String upass=request.getParameter("upass");
    if("zhangsan".equals(uname)
            && "123456".equals(upass)) {
        logger.info("成功");
        return "login";  //注册成功，跳转到 login.jsp
    }else{
        logger.info("失败");
        //在 register.jsp 页面上可以使用 EL 表达式取出 model 的 uname 值
        model.addAttribute("uname", uname);
        return "register";  //返回 register.jsp
    }
}
```

10.2.4 通过@PathVariable 接收 URL 中的请求参数

通过@PathVariable 获取 URL 中的参数，控制器类示例代码如下：

```
package controller;
import org.springframework.stereotype.Controller;
import org.springframework.ui.Model;
import org.springframework.web.bind.annotation.PathVariable;
import org.springframework.web.bind.annotation.RequestMapping;
import org.springframework.web.bind.annotation.RequestMethod;
@Controller
@RequestMapping("/user")
public class UserController {
    @RequestMapping(value="/register/{uname}/{upass}", method=Request Method. GET)
    //必须加 method 属性
    /**
     * 通过@PathVariable 获取 URL 中的参数
     */
    public String register(@PathVariable String uname,@PathVariable String
    upass, Model model) {
        if("zhangsan".equals(uname)
                && "123456".equals(upass))
            return "login";  //注册成功，跳转到 login.jsp
        else{
```

```
            //在 register.jsp 页面上可以使用 EL 表达式取出 model 的 uname 值
            model.addAttribute("uname", uname);
            return "register";  //返回 register.jsp
        }
    }
}
```

在访问“http://localhost:8080/ch2/user/register/zhangsan/123456”路径时，上述代码自动将 URL 中的模板变量{uname}和{upass}绑定到通过@PathVariable 注解的同名参数上，即 uname=zhangsan、upass=123456。

10.2.5　通过@RequestParam 接收请求参数

通过@RequestParam 接收请求参数适用于 get 和 post 提交请求方式，可以将 10.2.1 节中控制器类 UserController 中 register 方法的代码修改如下：

```
@RequestMapping("/register")
/**
 * 通过@RequestParam 接收请求参数
 */
public String register(@RequestParam String uname, @RequestParam String
upass, Model model) {
    if("zhangsan".equals(uname)
            && "123456".equals(upass)) {
        logger.info("成功");
        return "login";  //注册成功，跳转到 login.jsp
    }else{
        logger.info("失败");
        //在 register.jsp 页面上可以使用 EL 表达式取出 model 的 uname 值
        model.addAttribute("uname", uname);
        return "register";  //返回 register.jsp
    }
}
```

通过@RequestParam 接收请求参数与 10.2.2 节“通过处理方法的形参接收请求参数”的区别如下：当请求参数与接收参数名不一致时，“通过处理方法的形参接收请求参数”不会报 400 错误，而“通过@RequestParam 接收请求参数”会报 400 错误。

10.2.6　通过@ModelAttribute 接收请求参数

当@ModelAttribute 注解放在处理方法的形参上时，用于将多个请求参数封装到一个实体对象，从而简化数据绑定流程，而且自动暴露为模型数据，在视图页面展示时使用。而 10.2.1 节中只是将多个请求参数封装到一个实体对象，并不能暴露为模型数据（需要使用 model.addAttribute 语句才能暴露为模型数据，数据绑定与模型数据展示可参考第 12 章的内容）。

通过@ModelAttribute 注解接收请求参数适用于 get 和 post 提交请求方式，可以将 10.2.1 节中控制器类 UserController 中 register 方法的代码修改如下：

```
@RequestMapping("/register")
public String register(@ModelAttribute("user") UserForm user) {
    if("zhangsan".equals(user.getUname())
            && "123456".equals(user.getUpass())){
        logger.info("成功");
        return "login";  //注册成功，跳转到 login.jsp
    }else{
        logger.info("失败");
    //使用@ModelAttribute("user")与 model.addAttribute("user", user)的功能相同
    //在 register.jsp 页面上可以使用 EL 表达式${user.uname}取出 ModelAttribute 的 uname 值
        return "register";  //返回 register.jsp
    }
}
```

10.3 重定向与转发

重定向是将用户从当前处理请求定向到另一个视图（例如 JSP）或处理请求，以前的请求（request）中存放的信息全部失效，并进入一个新的 request 作用域；转发是将用户对当前处理的请求转发给另一个视图或处理请求，以前的 request 中存放的信息不会失效。

视频讲解

转发是服务器行为，重定向是客户端行为。

转发过程：客户浏览器发送 http 请求，Web 服务器接受此请求，调用内部的一个方法在容器内部完成请求处理和转发动作，将目标资源发送给客户；在这里转发的路径必须是同一个 Web 容器下的 URL，其不能转向到其他的 Web 路径上，中间传递的是自己的容器内的 request。在客户浏览器的地址栏中显示的仍然是其第一次访问的路径，也就是说客户是感觉不到服务器做了转发的。转发行为是浏览器只做了一次访问请求。

重定向过程：客户浏览器发送 http 请求，Web 服务器接受后发送 302 状态码响应及对应新的 location 给客户浏览器，客户浏览器发现是 302 响应，则自动再发送一个新的 http 请求，请求 URL 是新的 location 地址，服务器根据此请求寻找资源并发送给客户。在这里 location 可以重定向到任意 URL，既然是浏览器重新发出了请求，那么就没有什么 request 传递的概念了。在客户浏览器的地址栏中显示的是其重定向的路径，客户可以观察到地址的变化。重定向行为是浏览器做了至少两次的访问请求。

在 Spring MVC 框架中，控制器类中处理方法的 return 语句默认就是转发实现，只不过实现的是转发到视图。示例代码如下：

```
@RequestMapping("/register")
public String register() {
    return "register";  //转发到 register.jsp
```

```
}
```

在 Spring MVC 框架中，重定向与转发的示例代码如下：

```
package controller;
import org.springframework.stereotype.Controller;
import org.springframework.web.bind.annotation.RequestMapping;
@Controller
@RequestMapping("/index")
public class IndexController {
    @RequestMapping("/login")
    public String login() {
        //转发到一个请求方法（同一个控制器类中可以省略/index/）
        return "forward:/index/isLogin";
    }
    @RequestMapping("/isLogin")
    public String isLogin() {
        //重定向到一个请求方法
        return "redirect:/index/isRegister";
    }
    @RequestMapping("/isRegister")
    public String isRegister() {
        //转发到一个视图
        return "register";
    }
}
```

在 Spring MVC 框架中，不管是重定向或转发，都需要符合视图解析器的配置，如果直接重定向到一个不需要 DispatcherServlet 的资源，例如：

```
return "redirect:/html/my.html";
```

则需要使用 mvc:resources 配置：

```
<mvc:resources location="/html/" mapping="/html/**"></mvc:resources>
```

10.4　应用@Autowired 进行依赖注入

在前面学习的控制器中并没有体现 MVC 的 M 层，这是因为控制器既充当 C 层又充当 M 层。这样设计程序的系统结构很不合理，应该将 M 层从控制器中分离出来。Spring MVC 框架本身就是一个非常优秀的 MVC 框架，它具有依赖注入的优点，可以通过 org.springframework.beans.factory.annotation.Autowired 注解类型将依赖注入到一个属性（成员变量）或方法，例如：

视频讲解

```
@Autowired
public UserService userService;
```

在 Spring MVC 中，为了能被作为依赖注入，类必须使用 org.springframework.stereotype.Service 注解类型注明为@Service（一个服务）。另外，还需要在配置文件中使用<context:component-scan base-package="基本包"/>元素来扫描依赖基本包。下面将 10.2 节中“登录”和“注册”的业务逻辑处理分离出来，使用 Service 层实现。

首先创建 service 包，在该包中创建 UserService 接口和 UserServiceImpl 实现类。

UserService 接口的具体代码如下：

```
package service;
import pojo.UserForm;
public interface UserService {
    boolean login(UserForm user);
    boolean register(UserForm user);
}
```

UserServiceImpl 实现类的具体代码如下：

```
package service;
import org.springframework.stereotype.Service;
import pojo.UserForm;
//注解为一个服务
@Service
public class UserServiceImpl implements UserService{
    @Override
    public boolean login(UserForm user) {
        if("zhangsan".equals(user.getUname())
                && "123456".equals(user.getUpass()))
            return true;
        return false;
    }
    @Override
    public boolean register(UserForm user) {
        if("zhangsan".equals(user.getUname())
                && "123456".equals(user.getUpass()))
            return true;
        return false;
    }
}
```

然后在配置文件中添加一个<context:component-scan base-package="基本包"/>元素，具体代码如下：

```
<context:component-scan base-package="service"/>
```

最后修改控制器类 UserController，具体代码如下：

```
package controller;
import javax.servlet.http.HttpSession;
import org.apache.commons.logging.Log;
```

```
import org.apache.commons.logging.LogFactory;
import org.springframework.beans.factory.annotation.Autowired;
import org.springframework.stereotype.Controller;
import org.springframework.ui.Model;
import org.springframework.web.bind.annotation.RequestMapping;
import pojo.UserForm;
import service.UserService;
@Controller
@RequestMapping("/user")
public class UserController {
    //得到一个用来记录日志的对象，这样在打印信息的时候能够标记打印的是哪个类的信息
    private static final Log logger=LogFactory.getLog(UserController.class);
    //将服务依赖注入到属性 userService
    @Autowired
     public UserService userService;
    /**
     * 处理登录
     */
    @RequestMapping("/login")
    public String login(UserForm user, HttpSession session, Model model) {
        if(userService.login(user)){
            session.setAttribute("u", user);
            logger.info("成功");
            return "main";  //登录成功，跳转到 main.jsp
        }else{
            logger.info("失败");
            model.addAttribute("messageError", "用户名或密码错误");
            return "login";
        }
    }
    /**
     *处理注册
     */
    @RequestMapping("/register")
    public String register(@ModelAttribute("user") UserForm user) {
        if(userService.register(user)){
            logger.info("成功");
            return "login";  //注册成功，跳转到 login.jsp
        }else{
            logger.info("失败");
          //使用@ModelAttribute("user")与 model.addAttribute("user", user)
          的功能相同
          //在 register.jsp 页面上可以使用 EL 表达式${user.uname}取出 ModelAttribute
          的 uname 值
            return "register";  //返回 register.jsp
```

```
        }
    }
}
```

10.5 @ModelAttribute

视频讲解

通过 org.springframework.web.bind.annotation.ModelAttribute 注解类型可经常实现以下两个功能：

❶ 绑定请求参数到实体对象（表单的命令对象）

该用法如 10.2.6 节中的内容：

```
@RequestMapping("/register")
public String register(@ModelAttribute("user") UserForm user) {
    if("zhangsan".equals(user.getUname())
            && "123456".equals(user.getUpass())){
        return "login";
    }else{
        return "register";
    }
}
```

在上述代码中“@ModelAttribute("user") UserForm user”语句的功能有两个：一是将请求参数的输入封装到 user 对象中；二是创建 UserForm 实例，以“user”为键值存储在 Model 对象中，和“model.addAttribute("user", user)”语句的功能一样。如果没有指定键值，即“@ModelAttribute UserForm user”，那么在创建 UserForm 实例时以“userForm”为键值存储在 Model 对象中，和“model.addAttribute("userForm", user)”语句的功能一样。

❷ 注解一个非请求处理方法

被@ModelAttribute 注解的方法将在每次调用该控制器类的请求处理方法前被调用。这种特性可以用来控制登录权限，当然控制登录权限的方法有很多，例如拦截器、过滤器等。

使用该特性控制登录权限的示例代码如下：

```
package controller;
import javax.servlet.http.HttpSession;
import org.springframework.web.bind.annotation.ModelAttribute;
public class BaseController {
    @ModelAttribute
    public void isLogin(HttpSession session) throws Exception {
      if(session.getAttribute("user")==null){
          throw new Exception("没有权限");
      }
    }
```

```
}

package controller;
import org.springframework.stereotype.Controller;
import org.springframework.web.bind.annotation.RequestMapping;
@Controller
@RequestMapping("/admin")
public class ModelAttributeController extends BaseController{
    @RequestMapping("/add")
    public String add(){
        return "addSuccess";
    }
    @RequestMapping("/update")
    public String update(){
        return "updateSuccess";
    }
    @RequestMapping("/delete")
    public String delete(){
        return "deleteSuccess";
    }
}
```

在上述 ModelAttributeController 类中的 add、update、delete 请求处理方法执行时，首先执行父类 BaseController 中的 isLogin 方法判断登录权限，可以通过地址“http://localhost:8080/ch2/admin/add”测试登录权限。

10.6　本章小结

本章是整个 Spring MVC 框架的核心部分。通过本章的学习，读者务必要掌握如何编写基于注解的控制器类。

习题 10

1．在 Spring MVC 的控制器类中如何访问 Servlet API？
2．控制器接收请求参数的常见方式有哪几种？
3．如何编写基于注解的控制器类？
4．@ModelAttribute 可实现哪些功能？

第11章 类型转换和格式化

学习目的与要求

本章主要学习类型转换器和格式化转换器。通过本章的学习，读者应该理解类型转换器和格式化转换器的原理，掌握类型转换器和格式化转换器的用法。

本章主要内容

- Converter；
- Formatter。

在 Spring MVC 框架中需要收集用户请求参数，并将请求参数传递给应用的控制器组件。此时存在一个问题，即所有的请求参数类型只能是字符串数据类型，但 Java 是强类型语言，所以 Spring MVC 框架必须将这些字符串请求参数转换成相应的数据类型。

Spring MVC 框架不仅提供了强大的类型转换和格式化机制，而且开发者还可以方便地开发出自己的类型转换器和格式化转换器，完成字符串和各种数据类型之间的转换。这正是学习本章的目的所在。

11.1 类型转换的意义

本节以一个简单应用（JSP + Servlet）为示例来介绍类型转换的意义。如图 11.1 所示的添加商品页面用于收集用户输入的商品信息，商品信息包括商品名称（字符串类型 String）、商品价格（双精度浮点类型 double）、商品数量（整数类型 int）。

http://localhost:8080/ch11a/addGoods.jsp

商品名称：
商品价格：
商品数量：
提交

图 11.1 添加商品信息的收集页面

addGoods.jsp 页面的代码如下：

```
<body>
    <form action="addGoods" method="post">
        商品名称：<input type="text" name="goodsname"/><br>
        商品价格：<input type="text" name="goodsprice"/><br>
        商品数量：<input type="text" name="goodsnumber"/><br>
        <input type="submit" value="提交"/>
    </form>
</body>
```

希望页面收集到的数据提交到 addGoods 的 Servlet（AddGoodsServlet 类），该 Servlet 将这些请求信息封装成一个 Goods 类的值对象。

Goods 类的代码如下：

```
package pojo;
public class Goods {
    private String goodsname;
    private double goodsprice;
    private int goodsnumber;
    //无参数的构造方法
    public Goods(){}
    //有参数的构造方法
    public Goods(String goodsname, double goodsprice, int goodsnumber) {
        super();
        this.goodsname=goodsname;
        this.goodsprice=goodsprice;
        this.goodsnumber=goodsnumber;
    }
    //此处省略了 setter 和 getter 方法
    …
}
```

AddGoodsServlet 类的代码如下：

```
package servlet;
import java.io.IOException;
import javax.servlet.ServletException;
import javax.servlet.http.HttpServlet;
import javax.servlet.http.HttpServletRequest;
import javax.servlet.http.HttpServletResponse;
import domain.Goods;
public class AddGoodsServlet extends HttpServlet {
    public void doGet(HttpServletRequest request, HttpServletResponse response)
            throws ServletException, IOException {
        doPost(request, response);
    }
    public void doPost(HttpServletRequest request, HttpServletResponse response)
```

```
            throws ServletException, IOException {
        response.setContentType("text/html;charset=utf-8");
        //设置编码，防止乱码
        request.setCharacterEncoding("utf-8");
        //获取参数值
        String goodsname=request.getParameter("goodsname");
        String goodsprice=request.getParameter("goodsprice");
        String goodsnumber=request.getParameter("goodsnumber");
        //下面进行类型转换
        double newgoodsprice=Double.parseDouble(goodsprice);
        int newgoodsnumber=Integer.parseInt(goodsnumber);
        //将转换后的数据封装成 goods 值对象
        Goods goods = new Goods(goodsname, newgoodsprice, newgoodsnumber);
        //将 goods 值对象传递给数据访问层，进行添加操作，代码省略
        …
    }
}
```

对于上面这个应用而言，开发者需要自己在 Servlet 中进行类型转换，并将其封装成值对象。这些类型转换操作全部手工完成，异常烦琐。

对于 Spring MVC 框架而言，它必须将请求参数转换成值对象类中各属性对应的数据类型——这就是类型转换的意义。

11.2 Converter

Spring MVC 框架的 Converter<S, T>是一个可以将一种数据类型转换成另一种数据类型的接口，这里 S 表示源类型，T 表示目标类型。开发者在实际应用中使用框架内置的类型转换器基本上就够了，但有时需要编写具有特定功能的类型转换器。

视频讲解

11.2.1 内置的类型转换器

在 Spring MVC 框架中，对于常用的数据类型，开发者无须创建自己的类型转换器，因为 Spring MVC 框架有许多内置的类型转换器用于完成常用的类型转换。Spring MVC 框架提供的内置类型转换包括以下几种类型。

❶ 标量转换器

- StringToBooleanConverter：String 到 boolean 类型转换。
- ObjectToStringConverter：Object 到 String 转换，调用 toString 方法转换。
- StringToNumberConverterFactory：String 到数字转换（例如 Integer、Long 等）。
- NumberToNumberConverterFactory：数字子类型（基本类型）到数字类型（包装类型）转换。
- StringToCharacterConverter：String 到 Character 转换，取字符串中的第一个字符。

- NumberToCharacterConverter：数字子类型到 Character 转换。
- CharacterToNumberFactory：Character 到数字子类型转换。
- StringToEnumConverterFactory：String 到枚举类型转换，通过 Enum.valueOf 将字符串转换为需要的枚举类型。
- EnumToStringConverter：枚举类型到 String 转换，返回枚举对象的 name 值。
- StringToLocaleConverter：String 到 java.util.Locale 转换。
- PropertiesToStringConverter：java.util.Properties 到 String 转换，默认通过 ISO-8859-1 解码。
- StringToPropertiesConverter：String 到 java.util.Properties 转换，默认使用 ISO-8859-1 编码。

❷ 集合、数组相关转换器

- ArrayToCollectionConverter：任意数组到任意集合（List、Set）转换。
- CollectionToArrayConverter：任意集合到任意数组转换。
- ArrayToArrayConverter：任意数组到任意数组转换。
- CollectionToCollectionConverter：集合之间的类型转换。
- MapToMapConverter：Map 之间的类型转换。
- ArrayToStringConverter：任意数组到 String 转换。
- StringToArrayConverter：字符串到数组的转换，默认通过“,”分割，且去除字符串两边的空格（trim）。
- ArrayToObjectConverter：任意数组到 Object 的转换，如果目标类型和源类型兼容，直接返回源对象；否则返回数组的第一个元素并进行类型转换。
- ObjectToArrayConverter：Object 到单元素数组转换。
- CollectionToStringConverter：任意集合（List、Set）到 String 转换。
- StringToCollectionConverter：String 到集合（List、Set）转换，默认通过“,”分割，且去除字符串两边的空格（trim）。
- CollectionToObjectConverter：任意集合到任意 Object 的转换，如果目标类型和源类型兼容，直接返回源对象；否则返回集合的第一个元素并进行类型转换。
- ObjectToCollectionConverter：Object 到单元素集合的类型转换。

类型转换是在视图与控制器相互传递数据时发生的。Spring MVC 框架对于基本类型（例如 int、long、float、double、boolean 以及 char 等）已经做好了基本类型转换。例如，对于 11.1 节 addGoods.jsp 的提交请求，可以由以下处理方法来接收请求参数并处理：

```
package controller;
import org.springframework.stereotype.Controller;
import org.springframework.web.bind.annotation.RequestMapping;
@Controller
public class GoodsController {
    @RequestMapping("/addGoods")
    public String add(String goodsname, double goodsprice, int goodsnumber) {
        double total=goodsprice*goodsnumber;
```

```
        System.out.println(total);
        return "success";
    }
}
```

注意：在使用内置类型转换器时，请求参数输入值与接收参数类型要兼容，否则会报400 错误。请求参数类型与接收参数类型不兼容问题需要学习输入校验后才可解决。

11.2.2 自定义类型转换器

当 Spring MVC 框架内置的类型转换器不能满足需求时，开发者可以开发自己的类型转换器。例如有一个应用 ch11a 希望用户在页面表单中输入信息来创建商品信息。当输入"apple,10.58,200"时表示在程序中自动创建一个 new Goods，并将"apple"值自动赋给 goodsname 属性，将"10.58"值自动赋给 goodsprice 属性，将"200"值自动赋给 goodsnumber 属性。ch11a 应用与第 10 章中的 ch10 应用具有相同的 JAR 包、web.xml。

如果想实现上述应用，需要做以下 5 件事：

- 创建实体类；
- 创建控制器类；
- 创建自定义类型转换器类；
- 注册类型转换器；
- 创建相关视图。

按照上述步骤采用自定义类型转换器完成需求。

第 1 步　创建实体类

在 ch11a 的 src 目录下创建 pojo 包，并在该包中创建名为 GoodsModel 的实体类，代码如下：

```
package pojo;
public class GoodsModel {
    private String goodsname;
    private double goodsprice;
    private int goodsnumber;
    //省略 setter 和 getter 方法
}
```

第 2 步　创建控制器类

在 ch11a 的 src 目录下创建 controller 包，并在该包中创建名为 ConverterController 的控制器类，代码如下：

```
package controller;
import org.springframework.stereotype.Controller;
import org.springframework.ui.Model;
import org.springframework.web.bind.annotation.RequestMapping;
import org.springframework.web.bind.annotation.RequestParam;
```

```
import pojo.GoodsModel;
@Controller
@RequestMapping("/my")
public class ConverterController {
    @RequestMapping("/converter")
    /*使用@RequestParam("goods")接收请求参数,
    然后调用自定义类型转换器 GoodsConverter 将字符串值转换为 GoodsModel 的对象 gm
    */
    public String myConverter(@RequestParam("goods") GoodsModel gm, Model model){
        model.addAttribute("goods",gm);
        return "showGoods";
    }
}
```

第 3 步　创建自定义类型转换器类

自定义类型转换器类需要实现 Converter<S, T>接口，重写 convert(S)接口方法。convert(S)方法的功能是将源数据类型 S 转换成目标数据类型 T。在 ch11a 的 src 目录下创建 converter 包，并在该包中创建名为 GoodsConverter 的自定义类型转换器类，代码如下：

```
package converter;
import org.springframework.core.convert.converter.Converter;
import pojo.GoodsModel;
public class GoodsConverter implements Converter<String, GoodsModel>{
    @Override
    public GoodsModel convert(String source) {
        //创建一个 Goods 实例
        GoodsModel goods=new GoodsModel();
        //以“,”分隔
        String stringValues[]=source.split(",");
        if(stringValues!=null &&
                stringValues.length==3){
            //为 Goods 实例赋值
            goods.setGoodsname(stringValues[0]);
            goods.setGoodsprice(Double.parseDouble(stringValues[1]));
            goods.setGoodsnumber(Integer.parseInt(stringValues[2]));
            return goods;
        }else{
            throw new IllegalArgumentException(String.format("类型转换失败,
            需要格式‘apple,10.58,200’,但格式是[%s]", source));
        }
    }
}
```

第 4 步　注册类型转换器

在 ch11a 的 WEB-INF 目录下创建配置文件 springmvc-servlet.xml，并在配置文件中注册自定义类型转换器，配置文件代码如下：

```
<?xml version="1.0" encoding="UTF-8"?>
<beans xmlns="http://www.springframework.org/schema/beans"
    xmlns:xsi="http://www.w3.org/2001/XMLSchema-instance"
    xmlns:context="http://www.springframework.org/schema/context"
    xmlns:mvc="http://www.springframework.org/schema/mvc"
    xsi:schemaLocation="
    http://www.springframework.org/schema/beans
    http://www.springframework.org/schema/beans/spring-beans.xsd
    http://www.springframework.org/schema/context
    http://www.springframework.org/schema/context/spring-context.xsd
    http://www.springframework.org/schema/mvc
    http://www.springframework.org/schema/mvc/spring-mvc.xsd">
    <!-- 使用扫描机制扫描 controller 包 -->
    <context:component-scan base-package="controller"/>
    <!-- 注册类型转换器 GoodsConverter -->
    <bean id="conversionService" class="org.springframework.context. support.
    ConversionServiceFactoryBean">
        <property name="converters">
            <list>
                <bean class="converter.GoodsConverter"/>
            </list>
        </property>
    </bean>
    <mvc:annotation-driven conversion-service="conversionService"/>
    <!-- 配置视图解析器 -->
    <bean class="org.springframework.web.servlet.view.InternalResource ViewResolver"
        id="internalResourceViewResolver">
        <!-- 前缀 -->
        <property name="prefix" value="/WEB-INF/jsp/" />
        <!-- 后缀 -->
        <property name="suffix" value=".jsp" />
  </bean>
</beans>
```

第 5 步　创建相关视图

在 ch11a 应用的 WebContent 目录下创建信息采集页面 input.jsp，核心代码如下：

```
<form action="${pageContext.request.contextPath }/my/converter" method=
"post">
    请输入商品信息（格式为 apple,10.58,200）：
    <input type="text" name="goods"/><br>
    <input type="submit" value="提交"/>
</form>
```

在 ch11a 应用的/WEB-INF/jsp 目录下创建信息显示页面 showGoods.jsp，核心代码如下：

```
<body>
    您创建的商品信息如下：<br>
  <!-- 使用 EL 表达式取出 model 中 goods 的信息 -->
    商品名称为：${goods.goodsname }，
    商品价格为：${goods.goodsprice }，
    商品数量为：${goods.goodsnumber }。
</body>
```

最后，使用地址“http://localhost:8080/ch11a/input.jsp”测试应用。

11.3 Formatter

Spring MVC 框架的 Formatter<T>与 Converter<S, T>一样，也是一个可以将一种数据类型转换成另一种数据类型的接口。不同的是，Formatter<T>的源数据类型必须是 String 类型，而 Converter<S, T>的源数据类型是任意数据类型。

视频讲解

在 Web 应用中由 HTTP 发送的请求数据到控制器中都是以 String 类型获取，因此在 Web 应用中选择 Formatter<T>比选择 Converter<S, T>更加合理。

11.3.1 内置的格式化转换器

Spring MVC 提供了几个内置的格式化转换器，具体如下。

- NumberFormatter：实现 Number 与 String 之间的解析与格式化。
- CurrencyFormatter：实现 Number 与 String 之间的解析与格式化（带货币符号）。
- PercentFormatter：实现 Number 与 String 之间的解析与格式化（带百分数符号）。
- DateFormatter：实现 Date 与 String 之间的解析与格式化。

11.3.2 自定义格式化转换器

自定义格式化转换器就是编写一个实现 org.springframework.format.Formatter 接口的 Java 类。该接口声明如下：

```
public interface Formatter<T>
```

这里的 T 表示由字符串转换的目标数据类型。该接口有 parse 和 print 两个接口方法，自定义格式化转换器类必须覆盖它们。

```
public T parse(String s, java.util.Locale locale)
public String print(T object, java.util.Locale locale)
```

parse 方法的功能是利用指定的 Locale 将一个 String 类型转换成目标类型，print 方法与之相反，用于返回目标对象的字符串表示。

下面通过具体应用 ch11b 讲解自定义格式化转换器的用法，ch11b 应用与 ch11a 应用具有相同的 JAR 包、web.xml。

应用的具体要求如下：

（1）用户在页面表单中输入信息来创建商品，输入页面效果如图 11.2 所示。

http://localhost:8080/ch11b/inputb.jsp
商品名称：
商品价格：
商品数量：
商品日期：（yyyy-MM-dd）
提交

图 11.2　信息输入页面

（2）控制器使用实体 bean 类 GoodsModelb 接收页面提交的请求参数，GoodsModelb 类的属性如下。

```
private String goodsname;
private double goodsprice;
private int goodsnumber;
private Date goodsdate;
```

（3）GoodsModelb 实体类接收请求参数时，商品名称、价格和数量使用内置的类型转换器完成转换；商品日期需要用自定义的格式化转换器完成。

（4）用格式化转换器转换之后的数据显示在 showGoodsb.jsp 页面，效果如图 11.3 所示。

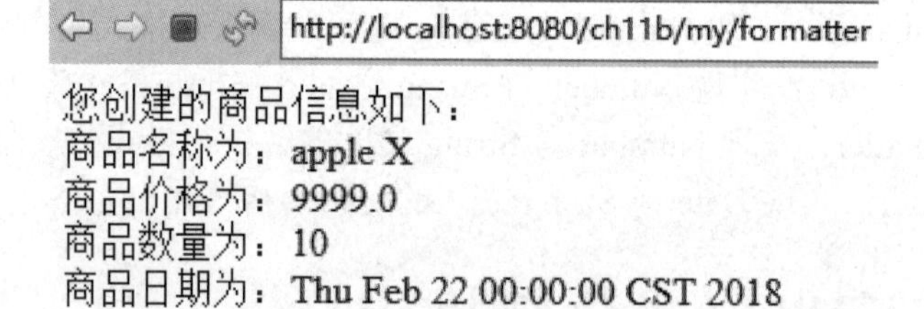

图 11.3　格式化后信息的显示页面

由图 11.3 可以看出，日期由字符串值“2018-02-22”格式化成 Date 类型。

如果想实现上述 ch11b 应用的需求，需要做以下 5 件事：

- 创建实体类；
- 创建控制器类；
- 创建自定义格式化转换器类；
- 注册格式化转换器；
- 创建相关视图。

按照上述步骤采用自定义格式化转换器完成需求。

第 1 步　创建实体类

在 ch11b 的 src 目录下创建 pojo 包，并在该包中创建名为 GoodsModelb 的实体类，代码如下：

```
package pojo;
import java.util.Date;
public class GoodsModelb {
    private String goodsname;
    private double goodsprice;
    private int goodsnumber;
    private Date goodsdate;
    //省略 setter 和 getter 方法
}
```

第 2 步　创建控制器类

在 ch11b 的 src 目录下创建 controller 包，并在该包中创建名为 FormatterController 的控制器类，代码如下：

```
package controller;
import org.springframework.stereotype.Controller;
import org.springframework.ui.Model;
import org.springframework.web.bind.annotation.RequestMapping;
import domain.GoodsModelb;
@Controller
@RequestMapping("/my")
public class FormatterController {
    @RequestMapping("/formatter")
    public String myConverter(GoodsModelb gm, Model model){
        model.addAttribute("goods",gm);
        return "showGoodsb";
    }
}
```

第 3 步　创建自定义格式化转换器类

在 ch11b 的 src 目录下创建 formatter 包，并在该包中创建名为 MyFormatter 的自定义格式化转换器类，代码如下：

```
package formatter;
import java.text.ParseException;
import java.text.SimpleDateFormat;
import java.util.Date;
import java.util.Locale;
import org.springframework.format.Formatter;
public class MyFormatter implements Formatter<Date>{
    SimpleDateFormat dateFormat=new SimpleDateFormat("yyyy-MM-dd");
    @Override
```

```
    public String print(Date object, Locale arg1) {
        return dateFormat.format(object);
    }
    @Override
    public Date parse(String source, Locale arg1) throws ParseException {
        return dateFormat.parse(source);  //Formatter只能对字符串转换
    }
}
```

第 4 步　注册格式化转换器

在 ch11b 的 WEB-INF 目录下创建配置文件 springmvc-servlet.xml，并在配置文件中注册格式化转换器，具体代码如下：

```
<?xml version="1.0" encoding="UTF-8"?>
<beans xmlns="http://www.springframework.org/schema/beans"
    xmlns:xsi="http://www.w3.org/2001/XMLSchema-instance"
    xmlns:context="http://www.springframework.org/schema/context"
    xmlns:mvc="http://www.springframework.org/schema/mvc"
    xsi:schemaLocation="
    http://www.springframework.org/schema/beans
    http://www.springframework.org/schema/beans/spring-beans.xsd
    http://www.springframework.org/schema/context
    http://www.springframework.org/schema/context/spring-context.xsd
    http://www.springframework.org/schema/mvc
    http://www.springframework.org/schema/mvc/spring-mvc.xsd">
    <!-- 使用扫描机制扫描 controller 包 -->
    <context:component-scan base-package="controller"/>
    <!-- 注册 MyFormatter -->
    <bean id="conversionService" class="org.springframework.format.
    support.FormattingConversionServiceFactoryBean">
     <property name="formatters">
         <set>
             <bean class="formatter.MyFormatter"/>
         </set>
     </property>
    </bean>
    <mvc:annotation-driven conversion-service="conversionService"/>
     <!-- 配置视图解析器 -->
     <bean  class="org.springframework.web.servlet.view.InternalResourceView
     Resolver"
           id="internalResourceViewResolver">
       <!-- 前缀 -->
       <property name="prefix" value="/WEB-INF/jsp/" />
       <!-- 后缀 -->
       <property name="suffix" value=".jsp" />
  </bean>
</beans>
```

第 5 步　创建相关视图

在 ch11b 应用的 WebContent 目录下创建信息输入页面 inputb.jsp，核心代码如下：

```
<form action="${pageContext.request.contextPath }/my/formatter" method=
"post">
    商品名称：<input type="text" name="goodsname"/><br>
    商品价格：<input type="text" name="goodsprice"/><br>
    商品数量：<input type="text" name="goodsnumber"/><br>
    商品日期：<input type="text" name="goodsdate"/>（yyyy-MM-dd）<br>
    <input type="submit" value="提交"/>
</form>
```

在 ch11b 应用的/WEB-INF/jsp 目录下创建信息显示页面 showGoodsb.jsp，核心代码如下：

```
<body>
    您创建的商品信息如下：<br>
  <!-- 使用 EL 表达式取出 Action 类的属性 goods 的值 -->
    商品名称为：${goods.goodsname }<br>
    商品价格为：${goods.goodsprice }<br>
    商品数量为：${goods.goodsnumber }<br>
    商品日期为：${goods.goodsdate }
</body>
```

最后通过地址“http://localhost:8080/ch11b/inputb.jsp”测试应用。

11.4　本章小结

本章重点讲解了自定义类型转换器和格式化转换器的实现与注册，但在实际应用中开发者很少自定义类型转换器和格式化类型转换器，一般都是使用内置的转换器。

习题 11

1. 在 MVC 框架中为什么要进行类型转换？
2. Converter 与 Formatter 的区别是什么？
3. 在 Spring MVC 框架中如何自定义类型转换器类？如何注册类型转换器？
4. 在 Spring MVC 框架中如何自定义格式化转换器类？如何注册格式化转换器？

第12章

数据绑定和表单标签库

学习目的与要求

本章主要讲解数据绑定和表单标签库。通过本章的学习，读者能够理解数据绑定的基本原理，掌握表单标签库的用法。

本章主要内容

- 数据绑定;
- 表单标签库;
- 数据绑定的应用;
- JSON 数据交互。

数据绑定是将用户参数输入值绑定到领域模型的一种特性，在 Spring MVC 的 Controller 和 View 参数数据传递中所有 HTTP 请求参数的类型均为字符串，如果模型需要绑定的类型为 double 或 int，则需要手动进行类型转换，而有了数据绑定后就不再需要手动将 HTTP 请求中的 String 类型转换为模型需要的类型。数据绑定的另一个好处是当输入验证失败时会重新生成一个 HTML 表单，无须重新填写输入字段。

在 Spring MVC 中，为了方便、高效地使用数据绑定，还需要学习表单标签库。

12.1 数据绑定

在 Spring MVC 框架中数据绑定有这样几层含义：绑定请求参数输入值到领域模型（如 10.2 节）、模型数据到视图的绑定（输入验证失败时）、模型数据到表单元素的绑定（如下列列表选项值由控制器初始化）。有关数据绑定的示例请读者参见 12.3 节“数据绑定的应用”。

12.2 表单标签库

视频讲解

表单标签库中包含了可以用在 JSP 页面中渲染 HTML 元素的标签。在

JSP 页面使用 Spring 表单标签库时，必须在 JSP 页面开头处声明 taglib 指令，指令代码如下：

```
<%@ taglib prefix="form" uri="http://www.springframework.org/tags/form" %>
```

在表单标签库中有 form、input、password、hidden、textarea、checkbox、checkboxes、radiobutton、radiobuttons、select、option、options、errors 等标签。

- form：渲染表单元素。
- input：渲染<input type="text"/>元素。
- password：渲染<input type="password"/>元素。
- hidden：渲染<input type="hidden"/>元素。
- textarea：渲染 textarea 元素。
- checkbox：渲染一个<input type="checkbox"/>元素。
- checkboxes：渲染多个<input type="checkbox"/>元素。
- radiobutton：渲染一个<input type="radio"/>元素。
- radiobuttons：渲染多个<input type="radio"/>元素。
- select：渲染一个选择元素。
- option：渲染一个选项元素。
- options：渲染多个选项元素。
- errors：在 span 元素中渲染字段错误。

12.2.1　表单标签

表单标签的语法格式如下：

```
<form:form modelAttribute="xxx" method="post" action="xxx">
    …
</form:form>
```

表单标签除了具有 HTML 表单元素属性以外，还具有 acceptCharset、commandName、cssClass、cssStyle、htmlEscape 和 modelAttribute 等属性。

- acceptCharset：定义服务器接受的字符编码列表。
- commandName：暴露表单对象的模型属性名称，默认为 command。
- cssClass：定义应用到 form 元素的 CSS 类。
- cssStyle：定义应用到 form 元素的 CSS 样式。
- htmlEscape：true 或 false，表示是否进行 HTML 转义。
- modelAttribute：暴露 form backing object 的模型属性名称，默认为 command。

其中，commandName 和 modelAttribute 属性的功能基本一致，属性值绑定一个 JavaBean 对象。假设控制器类 UserController 的方法 inputUser 是返回 userAdd.jsp 的请求处理方法，inputUser 方法的代码如下：

```
@RequestMapping(value="/input")
public String inputUser(Model model) {
    …
```

```
        model.addAttribute("user", new User());
        return "userAdd";
    }
```

userAdd.jsp 的表单标签代码如下：

```
<form:form modelAttribute="user" method="post" action="user/save">
    ...
</form:form>
```

注意：在 inputUser 方法中，如果没有 Model 属性 user，userAdd.jsp 页面就会抛出异常，因为表单标签无法找到在其 modelAttribute 属性中指定的 form backing object。

12.2.2 input 标签

input 标签的语法格式如下：

```
<form:input path="xxx"/>
```

该标签除了有 cssClass、cssStyle、htmlEscape 属性以外，还有一个最重要的属性——path。path 属性将文本框输入值绑定到 form backing object 的一个属性。示例代码如下：

```
<form:form modelAttribute="user" method="post" action="user/save">
    <form:input path="userName"/>
</form:form>
```

上述代码将输入值绑定到 user 对象的 userName 属性。

12.2.3 password 标签

password 标签的语法格式如下：

```
<form:password path="xxx"/>
```

该标签与 input 标签的用法完全一致，这里不再赘述。

12.2.4 hidden 标签

hidden 标签的语法格式如下：

```
<form:hidden path="xxx"/>
```

该标签与 input 标签的用法基本一致，只不过它不可显示，不支持 cssClass 和 cssStyle 属性。

12.2.5 textarea 标签

textarea 基本上就是一个支持多行输入的 input 元素，语法格式如下：

```
<form:textarea path="xxx"/>
```

该标签与 input 标签的用法完全一致，这里不再赘述。

12.2.6　checkbox 标签

checkbox 标签的语法格式如下：

```
<form:checkbox path="xxx" value="xxx"/>
```

多个 path 相同的 checkbox 标签，它们是一个选项组，允许多选，选项值绑定到一个数组属性。示例代码如下：

```
<form:checkbox path="friends" value="张三"/>张三
<form:checkbox path="friends" value="李四"/>李四
<form:checkbox path="friends" value="王五"/>王五
<form:checkbox path="friends" value="赵六"/>赵六
```

上述示例代码中复选框的值绑定到一个字符串数组属性 friends（String[] friends）。该标签的其他用法与 input 标签基本一致，这里不再赘述。

12.2.7　checkboxes 标签

checkboxes 标签渲染多个复选框，是一个选项组，等价于多个 path 相同的 checkbox 标签。它有 3 个非常重要的属性，即 items、itemLabel 和 itemValue。

- items：用于生成 input 元素的 Collection、Map 或 Array。
- itemLabel：items 属性中指定的集合对象的属性，为每个 input 元素提供 label。
- itemValue：items 属性中指定的集合对象的属性，为每个 input 元素提供 value。

checkboxes 标签的语法格式如下：

```
<form:checkboxes items="xxx" path="xxx"/>
```

示例代码如下：

```
<form:checkboxes items="${hobbys}" path="hobby" />
```

上述示例代码是将 model 属性 hobbys 的内容（集合元素）渲染为复选框。在 itemLabel 和 itemValue 省略的情况下，如果集合是数组，复选框的 label 和 value 相同；如果是 Map 集合，复选框的 label 是 Map 的值（value），复选框的 value 是 Map 的关键字（key）。

12.2.8　radiobutton 标签

radiobutton 标签的语法格式如下：

```
<form:radiobutton path="xxx" value="xxx"/>
```

多个 path 相同的 radiobutton 标签，它们是一个选项组，只允许单选。

12.2.9　radiobuttons 标签

radiobuttons 标签渲染多个 radio，是一个选项组，等价于多个 path 相同的 radiobutton 标签。radiobuttons 标签的语法格式如下：

```
<form:radiobuttons path="xxx" items="xxx"/>
```

该标签的 itemLabel 和 itemValue 属性与 checkboxes 标签的 itemLabel 和 itemValue 属性完全一样，但只允许单选。

12.2.10　select 标签

select 标签的选项可能来自其属性 items 指定的集合，或者来自一个嵌套的 option 标签或 options 标签。其语法格式如下：

```
<form:select path="xxx" items="xxx" />
```

或

```
<form:select path="xxx" items="xxx" >
    <option value="xxx">xxx</option>
</ form:select>
```

或

```
<form:select path="xxx">
    <form:options items="xxx"/>
</form:select>
```

该标签的 itemLabel 和 itemValue 属性与 checkboxes 标签的 itemLabel 和 itemValue 属性完全一样。

12.2.11　options 标签

options 标签生成一个 select 标签的选项列表，因此需要和 select 标签一同使用，具体用法参见 12.2.10 节“select 标签”。

12.2.12　errors 标签

errors 标签渲染一个或者多个 span 元素，每个 span 元素包含一个错误消息。它可以用于显示一个特定的错误消息，也可以显示所有错误消息。其语法格式如下：

```
<form:errors path="*"/>
```

或

```
<form:errors path="xxx"/>
```

其中，“*”表示显示所有错误消息；“xxx”表示显示由“xxx”指定的特定错误消息。

12.3　数据绑定的应用

为了让读者进一步学习数据绑定和表单标签，本节给出了一个应用范例 ch12。该应用中实现了 User 类属性和 JSP 页面中表单参数的绑定，同时在 JSP 页面中分别展示了 input、textarea、checkbox、checkboxs、select 等标签。

视频讲解

12.3.1　应用的相关配置

在 ch12 应用中需要使用 JSTL，因此不仅需要将 Spring MVC 的相关 JAR 包复制到应用的 WEN-INF/lib 目录下，还需要从 Tomcat 的 webapps\examples\WEB-INF\lib 目录下将 JSTL 的相关 JAR 包复制到应用的 WEN-INF/lib 目录下。ch12 的 JAR 包如图 12.1 所示。

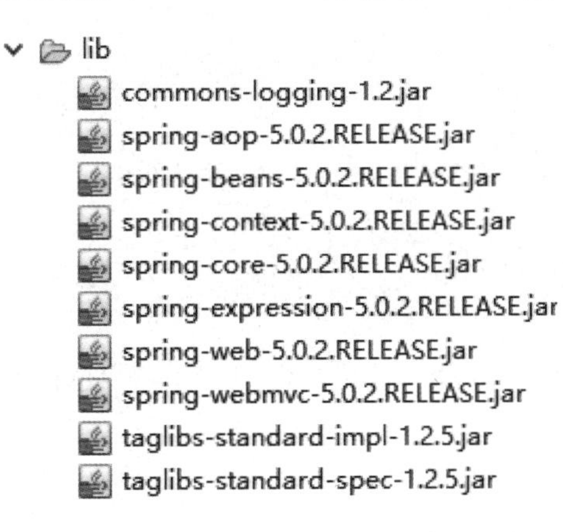

图 12.1　ch12 的 JAR 包

为了避免出现中文乱码问题，需要在 web.xml 文件中增加编码过滤器，同时将 JSP 页面编码设置为 UTF-8，form 表单的提交方式必须为 post。

web.xml 的代码如下：

```
<?xml version="1.0" encoding="UTF-8"?>
<web-app xmlns:xsi="http://www.w3.org/2001/XMLSchema-instance"
    xmlns="http://xmlns.jcp.org/xml/ns/javaee"
    xsi:schemaLocation="http://xmlns.jcp.org/xml/ns/javaee
    http://xmlns.jcp.org/xml/ns/javaee/web-app_3_1.xsd"
    id="WebApp_ID" version="3.1">
    <! -- 配置 DispatcherServlet -->
    <servlet>
```

```
        <servlet-name>springmvc</servlet-name>
        <servlet-class>org.springframework.web.servlet.DispatcherServlet
        </servlet-class>
        <load-on-startup>1</load-on-startup>
    </servlet>
    <servlet-mapping>
        <servlet-name>springmvc</servlet-name>
        <url-pattern>/</url-pattern>
    </servlet-mapping>
    <!-- 避免中文乱码 -->
    <filter>
        <filter-name>characterEncodingFilter</filter-name>
        <filter-class>org.springframework.web.filter.CharacterEncoding
        Filter</filter-class>
        <init-param>
            <param-name>encoding</param-name>
            <param-value>UTF-8</param-value>
        </init-param>
        <init-param>
            <param-name>forceEncoding</param-name>
            <param-value>true</param-value>
        </init-param>
    </filter>
    <filter-mapping>
        <filter-name>characterEncodingFilter</filter-name>
        <url-pattern>/*</url-pattern>
    </filter-mapping>
</web-app>
```

配置文件 springmvc-servlet.xml 与第 10 章中学习过的配置文件没有区别，这里不再赘述。

12.3.2 领域模型

应用中实现了 User 类属性和 JSP 页面中表单参数的绑定，User 类包含了和表单参数名对应的属性，以及属性的 set 和 get 方法。在 ch12 应用的 src 目录下创建 pojo 包，并在该包中创建 User 类。

User 类的代码如下：

```
package pojo;
public class User {
    private String userName;
    private String[] hobby;  //兴趣爱好
    private String[] friends;  //朋友
    private String carrer;
    private String houseRegister;
    private String remark;
```

```
    //省略 setter 和 getter 方法
}
```

12.3.3　Service 层

应用中使用了 Service 层，在 Service 层使用静态集合变量 users 模拟数据库存储用户信息，包括添加用户和查询用户两个功能方法。在 ch12 应用的 src 目录下创建 service 包，并在该包中创建 UserService 接口和 UserServiceImpl 实现类。

UserService 接口的代码如下：

```
package service;
import java.util.ArrayList;
import pojo.User;
public interface UserService {
    boolean addUser(User u);
    ArrayList<User> getUsers();
}
```

UserServiceImpl 实现类的代码如下：

```
package service;
import java.util.ArrayList;
import org.springframework.stereotype.Service;
import pojo.User;
@Service
public class UserServiceImpl implements UserService{
    //使用静态集合变量 users 模拟数据库
    private static ArrayList<User> users=new ArrayList<User>();
    @Override
    public boolean addUser(User u) {
        if(!"IT 民工".equals(u.getCarrer())){  //不允许添加 IT 民工
            users.add(u);
            return true;
        }
        return false;
    }
    @Override
    public ArrayList<User> getUsers() {
        return users;
    }
}
```

12.3.4　Controller 层

在 Controller 类 UserController 中定义了请求处理方法，包括处理 user/input 请求的 inputUser 方法以及处理 user/save 请求的 addUser 方法，其中在 addUser 方法中用到了重定向。在 UserController 类中，通过@Autowired 注解在 UserController 对象中主动注入 UserService 对象，实现对 user 对象的添加和查询等操作；通过 model 的 addAttribute 方法将 User 类对象、HashMap 类型的 hobbys 对象、String[]类型的 carrers 对象以及 String[]类型的 houseRegisters 对象传递给 View(userAdd.jsp)。在 ch12 应用的 src 目录下创建 controller 包，并在该包中创建 UserController 控制器类。

UserController 类的代码如下：

```
package controller;
import java.util.HashMap;
import java.util.List;
import org.apache.commons.logging.Log;
import org.apache.commons.logging.LogFactory;
import org.springframework.beans.factory.annotation.Autowired;
import org.springframework.stereotype.Controller;
import org.springframework.ui.Model;
import org.springframework.web.bind.annotation.ModelAttribute;
import org.springframework.web.bind.annotation.RequestMapping;
import pojo.User;
import service.UserService;
@Controller
@RequestMapping("/user")
public class UserController {
    //得到一个用来记录日志的对象，这样在打印信息的时候能够标记打印的是哪个类的信息
    private static final Log logger = LogFactory.getLog(UserController. class);
    @Autowired
    private UserService userService;
    @RequestMapping(value = "/input")
    public String inputUser(Model model) {
        HashMap<String, String> hobbys = new HashMap<String, String>();
        hobbys.put("篮球", "篮球");
        hobbys.put("乒乓球", "乒乓球");
        hobbys.put("电玩", "电玩");
        hobbys.put("游泳", "游泳");
        //如果model中没有user属性，userAdd.jsp会抛出异常，因为表单标签无法找到
        // modelAttribute属性指定的form backing object
        model.addAttribute("user", new User());
        model.addAttribute("hobbys", hobbys);
    model.addAttribute("carrers", new String[] { "教师", "学生", "coding搬
    运工", "IT民工", "其他" });
        model.addAttribute("houseRegisters", new String[] { "北京","上海",
        "广州", "深圳", "其他" });
        return "userAdd";
    }
    @RequestMapping(value = "/save")
```

```
    public String addUser(@ModelAttribute User user, Model model) {
        if (userService.addUser(user)) {
            logger.info("成功");
            return "redirect:/user/list";
        } else {
            logger.info("失败");
            HashMap<String, String> hobbys = new HashMap<String, String>();
            hobbys.put("篮球", "篮球");
            hobbys.put("乒乓球", "乒乓球");
            hobbys.put("电玩", "电玩");
            hobbys.put("游泳", "游泳");
            //这里不需要 model.addAttribute("user", new
            // User()), 因为@ModelAttribute 指定 form backing object
            model.addAttribute("hobbys", hobbys);
            model.addAttribute("carrers", new String[] { "教师", "学生",
            "coding 搬运工", "IT 民工", "其他" });
            model.addAttribute("houseRegisters", new String[] { "北京","
            上海","广州", "深圳", "其他" });
            return "userAdd";
        }
    }
    @RequestMapping(value = "/list")
    public String listUsers(Model model) {
        List<User> users = userService.getUsers();
        model.addAttribute("users", users);
        return "userList";
    }
}
```

12.3.5 View 层

View 层包含两个 JSP 页面，一个是信息输入页面 userAdd.jsp，一个是信息显示页面 userList.jsp。在 ch12 应用的 WEB-INF/jsp 目录下创建这两个 JSP 页面。

在 userAdd.jsp 页面中将 Map 类型的 hobbys 绑定到 checkboxes 上，将 String[]类型的 carrers 和 houseRegisters 绑定到 select 上，实现通过 option 标签对 select 添加选项，同时表单的 method 方法需指定为 post 来避免中文乱码问题。

在 userList.jsp 页面中使用 JSTL 标签遍历集合中的用户信息，对于 JSTL 的相关知识，请读者参见本书的相关内容。

userAdd.jsp 的代码如下：

```
<%@ page language="java" contentType="text/html; charset=UTF-8" pageEncoding=
"UTF-8"%>
<%@ taglib prefix="form" uri="http://www.springframework.org/tags/form" %>
<!DOCTYPE html PUBLIC "-//W3C//DTD HTML 4.01 Transitional//EN"
```

```
"http://www.w3.org/TR/html4/loose.dtd">
<html>
<head>
<meta http-equiv="Content-Type" content="text/html; charset=UTF-8">
<title>Insert title here</title>
</head>
<body>
<form:form modelAttribute="user" method="post" action=" ${pageContext.
request.contextPath }/user/save">
    <fieldset>
        <legend>添加一个用户</legend>
        <p>
            <label>用户名:</label>
            <form:input path="userName"/>
        </p>
        <p>
            <label>爱好:</label>
            <form:checkboxes items="${hobbys}" path="hobby" />
        </p>
        <p>
            <label>朋友:</label>
            <form:checkbox path="friends" value="张三"/>张三
            <form:checkbox path="friends" value="李四"/>李四
            <form:checkbox path="friends" value="王五"/>王五
            <form:checkbox path="friends" value="赵六"/>赵六
        </p>
        <p>
            <label>职业:</label>
            <form:select path="carrer">
                <option/>请选择职业
                <form:options items="${carrers }"/>
            </form:select>
        </p>
        <p>
            <label>户籍:</label>
            <form:select path="houseRegister">
                <option/>请选择户籍
                <form:options items="${houseRegisters }"/>
            </form:select>
        </p>
        <p>
            <label>个人描述:</label>
            <form:textarea path="remark" rows="5"/>
        </p>
        <p id="buttons">
```

```
            <input id="reset" type="reset">
            <input id="submit" type="submit" value="添加">
        </p>
    </fieldset>
</form:form>
</body>
</html>
```

userList.jsp 的代码如下：

```
<%@ page language="java" contentType="text/html; charset=UTF-8"
pageEncoding="UTF-8"%>
<%@ taglib uri="http://java.sun.com/jsp/jstl/core" prefix="c" %>
<!DOCTYPE html PUBLIC "-//W3C//DTD HTML 4.01 Transitional//EN"
"http://www.w3.org/TR/html4/loose.dtd">
<html>
<head>
<meta http-equiv="Content-Type" content="text/html; charset=UTF-8">
<title>Insert title here</title>
</head>
<body>
    <h1>用户列表</h1>
    <a href="<c:url value="${pageContext.request.contextPath }/user/
    input"/>">继续添加</a>
    <table>
        <tr>
            <th>用户名</th>
            <th>兴趣爱好</th>
            <th>朋友</th>
            <th>职业</th>
            <th>户籍</th>
            <th>个人描述</th>
        </tr>
        <!-- JSTL 标签请参考本书的相关内容 -->
        <c:forEach items="${users}" var="user">
            <tr>
                <td>${user.userName }</td>
                <td>
                  <c:forEach items="${user.hobby }" var="hobby">
                      ${hobby } 
                  </c:forEach>
                </td>
                <td>
                  <c:forEach items="${user.friends }" var="friend">
                      ${friend } 
                  </c:forEach>
```

```
                </td>
                <td>${user.carrer }</td>
                <td>${user.houseRegister }</td>
                <td>${user.remark }</td>
            </tr>
        </c:forEach>
    </table>
</body>
</html>
```

12.3.6 测试应用

通过地址“http://localhost:8080/ch12/user/input”测试应用，添加用户信息页面效果如图 12.2 所示。

如果在图 12.2 中职业选择“IT 民工”，添加失败。失败后还回到添加页面，输入过的信息不再输入，自动回填（必须结合 form 标签）。自动回填是数据绑定的一个优点。失败页面如图 12.3 所示。

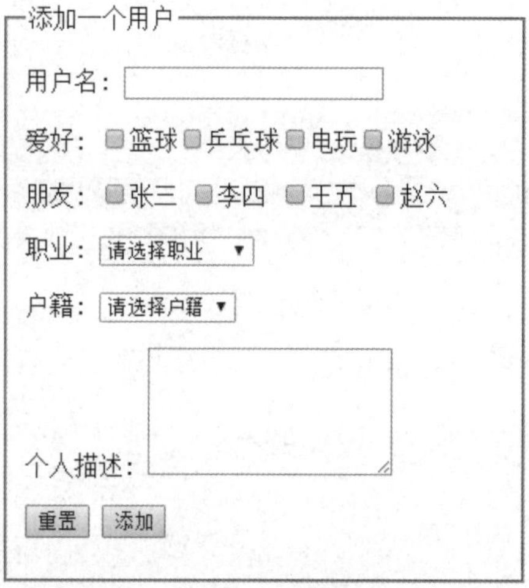

添加一个用户
用户名：
爱好： 篮球 乒乓球 电玩 游泳
朋友： 张三 李四 王五 赵六
职业： 请选择职业
户籍： 请选择户籍
个人描述：
重置 添加

图 12.2 添加用户信息页面

添加一个用户
用户名： 陈恒
爱好： 游泳 篮球 电玩 乒乓球
朋友： 张三 李四 王五 赵六
职业： IT民工
户籍： 其他
一个IT民工
个人描述：
重置 添加

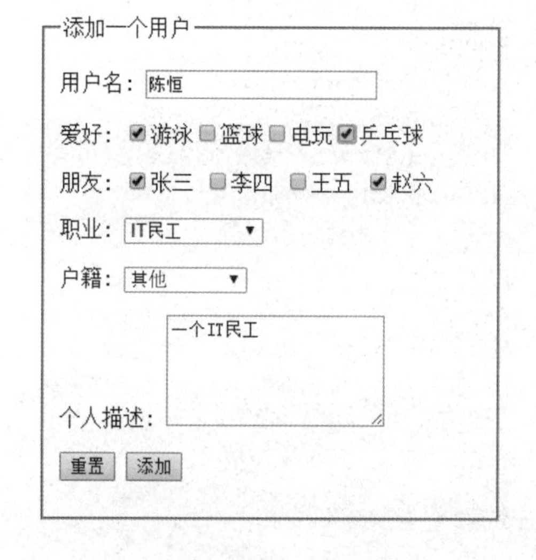

图 12.3 添加用户信息失败页面

在图 12.2 中输入正确信息，添加成功后重定向到信息显示页面，效果如图 12.4 所示。

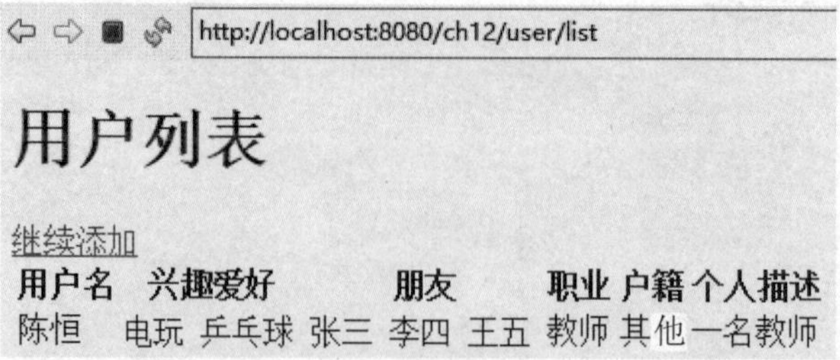

http://localhost:8080/ch12/user/list
用户列表
继续添加
用户名 兴趣爱好 朋友 职业 户籍 个人描述
陈恒 电玩 乒乓球 张三 李四 王五 教师 其他 一名教师

图 12.4 信息显示页面

12.4 JSON 数据交互

Spring MVC 在数据绑定的过程中需要对传递数据的格式和类型进行转换，它既可以转换 String 等类型的数据，也可以转换 JSON 等其他类型的数据。本节将针对 Spring MVC 中 JSON 类型的数据交互进行讲解。

视频讲解

12.4.1 JSON 概述

JSON（JavaScript Object Notation，JS 对象标记）是一种轻量级的数据交换格式。与 XML 一样，JSON 也是基于纯文本的数据格式。它有对象结构和数组结构两种数据结构。

❶ 对象结构

对象结构以“{”开始、以“}”结束，中间部分由 0 个或多个以英文“,”分隔的 key/value 对构成，key 和 value 之间以英文“:”分隔。对象结构的语法结构如下：

```
{
    key1:value1,
    key2:value2,
    …
}
```

其中，key 必须为 String 类型，value 可以是 String、Number、Object、Array 等数据类型。例如，一个 person 对象包含姓名、密码、年龄等信息，使用 JSON 的表示形式如下：

```
{
    "pname":"陈恒",
    "password":"123456",
    "page":40
}
```

❷ 数组结构

数组结构以“[”开始、以“]”结束，中间部分由 0 个或多个以英文“,”分隔的值的列表组成。数组结构的语法结构如下：

```
[
    value1,
    value2,
    …
]
```

上述两种（对象、数组）数据结构也可以分别组合构成更加复杂的数据结构。例如，一个 student 对象包含 sno、sname、hobby 和 college 对象，其 JSON 的表示形式如下：

```
{
    "sno":"201802228888",
    "sname":"陈恒",
```

```
    "hobby":["篮球","足球"],
    "college":{
        "cname":"清华大学",
        "city":"北京"
    }
}
```

12.4.2 JSON 数据转换

为实现浏览器与控制器类之间的 JSON 数据交互，Spring MVC 提供了 MappingJackson2HttpMessageConverter 实现类默认处理 JSON 格式请求响应。该实现类利用 Jackson 开源包读写 JSON 数据，将 Java 对象转换为 JSON 对象和 XML 文档，同时也可以将 JSON 对象和 XML 文档转换为 Java 对象。

Jackson 开源包及其描述如下。

- jackson-annotations-2.9.4.jar：JSON 转换注解包。
- jackson-core-2.9.4.jar：JSON 转换核心包。
- jackson-databind-2.9.4.jar：JSON 转换的数据绑定包。

以上 3 个 Jackson 的开源包在编写本书时最新版本是 2.9.4，读者可通过地址“http://mvnrepository.com/artifact/com.fasterxml.jackson.core”下载得到。

在使用注解开发时需要用到两个重要的 JSON 格式转换注解，分别是@RequestBody 和@ResponseBody。

- @RequestBody：用于将请求体中的数据绑定到方法的形参中，该注解应用在方法的形参上。
- @ResponseBody：用于直接返回 return 对象，该注解应用在方法上。

下面通过一个案例来演示如何进行 JSON 数据交互，具体步骤如下。

❶ 创建应用并导入相关 JAR 包

创建 ch12b 应用，然后将 Spring MVC 的相关 JAR 包、JSON 转换包添加到 ch12b 的 lib 目录下。ch12b 的 lib 目录如图 12.5 所示。

图 12.5　ch12b 的相关 JAR 包

❷ **配置 web.xml**

在 web.xml 文件中对 Spring MVC 的前端控制器等信息进行配置，其代码与第 10 章中 ch10 的相同，这里不再赘述。

❸ **配置 Spring MVC 的核心配置文件**

在 WEB-INF 目录下创建 Spring MVC 的核心配置文件 springmvc-servlet.xml，代码如下：

```
<?xml version="1.0" encoding="UTF-8"?>
<beans xmlns="http://www.springframework.org/schema/beans"
    xmlns:xsi="http://www.w3.org/2001/XMLSchema-instance"
    xmlns:context="http://www.springframework.org/schema/context"
    xmlns:mvc="http://www.springframework.org/schema/mvc"
    xsi:schemaLocation="
    http://www.springframework.org/schema/beans
    http://www.springframework.org/schema/beans/spring-beans.xsd
    http://www.springframework.org/schema/context
    http://www.springframework.org/schema/context/spring-context.xsd
    http://www.springframework.org/schema/mvc
    http://www.springframework.org/schema/mvc/spring-mvc.xsd">
    <!-- 使用扫描机制扫描控制器类 -->
    <context:component-scan base-package="controller"/>
    <mvc:annotation-driven />
    <!-- annotation-driven 用于简化开发的配置，
    注解 DefaultAnnotationHandlerMapping 和 AnnotationMethodHandlerAdapter -->
    <!-- 使用 resources 过滤掉不需要 dispatcher servlet 的资源。
    在使用 resources 时必须使用 annotation-driven，否则 resources 元素会阻止任意控
    制器被调用
   -->
    <!-- 配置静态资源，允许 js 目录下的所有文件可见 -->
    <mvc:resources location="/js/" mapping="/js/**"></mvc:resources>
     <!-- 配置视图解析器 -->
     <bean  class="org.springframework.web.servlet.view.InternalResourceView
     Resolver"
            id="internalResourceViewResolver">
        <!-- 前缀 -->
        <property name="prefix" value="/WEB-INF/jsp/" />
        <!-- 后缀 -->
        <property name="suffix" value=".jsp" />
   </bean>
</beans>
```

❹ **创建 POJO 类**

在 src 目录下创建 pojo 包，并在该包中创建 POJO 类 Person，代码如下：

```
package pojo;
```

```
public class Person {
    private String pname;
    private String password;
    private Integer page;
    //省略 setter 和 getter 方法
}
```

❺ 创建 JSP 页面测试 JSON 数据交互

在 WebContent 目录下创建页面 index.jsp 来测试 JSON 数据交互，代码如下：

```
<%@ page language="java" contentType="text/html; charset=UTF-8"
pageEncoding="UTF-8"%>
<!DOCTYPE html PUBLIC "-//W3C//DTD HTML 4.01 Transitional//EN"
"http://www.w3.org/TR/html4/loose.dtd">
<html>
<head>
<meta http-equiv="Content-Type" content="text/html; charset=UTF-8">
<title>Insert title here</title>
<script type="text/javascript" src="${pageContext.request.contextPath }/
js/jquery-3.2.1.min.js"></script>
<script type="text/javascript">
    function testJson() {
        //获取输入的值 pname 为 id
        var pname=$("#pname").val();
        var password=$("#password").val();
        var page=$("#page").val();
        $.ajax({
            //请求路径
            url : "${pageContext.request.contextPath }/testJson",
            //请求类型
            type : "post",
            //data 表示发送的数据
            data : JSON.stringify({pname:pname,password:password,page:page}),
            //定义发送请求的数据格式为 JSON 字符串
            contentType : "application/json;charset=utf-8",
            //定义回调响应的数据格式为 JSON 字符串，该属性可以省略
            dataType : "json",
            //成功响应的结果
            success : function(data){
                if(data!=null){
        alert("输入的用户名:" + data.pname + ", 密码: " + data.password + ",
        年龄: " + data.page);
                }
            }
```

```
            });
        }
</script>
</head>
<body>
    <form action="">
        用户名：<input type="text" name="pname" id="pname"/><br>
        密码：<input type="password" name="password" id="password"/><br>
        年龄：<input type="text" name="page" id="page"/><br>
        <input type="button" value="测试" onclick="testJson()"/>
    </form>
</body>
</html>
```

在 index.jsp 页面中编写了一个测试 JSON 交互的表单，当单击“测试”按钮时执行页面中的 testJson()函数。在该函数中使用了 jQuery 的 AJAX 方式将 JSON 格式的数据传递给以“/testJson”结尾的请求中。

因为在 index.jsp 中使用的是 jQuery 的 AJAX 进行的 JSON 数据提交和响应，所以还需要引入 jquery.js 文件。本例引入了 WebContent 目录下 js 文件夹中的 jquery-3.2.1.min.js，读者可以在源程序中找到该文件。

❻ 创建控制器类

在 src 目录下创建 controller 包，并在该包中创建一个用于用户操作的控制器类 TestController，代码如下：

```
package controller;
import org.springframework.stereotype.Controller;
import org.springframework.web.bind.annotation.RequestBody;
import org.springframework.web.bind.annotation.RequestMapping;
import org.springframework.web.bind.annotation.ResponseBody;
import pojo.Person;
@Controller
public class TestController {
    /**
     * 接收页面请求的 JSON 数据，并返回 JSON 格式的结果
     */
    @RequestMapping("/testJson")
    @ResponseBody
    public Person testJson(@RequestBody Person user) {
        //打印接收的 JSON 格式数据
        System.out.println("pname=" + user.getPname() +
                ", password=" + user.getPassword() + ",page=" +
                user.getPage());
        //返回 JSON 格式的响应
        return user;
```

```
        }
}
```

在上述控制器类中编写了接收和响应 JSON 格式数据的 testJson 方法，方法中的 @RequestBody 注解用于将前端请求体中的 JSON 格式数据绑定到形参 user 上，@ResponseBody 注解用于直接返回 Person 对象（当返回 POJO 对象时默认转换为 JSON 格式数据进行响应）。

❼ 运行 index.jsp 页面，测试程序

将 ch12b 应用发布到 Tomcat 服务器并启动服务器，在浏览器中访问地址“http://localhost:8080/ch12b/index.jsp”，运行效果如图 12.6 所示。

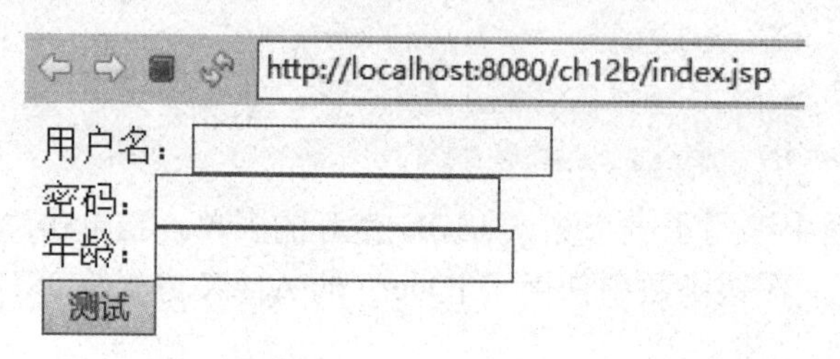

图 12.6　index.jsp 测试页面

在图 12.6 所示的输入框中输入信息后单击“测试”按钮，当程序正确执行时页面将弹出显示输入信息的对话框，如图 12.7 所示。

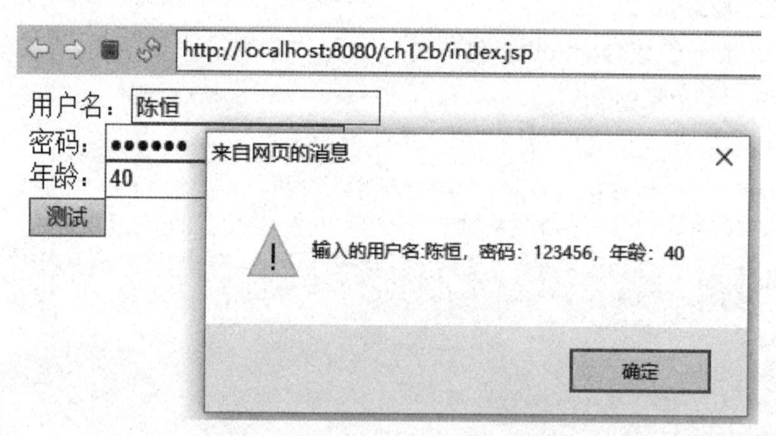

图 12.7　index.jsp 正确执行的效果

同时，Eclipse 的控制台将打印出相应数据，如图 12.8 所示。

图 12.8　Eclipse 运行结果

从图 12.7 和图 12.8 所示的结果可以看出，编写的代码可以将 JSON 格式的请求数据转换为方法中的 Java 对象，也可以将 Java 对象转换为 JSON 格式的响应数据。

12.5　本章小结

本章介绍了 Spring MVC 的数据绑定和表单标签，包括数据绑定的原理以及如何使用表单标签，然后给出了一个数据绑定应用示例，大致演示了数据绑定在实际开发中的使用，最后讲解了 JSON 数据交互的使用，使读者了解 JSON 数据的组织结构。

习题 12

1. 举例说明数据绑定的优点。
2. Spring MVC 有哪些表单标签？其中可以绑定集合数据的标签有哪些？

第 13 章

拦截器

学习目的与要求

本章主要介绍了拦截器的概念、原理以及实际应用。通过本章的学习，读者能够理解拦截器的原理，掌握拦截器的实际应用。

本章主要内容

- 拦截器的定义;
- 拦截器的配置;
- 拦截器的执行流程。

在开发一个网站时可能有这样的需求：某些页面只希望几个特定的用户浏览。对于这样的访问权限控制，应该如何实现呢？拦截器就可以实现上述需求。在 Struts 2 框架中，拦截器是其重要的组成部分，Spring MVC 框架也提供了拦截器功能。本章将针对 Spring MVC 中拦截器的使用进行详细讲解。

13.1 拦截器概述

Spring MVC 的拦截器（Interceptor）与 Java Servlet 的过滤器（Filter）类似，它主要用于拦截用户的请求并做相应的处理，通常应用在权限验证、记录请求信息的日志、判断用户是否登录等功能上。

视频讲解

13.1.1 拦截器的定义

在 Spring MVC 框架中定义一个拦截器需要对拦截器进行定义和配置，定义一个拦截器可以通过两种方式：一种是通过实现 HandlerInterceptor 接口或继承 HandlerInterceptor 接口的实现类来定义；另一种是通过实现 WebRequestInterceptor 接口或继承 WebRequestInterceptor 接口的实现类来定义。本章以实现 HandlerInterceptor 接口的定义方式为例讲解自定义拦截器的使用方法。示例代码如下：

```
package interceptor;
import javax.servlet.http.HttpServletRequest;
import javax.servlet.http.HttpServletResponse;
import org.springframework.web.servlet.HandlerInterceptor;
import org.springframework.web.servlet.ModelAndView;
public class TestInterceptor implements HandlerInterceptor{
    @Override
    public boolean preHandle(HttpServletRequest request, HttpServlet
    Response response, Object handler)
            throws Exception {
        System.out.println("preHandle 方法在控制器的处理请求方法前执行");
        /**返回 true 表示继续向下执行，返回 false 表示中断后续操作*/
        return true;
    }
    @Override
    public void postHandle(HttpServletRequest request, HttpServletResponse
    response, Object handler,
            ModelAndView modelAndView) throws Exception {
        System.out.println("postHandle 方法在控制器的处理请求方法调用之后，解析
        视图之前执行");
    }
    @Override
    public void afterCompletion(HttpServletRequest request, HttpServlet
    Response response, Object handler, Exception ex)
            throws Exception {
        System.out.println("afterCompletion 方法在控制器的处理请求方法执行完成
        后执行，即视图渲染结束之后执行");
    }
}
```

在上述拦截器的定义中实现了 HandlerInterceptor 接口，并实现了接口中的 3 个方法。有关这 3 个方法的描述如下。

- preHandle 方法：该方法在控制器的处理请求方法前执行，其返回值表示是否中断后续操作，返回 true 表示继续向下执行，返回 false 表示中断后续操作。
- postHandle 方法：该方法在控制器的处理请求方法调用之后、解析视图之前执行，可以通过此方法对请求域中的模型和视图做进一步的修改。
- afterCompletion 方法：该方法在控制器的处理请求方法执行完成后执行，即视图渲染结束后执行，可以通过此方法实现一些资源清理、记录日志信息等工作。

13.1.2 拦截器的配置

让自定义的拦截器生效需要在 Spring MVC 的配置文件中进行配置，配置示例代码

如下：

```
<!-- 配置拦截器 -->
<mvc:interceptors>
    <!-- 配置一个全局拦截器，拦截所有请求 -->
    <bean class="interceptor.TestInterceptor"/>
    <mvc:interceptor>
        <!-- 配置拦截器作用的路径 -->
        <mvc:mapping path="/**"/>
        <!-- 配置不需要拦截作用的路径 -->
        <mvc:exclude-mapping path=""/>
        <!-- 定义在<mvc:interceptor>元素中，表示匹配指定路径的请求才进行拦截 -->
        <bean class="interceptor.Interceptor1"/>
    </mvc:interceptor>
    <mvc:interceptor>
        <!-- 配置拦截器作用的路径 -->
        <mvc:mapping path="/gotoTest"/>
        <!-- 定义在<mvc:interceptor>元素中，表示匹配指定路径的请求才进行拦截 -->
        <bean class="interceptor.Interceptor2"/>
    </mvc:interceptor>
</mvc:interceptors>
```

在上述示例代码中，<mvc:interceptors>元素用于配置一组拦截器，其子元素<bean>定义的是全局拦截器，即拦截所有的请求。<mvc:interceptor>元素中定义的是指定路径的拦截器，其子元素<mvc:mapping>用于配置拦截器作用的路径，该路径在其属性 path 中定义。如上述示例代码中，path 的属性值“/**”表示拦截所有路径，“/gotoTest”表示拦截所有以“/gotoTest”结尾的路径。如果在请求路径中包含不需要拦截的内容，可以通过<mvc:exclude-mapping>子元素进行配置。

需要注意的是，<mvc:interceptor>元素的子元素必须按照<mvc:mapping.../>、<mvc:exclude-mapping .../>、<bean .../>的顺序配置。

13.2 拦截器的执行流程

13.2.1 单个拦截器的执行流程

视频讲解

在配置文件中如果只定义了一个拦截器，程序将首先执行拦截器类中的 preHandle 方法，如果该方法返回 true，程序将继续执行控制器中处理请求的方法，否则中断执行。如果 preHandle 方法返回 true，并且控制器中处理请求的方法执行后、返回视图前将执行 postHandle 方法，返回视图后才执行 afterCompletion 方法。下面通过一个应用 ch13 演示拦截器的执行流程，具体步骤如下：

❶ 创建应用

创建一个名为 ch13 的 Web 应用，并将 Spring MVC 相关的 JAR 包复制到 lib 目录中。

❷ 创建 web.xml

在 WEB-INF 目录下创建 web.xml 文件，该文件中的配置信息如下：

```
<?xml version="1.0" encoding="UTF-8"?>
<web-app  xmlns:xsi="http://www.w3.org/2001/XMLSchema-instance"  xmlns=
"http://xmlns.jcp.org/xml/ns/javaee"  xsi:schemaLocation="http://xmlns.jcp.
org/xml/ns/javaee http://xmlns.jcp.org/xml/ns/javaee/web-app_3_1.xsd" id=
"WebApp_ID" version="3.1">
  <servlet>
    <servlet-name>springmvc</servlet-name>
    <servlet-class>org.springframework.web.servlet.DispatcherServlet
    </servlet-class>
    <load-on-startup>1</load-on-startup>
  </servlet>
  <servlet-mapping>
    <servlet-name>springmvc</servlet-name>
    <url-pattern>/</url-pattern>
  </servlet-mapping>
</web-app>
```

❸ 创建控制器类

在 src 目录下创建一个名为 controller 的包，并在该包中创建控制器类 Interceptor Controller，代码如下：

```
package controller;
import org.springframework.stereotype.Controller;
import org.springframework.web.bind.annotation.RequestMapping;
@Controller
public class InterceptorController {
    @RequestMapping("/gotoTest")
    public String gotoTest() {
        System.out.println("正在测试拦截器，执行控制器的处理请求方法中");
        return "test";
    }
}
```

❹ 创建拦截器类

在 src 目录下创建一个名为 interceptor 的包，并在该包中创建拦截器类 TestInterceptor，代码与 13.1.1 节中的示例代码相同。

❺ 创建配置文件 springmvc-servlet.xml

在 WEB-INF 目录下创建配置文件 springmvc-servlet.xml，代码如下：

```
<?xml version="1.0" encoding="UTF-8"?>
<beans xmlns="http://www.springframework.org/schema/beans"
    xmlns:xsi="http://www.w3.org/2001/XMLSchema-instance"
    xmlns:context="http://www.springframework.org/schema/context"
    xmlns:mvc="http://www.springframework.org/schema/mvc"
    xsi:schemaLocation="
    http://www.springframework.org/schema/beans
    http://www.springframework.org/schema/beans/spring-beans.xsd
    http://www.springframework.org/schema/context
    http://www.springframework.org/schema/context/spring-context.xsd
    http://www.springframework.org/schema/mvc
    http://www.springframework.org/schema/mvc/spring-mvc.xsd">
    <!-- 使用扫描机制扫描控制器类 -->
    <context:component-scan base-package="controller"/>
    <!-- 配置视图解析器 -->
    <bean   class="org.springframework.web.servlet.view.InternalResource
    ViewResolver"
           id="internalResourceViewResolver">
      <!-- 前缀 -->
      <property name="prefix" value="/WEB-INF/jsp/" />
      <!-- 后缀 -->
      <property name="suffix" value=".jsp" />
    </bean>
    <mvc:interceptors>
        <!-- 配置一个全局拦截器，拦截所有请求 -->
        <bean class="interceptor.TestInterceptor"/>
    </mvc:interceptors>
</beans>
```

❻ 创建视图 JSP 文件

在 WEB-INF 目录下创建一个 jsp 文件夹，并在该文件夹中创建一个 JSP 文件 test.jsp，代码如下：

```
<%@ page language="java" contentType="text/html; charset=UTF-8"
pageEncoding="UTF-8"%>
<!DOCTYPE html PUBLIC "-//W3C//DTD HTML 4.01 Transitional//EN"
"http://www.w3.org/TR/html4/loose.dtd">
<html>
<head>
<meta http-equiv="Content-Type" content="text/html; charset=UTF-8">
```

```
<title>Insert title here</title>
</head>
<body>
    视图
    <%System.out.println("视图渲染结束。"); %>
</body>
</html>
```

❼ 测试拦截器

首先将 ch13 应用发布到 Tomcat 服务器，并启动 Tomcat 服务器，然后通过地址“http://localhost:8080/ch13/gotoTest”测试拦截器。程序正确执行后控制台的输出结果如图 13.1 所示。

```
Servers  Console
Tomcat v9.0 Server at localhost [Apache Tomcat] C:\Program Files\Java\jre1.8.0_152\bin\javaw.e
preHandle方法在控制器的处理请求方法前执行
正在测试拦截器，执行控制器的处理请求方法中
postHandle方法在控制器的处理请求方法调用之后，解析视图之前执行
视图渲染结束。
afterCompletion方法在控制器的处理请求方法执行完成后执行，即视图渲染结束之后执行
```

图 13.1　单个拦截器的执行过程

13.2.2　多个拦截器的执行流程

在 Web 应用中通常需要有多个拦截器同时工作，这时它们的 preHandle 方法将按照配置文件中拦截器的配置顺序执行，而它们的 postHandle 方法和 afterCompletion 方法则按照配置顺序的反序执行。

下面通过修改 13.2.1 小节的 ch13 应用来演示多个拦截器的执行流程，具体步骤如下：

❶ 创建多个拦截器

在 ch13 应用的 interceptor 包中创建两个拦截器类 Interceptor1 和 Interceptor2。

Interceptor1 类的代码如下：

```
package interceptor;
import javax.servlet.http.HttpServletRequest;
import javax.servlet.http.HttpServletResponse;
import org.springframework.web.servlet.HandlerInterceptor;
import org.springframework.web.servlet.ModelAndView;
public class Interceptor1 implements HandlerInterceptor{
    @Override
    public boolean preHandle(HttpServletRequest request, HttpServlet
```

```
    Response response, Object handler)
            throws Exception {
        System.out.println("Interceptor1 preHandle 方法执行");
        /**返回 true 表示继续向下执行，返回 false 表示中断后续的操作*/
        return true;
    }
    @Override
    public void postHandle(HttpServletRequest request, HttpServletResponse
    response, Object handler,
            ModelAndView modelAndView) throws Exception {
        System.out.println("Interceptor1 postHandle 方法执行");
    }
    @Override
    public void afterCompletion(HttpServletRequest request, HttpServlet
    Response response, Object handler, Exception ex)
            throws Exception {
        System.out.println("Interceptor1 afterCompletion 方法执行");
    }
}
```

Interceptor2 类的代码如下：

```
package interceptor;
import javax.servlet.http.HttpServletRequest;
import javax.servlet.http.HttpServletResponse;
import org.springframework.web.servlet.HandlerInterceptor;
import org.springframework.web.servlet.ModelAndView;
public class Interceptor2 implements HandlerInterceptor{
    @Override
    public boolean preHandle(HttpServletRequest request, HttpServlet
    Response response, Object handler)
            throws Exception {
        System.out.println("Interceptor2 preHandle 方法执行");
        /**返回 true 表示继续向下执行，返回 false 表示中断后续的操作*/
        return true;
    }
    @Override
    public void postHandle(HttpServletRequest request, HttpServletResponse
    response, Object handler,
            ModelAndView modelAndView) throws Exception {
        System.out.println("Interceptor2 postHandle 方法执行");
    }
    @Override
    public void afterCompletion(HttpServletRequest request, HttpServlet
    Response response, Object handler, Exception ex)
            throws Exception {
```

```
        System.out.println("Interceptor2 afterCompletion 方法执行");
    }
}
```

❷ **配置拦截器**

在配置文件 springmvc-servlet.xml 中的<mvc:interceptors>元素内配置两个拦截器 Interceptor1 和 Interceptor2，配置代码如下：

```
<mvc:interceptors>
        <!-- 配置一个全局拦截器，拦截所有请求 -->
        <!--  <bean class="interceptor.TestInterceptor"/>-->
        <mvc:interceptor>
            <!-- 配置拦截器作用的路径 -->
            <mvc:mapping path="/**"/>
            <!-- 定义在<mvc:interceptor>元素中，表示匹配指定路径的请求才进行拦截 -->
            <bean class="interceptor.Interceptor1"/>
        </mvc:interceptor>
        <mvc:interceptor>
            <!-- 配置拦截器作用的路径 -->
            <mvc:mapping path="/gotoTest"/>
            <!-- 定义在<mvc:interceptor>元素中，表示匹配指定路径的请求才进行拦截 -->
            <bean class="interceptor.Interceptor2"/>
        </mvc:interceptor>
    </mvc:interceptors>
```

❸ **测试多个拦截器**

首先将 ch13 应用发布到 Tomcat 服务器并启动 Tomcat 服务器，然后通过地址 “http://localhost:8080/ch13/gotoTest” 测试拦截器。程序正确执行后控制台的输出结果如图 13.2 所示。

```
Servers  Console
Tomcat v9.0 Server at localhost [Apache Tomcat] C:\Prc
Interceptor1 preHandle方法执行
Interceptor2 preHandle方法执行
正在测试拦截器，执行控制器的处理请求方法中
Interceptor2 postHandle方法执行
Interceptor1 postHandle方法执行
视图渲染结束。
Interceptor2 afterCompletion方法执行
Interceptor1 afterCompletion方法执行
```

图 13.2　多个拦截器的执行过程

13.3　应用案例——用户登录权限验证

本节将通过拦截器来完成一个用户登录权限验证的 Web 应用 ch13b，具体要求如下：

只有成功登录的用户才能访问系统的主页面 main.jsp，如果没有成功登录而直接访问主页面，则拦截器将请求拦截，并转发到登录页面 login.jsp。当成功登录的用户在系统主页面中单击“退出”链接时回到登录页面。具体实现步骤如下：

视频讲解

❶ 创建应用

创建 Web 应用 ch13b，并将 Spring MVC 相关的 JAR 包复制到 lib 目录中。

❷ 创建 POJO 类

在 ch13b 的 src 目录中创建 pojo 包，并在该包中创建 User 类，具体代码如下：

```
package pojo;
public class User {
    private String uname;
    private String upwd;
    //省略 setter 和 getter 方法
}
```

❸ 创建控制器类

在 ch13b 的 src 目录中创建 controller 包，并在该包中创建控制器类 UserController，具体代码如下：

```
package controller;
import javax.servlet.http.HttpSession;
import org.springframework.stereotype.Controller;
import org.springframework.ui.Model;
import org.springframework.web.bind.annotation.RequestMapping;
import pojo.User;
@Controller
public class UserController {
    /**
     * 登录页面初始化
     */
    @RequestMapping("/toLogin")
    public String initLogin() {
        return "login";
    }
    /**
     * 处理登录功能
     */
    @RequestMapping("/login")
    public String login(User user, Model model,HttpSession session) {
        System.out.println(user.getUname());
        if("chenheng".equals(user.getUname()) &&
                "123456".equals(user.getUpwd())) {
            //登录成功，将用户信息保存到 session 对象中
            session.setAttribute("user", user);
            //重定向到主页面的跳转方法
```

```
            return "redirect:main";
        }
        model.addAttribute("msg","用户名或密码错误，请重新登录！");
        return "login";
    }
    /**
     * 跳转到主页面
     */
    @RequestMapping("/main")
    public String toMain() {
        return "main";
    }
    /**
     * 退出登录
     */
    @RequestMapping("/logout")
    public String logout(HttpSession session) {
        //清除 session
        session.invalidate();
        return "login";
    }
}
```

❹ 创建拦截器类

在 ch13b 的 src 目录中创建 interceptor 包，并在该包中创建拦截器类 LoginInterceptor，具体代码如下：

```
package interceptor;
import javax.servlet.http.HttpServletRequest;
import javax.servlet.http.HttpServletResponse;
import javax.servlet.http.HttpSession;
import org.springframework.web.servlet.HandlerInterceptor;
public class LoginInterceptor implements HandlerInterceptor{
    @Override
    public boolean preHandle(HttpServletRequest request, HttpServlet
    Response response, Object handler)
            throws Exception {
        //获取请求的 URL
        String url=request.getRequestURI();
        //login.jsp 或登录请求放行，不拦截
        if(url.indexOf("/toLogin")>=0||url.indexOf("/login") >= 0) {
            return true;
        }
        //获取 session
        HttpSession session=request.getSession();
        Object obj=session.getAttribute("user");
```

```
        if(obj!=null)
            return true;
        //没有登录且不是登录页面，转发到登录页面，并给出提示错误信息
        request.setAttribute("msg", "还没登录，请先登录！");
        request.getRequestDispatcher("/WEB-INF/jsp/login.jsp").forward
        (request, response);
        return false;
    }
}
```

❺ 配置拦截器

在 WEB-INF 目录下创建配置文件 springmvc-servlet.xml 和 web.xml。web.xml 的代码和 ch13 一样，这里不再赘述。在 springmvc-servlet.xml 文件中配置拦截器 LoginInterceptor，具体代码如下：

```
<?xml version="1.0" encoding="UTF-8"?>
<beans xmlns="http://www.springframework.org/schema/beans"
    xmlns:xsi="http://www.w3.org/2001/XMLSchema-instance"
    xmlns:context="http://www.springframework.org/schema/context"
    xmlns:mvc="http://www.springframework.org/schema/mvc"
    xsi:schemaLocation="
    http://www.springframework.org/schema/beans
    http://www.springframework.org/schema/beans/spring-beans.xsd
    http://www.springframework.org/schema/context
    http://www.springframework.org/schema/context/spring-context.xsd
    http://www.springframework.org/schema/mvc
    http://www.springframework.org/schema/mvc/spring-mvc.xsd">
    <!-- 使用扫描机制扫描控制器类 -->
    <context:component-scan base-package="controller"/>
    <!-- 配置视图解析器 -->
    <bean class="org.springframework.web.servlet.view.InternalResourceViewResolver"
          id="internalResourceViewResolver">
      <!-- 前缀 -->
      <property name="prefix" value="/WEB-INF/jsp/" />
      <!-- 后缀 -->
      <property name="suffix" value=".jsp" />
    </bean>
    <mvc:interceptors>
        <mvc:interceptor>
            <!-- 配置拦截器作用的路径 -->
            <mvc:mapping path="/**"/>
            <bean class="interceptor.LoginInterceptor"/>
        </mvc:interceptor>
    </mvc:interceptors>
</beans>
```

❻ 创建视图 JSP 页面

在 WEB-INF 目录下创建文件夹 jsp，并在该文件夹中创建 login.jsp 和 main.jsp。

login.jsp 的代码如下：

```
<%@ page language="java" contentType="text/html; charset=UTF-8"
pageEncoding="UTF-8"%>
<!DOCTYPE html PUBLIC "-//W3C//DTD HTML 4.01 Transitional//EN"
"http://www.w3.org/TR/html4/loose.dtd">
<html>
<head>
<meta http-equiv="Content-Type" content="text/html; charset=UTF-8">
<title>Insert title here</title>
</head>
<body>
    ${msg}
    <form action="${pageContext.request.contextPath }/login" method="post">
        用户名：<input type="text" name="uname"/><br>
        密码：<input type="password" name="upwd"/><br>
        <input type="submit" value="登录"/>
    </form>
</body>
</html>
```

main.jsp 的代码如下：

```
<%@ page language="java" contentType="text/html; charset=UTF-8"
pageEncoding="UTF-8"%>
<!DOCTYPE html PUBLIC "-//W3C//DTD HTML 4.01 Transitional//EN"
"http://www.w3.org/TR/html4/loose.dtd">
<html>
<head>
<meta http-equiv="Content-Type" content="text/html; charset=UTF-8">
<title>Insert title here</title>
</head>
<body>
    当前用户：${user.uname }<br>
    <a href="${pageContext.request.contextPath }/logout">退出</a>
</body>
</html>
```

❼ 发布并测试应用

首先将 ch13b 应用发布到 Tomcat 服务器并启动 Tomcat 服务器，然后通过地址“http://localhost:8080/ch13b/main”测试应用，运行效果如图 13.3 所示。

图 13.3　没有登录直接访问主页面的效果

从图 13.3 可以看出，当用户没有登录而直接访问系统主页面时请求将被登录拦截器拦截，返回到登录页面，并提示信息。如果用户在用户名框中输入“chenheng”，在密码框中输入“123456”，单击“登录”按钮后浏览器的显示结果如图 13.4 所示。如果输入的用户名或密码错误，浏览器的显示结果如图 13.5 所示。

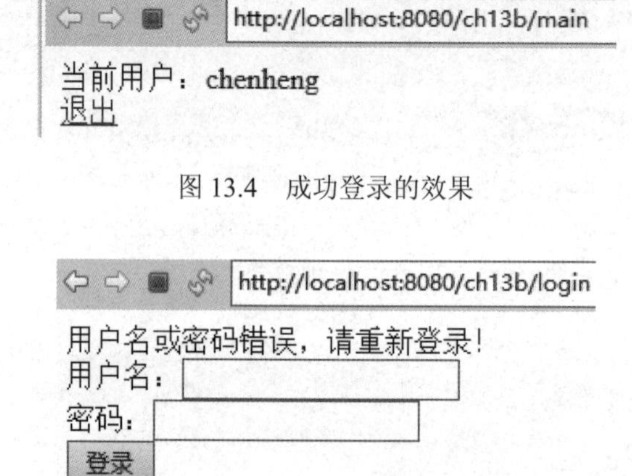

图 13.4　成功登录的效果

图 13.5　用户名或密码错误

当单击图 13.4 中的“退出”链接后，系统将从主页面返回到登录页面。

13.4　本章小结

本章首先讲解了在 Spring MVC 应用中如何定义和配置拦截器，然后详细讲解了拦截器的执行流程，包括单个和多个拦截器的执行流程，最后通过用户登录权限验证的应用案例演示了拦截器的实际应用。

习题 13

1．在 Spring MVC 框架中如何自定义拦截器？如何配置自定义拦截器？

2．请简述单个拦截器和多个拦截器的执行流程。

第 14 章

数据验证

学习目的与要求

本章重点讲解 Spring MVC 框架的输入验证体系。通过本章的学习，读者能够理解输入验证的流程，能够利用 Spring 的自带验证框架和 JSR 303（Java 验证规范）对数据进行验证。

本章主要内容

- 数据验证概述;
- Spring 验证;
- JSR 303 验证。

用户的输入一般是随意的，为了保证数据的合法性，数据验证是所有 Web 应用必须处理的问题。在 Spring MVC 框架中有两种方法可以验证输入数据，一种是利用 Spring 自带的验证框架，另一种是利用 JSR 303 实现。

14.1 数据验证概述

数据验证分为客户端验证和服务器端验证，客户端验证主要是过滤正常用户的误操作，通过 JavaScript 代码完成；服务器端验证是整个应用阻止非法数据的最后防线，通过在应用中编程实现。

视频讲解

14.1.1 客户端验证

在大多数情况下，使用 JavaScript 进行客户端验证的步骤如下：

（1）编写验证函数。

（2）在提交表单的事件中调用验证函数。

（3）根据验证函数来判断是否进行表单提交。

客户端验证可以过滤用户的误操作，是第一道防线，一般使用 JavaScript 代码实现。但仅有客户端验证是不够的，攻击者还可以绕过客户端验证直接进行非法输入，这样可能会引起系统异常，为了确保数据的合法性，防止用户通过非正常手段提交错误信息，必须加上服务器端验证。

14.1.2 服务器端验证

Spring MVC 的 Converter 和 Formatter 在进行类型转换时是将输入数据转换成领域对象的属性值（一种 Java 类型），一旦成功，服务器端验证器就会介入。也就是说，在 Spring MVC 框架中先进行数据类型转换，再进行服务器端验证。

服务器端验证对于系统的安全性、完整性、健壮性起到了至关重要的作用。在 Spring MVC 框架中可以利用 Spring 自带的验证框架验证数据，也可以利用 JSR 303 实现数据验证。

14.2 Spring 验证器

14.2.1 Validator 接口

创建自定义 Spring 验证器需要实现 org.springframework.validation.Validator 接口，该接口有两个接口方法：

```
boolean supports(Class<?> klass)
void validate(Object object, Errors errors)
```

当 supports 方法返回 true 时，验证器可以处理指定的 Class。validate 方法的功能是验证目标对象 object，并将验证错误消息存入 Errors 对象。

往 Errors 对象存入错误消息的方法是 reject 或 rejectValue，这两个方法的部分重载方法如下：

```
void reject(String errorCode)
void reject(String errorCode, String defaultMessage)
void rejectValue(String field, String errorCode)
void rejectValue(String field, String errorCode, String defaultMessage)
```

在一般情况下只需要给 reject 或 rejectValue 方法一个错误代码，Spring MVC 框架就会在消息属性文件中查找错误代码，获取相应错误消息。具体示例如下：

```
if(goods.getGprice() > 100 || goods.getGprice() < 0){
    errors.rejectValue("gprice", "gprice.invalid");  //gprice.invalid 为错误代码
}
```

14.2.2　ValidationUtils 类

org.springframework.validation.ValidationUtils 是一个工具类，该类中有几个方法可以帮助用户判定值是否为空。

例如：

```
if(goods.getGname()==null || goods.getGname().isEmpty()){
    errors.rejectValue("gname", "goods.gname.required")
}
```

上述 if 语句可以使用 ValidationUtils 类的 rejectIfEmpty 方法，代码如下：

```
//errors 为 Errors 对象
//gname 为 goods 对象的属性
ValidationUtils.rejectIfEmpty(errors, "gname", "goods.gname.required");
```

再如：

```
if(goods.getGname()==null || goods.getGname().trim().isEmpty()){
    errors.rejectValue("gname", "goods.gname.required")
}
```

上述 if 语句可以编写成：

```
//gname 为 goods 对象的属性
ValidationUtils.rejectIfEmptyOrWhitespace(errors, "gname", "goods.gname.required");
```

14.2.3　验证示例

本节使用一个应用 ch14a 讲解 Spring 验证器的编写及使用。该应用中有一个数据输入页面 addGoods.jsp，效果如图 14.1 所示；有一个数据显示页面 goodsList.jsp，效果如图 14.2 所示。

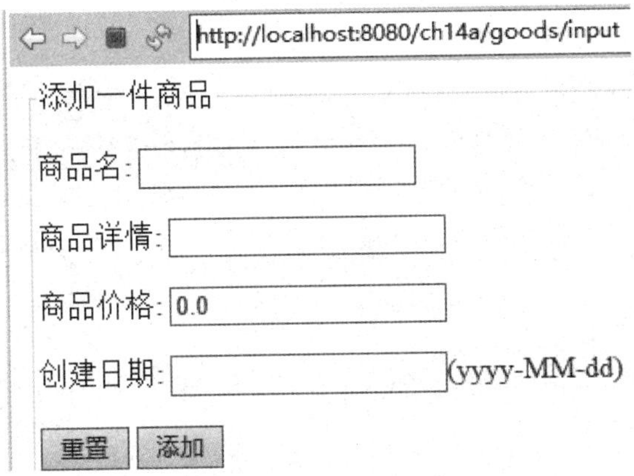

http://localhost:8080/ch14a/goods/input

添加一件商品

商品名：

商品详情：

商品价格：0.0

创建日期：(yyyy-MM-dd)

重置　添加

图 14.1　数据输入页面

http://localhost:8080/ch14a/goods/save

商品名 商品详情 商品价格 创建日期
鸡蛋 好鸡蛋 8.8 Sat Feb 03 00:00:00 CST 2018

图 14.2　数据显示页面

编写一个实现 org.springframework.validation.Validator 接口的验证器类 GoodsValidator，验证要求如下：

（1）商品名和商品详情不能为空。

（2）商品价格在 0 到 100。

（3）创建日期不能在系统日期之后。

根据上述要求，按照如下步骤完成 ch14a 应用。

❶ 创建应用

创建 ch14a 应用，并导入 Spring MVC 的相关 JAR 包。另外，需要使用 JSTL 标签显示数据，所以需要导入 JSTL 的 JAR 包。ch14a 需要的 JAR 包如图 14.3 所示。

图 14.3　ch14a 需要的 JAR 包

❷ 创建数据输入页面

在 WEB-INF 目录下创建文件夹 jsp，并在该文件夹中创建数据输入页面 addGoods.jsp。核心代码如下：

```
<body>
    <form:form modelAttribute="goods" action="${pageContext.request.
    contextPath }/goods/save" method="post">
        <fieldset>
       <legend>添加一件商品</legend>
       <p>
           <label>商品名:</label>
           <form:input path="gname"/>
       </p>
       <p>
           <label>商品详情:</label>
```

```
            <form:input path="gdescription"/>
        </p>
        <p>
            <label>商品价格:</label>
            <form:input path="gprice"/>
        </p>
        <p>
            <label>创建日期:</label>
            <form:input path="gdate"/>(yyyy-MM-dd)
        </p>
        <p id="buttons">
            <input id="reset" type="reset">
            <input id="submit" type="submit" value="添加">
        </p>
    </fieldset>
    <!-- 取出所有验证错误 -->
    <form:errors path="*"/>
    </form:form>
</body>
```

❸ 编写模型类

在 src 目录下创建 pojo 包，并在该包中定义领域模型类 Goods，封装输入参数。在该类中使用@DateTimeFormat(pattern="yyyy-MM-dd")格式化创建日期。模型类 Goods 的具体代码如下：

```
package pojo;
import java.util.Date;
import org.springframework.format.annotation.DateTimeFormat;
public class Goods {
    private String gname;
    private String gdescription;
    private double gprice;
    //日期格式化（需要在配置文件中配置 FormattingConversionServiceFactoryBean）
    @DateTimeFormat(pattern="yyyy-MM-dd")
    private Date gdate;
    //省略 setter 和 getter 方法
}
```

❹ 编写验证器类

在 src 目录下创建 validator 包，并在该包中编写实现 org.springframework.validation.Validator 接口的验证器类 GoodsValidator，使用@Component 注解将 GoodsValidator 类声明为验证组件。具体代码如下：

```
package validator;
import java.util.Date;
import org.springframework.stereotype.Component;
```

```
import org.springframework.validation.Errors;
import org.springframework.validation.ValidationUtils;
import org.springframework.validation.Validator;
import pojo.Goods;
@Component
public class GoodsValidator implements Validator{
    @Override
    public boolean supports(Class<?> klass) {
        //要验证的model，返回值为false则不验证
       return Goods.class.isAssignableFrom(klass);
    }
    @Override
    public void validate(Object object, Errors errors) {
        Goods goods=(Goods)object;  //object要验证的对象
        //goods.gname.required是错误消息属性文件中的编码（国际化后对应的是国际化
        的信息）
        ValidationUtils.rejectIfEmpty(errors, "gname", "goods.gname.required");
        ValidationUtils.rejectIfEmpty(errors, "gdescription", "goods.
        gdescription.required");
        if(goods.getGprice()>100 || goods.getGprice()<0){
            errors.rejectValue("gprice", "gprice.invalid");
        }
        Date goodsDate=goods.getGdate();
        //在系统时间之后
        if(goodsDate!=null && goodsDate.after(new Date())){
            errors.rejectValue("gdate", "gdate.invalid");
        }
    }
}
```

❺ 编写错误消息属性文件

在 WEB-INF 目录下创建文件夹 resource，并在该文件夹中编写属性文件 errorMessages.properties。文件内容如下：

```
goods.gname.required=请输入商品名称。
goods.gdescription.required=请输入商品详情。
gprice.invalid=价格为0～100。
gdate.invalid=创建日期不能在系统日期之后。
```

Unicode 编码（Eclipse 带有将汉字转换成 Unicode 编码的功能）的属性文件内容如下：

```
goods.gname.required=\u8BF7\u8F93\u5165\u5546\u54C1\u540D\u79F0\u3002
goods.gdescription.required=\u8BF7\u8F93\u5165\u5546\u54C1\u8BE6\u60C5\
u3002
gprice.invalid=\u4EF7\u683C\u57280-100\u4E4B\u95F4\u3002
gdate.invalid=\u521B\u5EFA\u65E5\u671F\u4E0D\u80FD\u5728\u7CFB\u7EDF\u6
5E5\u671F\u4E4B\u540E\u3002
```

在属性文件创建完成后需要告诉 Spring MVC 从该文件中获取错误消息，则需要在配置文件中声明一个 messageSource bean，具体代码如下：

```
<!-- 配置消息属性文件 -->
<bean id="messageSource"
    class="org.springframework.context.support.ReloadableResourceBundleM
    essageSource">
    <property name="basename" value="/WEB-INF/resource/errorMessages"/>
</bean>
```

❻ 编写 Service 层

在 src 目录下创建 service 包，并在该包中编写一个 GoodsService 接口和 GoodsServiceImpl 实现类。具体代码如下：

```
package service;
import java.util.ArrayList;
import pojo.Goods;
public interface GoodsService {
    boolean save(Goods g);
    ArrayList<Goods> getGoods();
}
package service;
import java.util.ArrayList;
import org.springframework.stereotype.Service;
import domain.Goods;
@Service
public class GoodsServiceImpl implements GoodsService{
    //使用静态集合变量 goods 模拟数据库
    private static ArrayList<Goods> goods = new ArrayList<Goods>();
    @Override
    public boolean save(Goods g) {
        goods.add(g);
        return true;
    }
    @Override
    public ArrayList<Goods> getGoods() {
        return goods;
    }
}
```

❼ 编写控制器类

在 src 目录下创建 controller 包，并在该包中编写控制器类 GoodsController，在该类中使用@Resource 注解注入自定义验证器。另外，在控制器类中包含两个处理请求的方法，具体代码如下：

```
package controller;
```

```
import javax.annotation.Resource;
import org.apache.commons.logging.Log;
import org.apache.commons.logging.LogFactory;
import org.springframework.beans.factory.annotation.Autowired;
import org.springframework.stereotype.Controller;
import org.springframework.ui.Model;
import org.springframework.validation.BindingResult;
import org.springframework.validation.Validator;
import org.springframework.web.bind.annotation.ModelAttribute;
import org.springframework.web.bind.annotation.RequestMapping;
import pojo.Goods;
import service.GoodsService;
@Controller
@RequestMapping("/goods")
public class GoodsController {
    //得到一个用来记录日志的对象，这样在打印信息的时候能够标记打印的是哪个类的信息
    private static final Log logger=LogFactory.getLog (GoodsController.
    class);
    @Autowired
    private GoodsService goodsService;
    //注解验证器相当于“GoodsValidator validator=new GoodsValidator();”
    @Resource
    private Validator validator;
    @RequestMapping("/input")
    public String input(Model model){
        //如果model中没有goods属性，addGoods.jsp会抛出异常
        //因为表单标签无法找到modelAttribute属性指定的form backing object
        model.addAttribute("goods", new Goods());
        return "addGoods";
    }
    @RequestMapping("/save")
    public String save(@ModelAttribute Goods goods, BindingResult result,
    Model model){
        this.validator.validate(goods, result);  //添加验证
        if (result.hasErrors()) {
            return "addGoods";
        }
        goodsService.save(goods);
        logger.info("添加成功");
        model.addAttribute("goodsList", goodsService.getGoods());
        return "goodsList";
    }
}
```

❽ 编写配置文件

在 WEB-INF 目录下编写配置文件 springmvc-servlet.xml，具体代码如下：

```
<?xml version="1.0" encoding="UTF-8"?>
<beans xmlns="http://www.springframework.org/schema/beans"
   xmlns:xsi="http://www.w3.org/2001/XMLSchema-instance"
   xmlns:p="http://www.springframework.org/schema/p"
   xmlns:context="http://www.springframework.org/schema/context"
   xmlns:mvc="http://www.springframework.org/schema/mvc"
   xsi:schemaLocation="
   http://www.springframework.org/schema/beans
   http://www.springframework.org/schema/beans/spring-beans.xsd
   http://www.springframework.org/schema/context
   http://www.springframework.org/schema/context/spring-context.xsd
   http://www.springframework.org/schema/mvc
   http://www.springframework.org/schema/mvc/spring-mvc.xsd">
   <!-- 使用扫描机制扫描包 -->
   <context:component-scan base-package="controller"/>
   <context:component-scan base-package="service"/>
   <context:component-scan base-package="validator"/>
   <!-- 注册格式化转换器，因为用到日期转换 -->
   <bean id="conversionService" class="org.springframework.format.
   support.FormattingConversionServiceFactoryBean">
   </bean>
   <mvc:annotation-driven conversion-service="conversionService"/>
   <!-- 配置消息属性文件 -->
   <bean id="messageSource" class="org.springframework.context.support.
   ReloadableResourceBundleMessageSource">
    <property name="basename" value="/WEB-INF/resource/errorMessages"/>
   </bean>
    <!-- 配置视图解析器 -->
    <bean  class="org.springframework.web.servlet.view.InternalResource
    ViewResolver"
         id="internalResourceViewResolver">
      <!-- 前缀 -->
      <property name="prefix" value="/WEB-INF/jsp/" />
      <!-- 后缀 -->
      <property name="suffix" value=".jsp" />
  </bean>
</beans>
```

❾ 创建数据显示页面

在 WEB-INF/jsp 目录下创建数据显示页面 goodsList.jsp。核心代码如下：

```
<body>
    <table>
        <tr>
            <td>商品名</td>
            <td>商品详情</td>
```

```
            <td>商品价格</td>
            <td>创建日期</td>
        </tr>
        <c:forEach items="${goodsList }" var="goods">
            <tr>
                <td>${goods.gname }</td>
                <td>${goods.gdescription }</td>
                <td>${goods.gprice }</td>
                <td>${goods.gdate }</td>
            </tr>
        </c:forEach>
    </table>
</body>
```

⑩ 创建 web.xml 文件

在 WEB-INF 目录下创建 web.xml 文件，在该文件中配置 Spring MVC 的核心控制器 DispatcherServlet 和字符编码过滤器，具体代码如下：

```
<?xml version="1.0" encoding="UTF-8"?>
<web-app  xmlns:xsi="http://www.w3.org/2001/XMLSchema-instance"  xmlns=
"http:// xmlns. jcp.org/xml/ns/javaee" xsi:schemaLocation="http://xmlns.
jcp.org/xml/ns/javaee  http://xmlns.jcp.org/xml/ns/javaee/web-app_3_1.xsd"
id="WebApp_ID" version="3.1">
  <servlet>
    <servlet-name>springmvc</servlet-name>
    <servlet-class>org.springframework.web.servlet.DispatcherServlet
    </servlet-class>
    <load-on-startup>1</load-on-startup>
  </servlet>
  <servlet-mapping>
    <servlet-name>springmvc</servlet-name>
    <url-pattern>/</url-pattern>
  </servlet-mapping>
  <!-- 避免中文乱码 -->
  <filter>
    <filter-name>characterEncodingFilter</filter-name>
    <filter-class>org.springframework.web.filter.CharacterEncodingFilter
    </filter-class>
    <init-param>
      <param-name>encoding</param-name>
      <param-value>UTF-8</param-value>
    </init-param>
    <init-param>
      <param-name>forceEncoding</param-name>
      <param-value>true</param-value>
    </init-param>
  </filter>
```

```
    <filter-mapping>
      <filter-name>characterEncodingFilter</filter-name>
      <url-pattern>/*</url-pattern>
    </filter-mapping>
</web-app>
```

⑪ 测试应用

发布 ch14a 应用并启动 Tomcat 服务器，然后通过地址“http://localhost:8080/ch14a/goods/input”测试应用。

14.3 JSR 303 验证

对于 JSR 303 验证，目前有两个实现，一个是 Hibernate Validator，一个是 Apache BVal。本书采用的是 Hibernate Validator，注意它和 Hibernate 无关，只是使用它进行数据验证。

视频讲解

14.3.1 JSR 303 验证配置

❶ 下载与安装 Hibernate Validator

用户可以通过地址“https://sourceforge.net/projects/hibernate/files/hibernate-validator/”下载 Hibernate Validator，本书选择的是 hibernate-validator-5.4.0.Final-dist.zip。

首先将下载的压缩包解压，然后将\hibernate-validator-5.4.0.Final\dist 目录下的 hibernate-validator-5.4.0.Final.jar 和\hibernate-validator-5.4.0.Final\dist\lib\required 目录下的 classmate-1.3.1.jar、javax.el-3.0.1-b08.jar、jboss-logging-3.3.0.Final.jar、validation-api-1.1.0.Final.ja 复制到应用的\WEB-INF\lib 目录下。

❷ 配置属性文件与验证器

如果将验证错误消息放在属性文件中，那么需要在配置文件中配置属性文件，并将属性文件与 Hibernate Validator 关联，具体配置代码如下：

```
<!-- 配置消息属性文件 -->
<bean id="messageSource"
    class="org.springframework.context.support.ReloadableResourceBundle
    MessageSource">
    <!-- 资源文件名-->
    <property name="basenames">
     <list>
       <value>/WEB-INF/resource/errorMessages</value>
     </list>
    </property>
    <!-- 资源文件编码格式 -->
    <property name="fileEncodings" value="utf-8" />
    <!-- 对资源文件内容缓存的时间，单位为秒 -->
```

```
    <property name="cacheSeconds" value="120" />
</bean>
<!-- 注册校验器 -->
<bean id="validator"
    class="org.springframework.validation.beanvalidation.LocalValidator
    FactoryBean">
    <!-- hibernate 校验器 -->
    <property name="providerClass" value="org.hibernate.validator.
    HibernateValidator" />
    <!-- 指定校验使用的资源文件，在文件中配置校验错误信息，如果不指定则默认使用
    classpath 下的 ValidationMessages.properties -->
    <property name="validationMessageSource" ref="messageSource" />
</bean>
<!-- 开启 Spring 的 Valid 功能 -->
<mvc:annotation-driven conversion-service="conversionService" validator=
"validator"/>
```

14.3.2 标注类型

JSR 303 不需要编写验证器，但需要利用它的标注类型在领域模型的属性上嵌入约束。

❶ 空检查

- @Null：验证对象是否为 null。
- @NotNull：验证对象是否不为 null，无法检查长度为 0 的字符串。
- @NotBlank：检查约束字符串是不是 null，以及被 trim 后的长度是否大于 0，只针对字符串，且会去掉前后空格。
- @NotEmpty：检查约束元素是否为 null 或者是 empty。

示例如下：

```
@NotBlank(message="{goods.gname.required}")  //goods.gname.required为属性
文件的错误代码
private String gname;
```

❷ boolean 检查

- @AssertTrue：验证 boolean 属性是否为 true。
- @AssertFalse：验证 boolean 属性是否为 false。

示例如下：

```
@AssertTrue
private boolean isLogin;
```

❸ 长度检查

- @Size(min=, max=)：验证对象（Array、Collection、Map、String）长度是否在给定的范围之内。
- @Length(min=, max=)：验证字符串长度是否在给定的范围之内。

示例如下：

```
@Length(min=1,max=100)
private String gdescription;
```

❹ **日期检查**

- @Past：验证 Date 和 Calendar 对象是否在当前时间之前。
- @Future：验证 Date 和 Calendar 对象是否在当前时间之后。
- @Pattern：验证 String 对象是否符合正则表达式的规则。

示例如下：

```
@Past(message="{gdate.invalid}")
private Date gdate;
```

❺ **数值检查**

- @Min：验证 Number 和 String 对象是否大于指定的值。
- @Max：验证 Number 和 String 对象是否小于指定的值。
- @DecimalMax：被标注的值必须不大于约束中指定的最大值，这个约束的参数是一个通过 BigDecimal 定义的最大值的字符串表示，小数存在精度。
- @DecimalMin：被标注的值必须不小于约束中指定的最小值，这个约束的参数是一个通过 BigDecimal 定义的最小值的字符串表示，小数存在精度。
- @Digits：验证 Number 和 String 的构成是否合法。
- @Digits(integer=,fraction=)：验证字符串是否符合指定格式的数字，integer 指定整数精度，fraction 指定小数精度。
- @Range(min=, max=)：检查数字是否介于 min 和 max 之间。
- @Valid：对关联对象进行校验，如果关联对象是个集合或者数组，那么对其中的元素进行校验，如果是一个 map，则对其中的值部分进行校验。
- @CreditCardNumber：信用卡验证。
- @Email：验证是否为邮件地址，如果为 null，不进行验证，通过验证。

示例如下：

```
@Range(min=0,max=100,message="{gprice.invalid}")
private double gprice;
```

14.3.3　验证示例

创建 ch14b 应用，该应用实现的功能与 14.2.3 节中的 ch14a 应用相同。ch14b 所需的 JAR 包如图 14.4 所示。

在 ch14b 应用中不需要创建验证器类 GoodsValidator。另外，Service 层、View 层以及错误消息属性文件都与 ch14a 应用的相同。与 ch14a 应用的实现不同的是模型类、控制器类和 Spring MVC 的核心配置文件，具体如下：

图 14.4　ch14b 的 JAR 包

❶ 模型类

在模型类 Goods 中利用 JSR 303 的标注类型对属性进行验证，具体代码如下：

```
package pojo;
import java.util.Date;
import javax.validation.constraints.Past;
import org.hibernate.validator.constraints.NotBlank;
import org.hibernate.validator.constraints.Range;
import org.springframework.format.annotation.DateTimeFormat;
public class Goods {
    //goods.gname.required 错误消息 key（国际化后对应的就是国际化的信息）
    @NotBlank(message="{goods.gname.required}")
    private String gname;
    @NotBlank(message="{goods.gdescription.required}")
    private String gdescription;
    @Range(min=0,max=100,message="{gprice.invalid}")
    private double gprice;
    //日期格式化（需要在配置文件中配置 FormattingConversionServiceFactoryBean）
    @DateTimeFormat(pattern="yyyy-MM-dd")
    @Past(message="{gdate.invalid}")
    private Date gdate;
    //省略 setter 和 getter 方法
}
```

❷ 控制器类

在控制器类 GoodsController 中使用@Valid 对模型对象进行验证，具体代码如下：

```
package controller;
import javax.validation.Valid;
```

```
import org.apache.commons.logging.Log;
import org.apache.commons.logging.LogFactory;
import org.springframework.beans.factory.annotation.Autowired;
import org.springframework.stereotype.Controller;
import org.springframework.ui.Model;
import org.springframework.validation.BindingResult;
import org.springframework.web.bind.annotation.ModelAttribute;
import org.springframework.web.bind.annotation.RequestMapping;
import pojo.Goods;
import service.GoodsService;
@Controller
@RequestMapping("/goods")
public class GoodsController {
    //得到一个用来记录日志的对象，这样在打印信息的时候能够标记打印的是哪个类的信息
    private static final Log logger=LogFactory.getLog (Goods Controller.
    class);
    @Autowired
    private GoodsService goodsService;
    @RequestMapping("/input")
    public String input(Model model){
        //如果model中没有goods属性，addGoods.jsp会抛出异常，因为表单标签无法找到
        // modelAttribute属性指定的form backing object
        model.addAttribute("goods", new Goods());
        return "addGoods";
    }
    @RequestMapping("/save")
    public String save(@Valid @ModelAttribute Goods goods, BindingResult
    result, Model model){
        if(result.hasErrors()){
            return "addGoods";
        }
        goodsService.save(goods);
        logger.info("添加成功");
        model.addAttribute("goodsList", goodsService.getGoods());
        return "goodsList";
    }
}
```

❸ 配置文件

配置文件 springmvc-servlet.xml 的代码如下：

```
<?xml version="1.0" encoding="UTF-8"?>
<beans xmlns="http://www.springframework.org/schema/beans"
   xmlns:xsi="http://www.w3.org/2001/XMLSchema-instance"
   xmlns:p="http://www.springframework.org/schema/p"
   xmlns:context="http://www.springframework.org/schema/context"
```

```
xmlns:mvc="http://www.springframework.org/schema/mvc"
xsi:schemaLocation="
http://www.springframework.org/schema/beans
http://www.springframework.org/schema/beans/spring-beans.xsd
http://www.springframework.org/schema/context
http://www.springframework.org/schema/context/spring-context.xsd
http://www.springframework.org/schema/mvc
http://www.springframework.org/schema/mvc/spring-mvc.xsd">
<!-- 使用扫描机制扫描包 -->
<context:component-scan base-package="controller"/>
<context:component-scan base-package="service"/>
 <!-- 配置消息属性文件 -->
<bean id="messageSource"
   class="org.springframework.context.support.ReloadableResource
   BundleMessageSource">
   <!-- 资源文件名 -->
   <property name="basenames">
    <list>
      <value>/WEB-INF/resource/errorMessages</value>
    </list>
   </property>
   <!-- 资源文件编码格式 -->
   <property name="fileEncodings" value="utf-8" />
   <!-- 对资源文件内容缓存的时间，单位为秒 -->
   <property name="cacheSeconds" value="120" />
</bean>
<!-- 注册校验器 -->
<bean id="validator"
   class="org.springframework.validation.beanvalidation.LocalValidator
   FactoryBean">
   <!-- hibernate 校验器-->
   <property name="providerClass" value="org.hibernate.validator.
   HibernateValidator" />
   <!-- 指定校验使用的资源文件，在文件中配置校验错误信息，如果不指定则默认使用
   classpath 下的 ValidationMessages.properties -->
   <property name="validationMessageSource" ref="messageSource" />
</bean>
<!-- 开启 Spring 的 Valid 功能 -->
<mvc:annotation-driven conversion-service="conversionService" validator=
"validator"/>
 <!-- 注册格式化转换器 -->
<bean id="conversionService" class="org.springframework.format.
support.FormattingConversionServiceFactoryBean">
</bean>
 <!-- 配置视图解析器 -->
```

```
    <bean class="org.springframework.web.servlet.view.InternalResource
    ViewResolver"
            id="internalResourceViewResolver">
        <!-- 前缀 -->
        <property name="prefix" value="/WEB-INF/jsp/" />
        <!-- 后缀 -->
        <property name="suffix" value=".jsp" />
  </bean>
</beans>
```

❹ 测试应用

通过地址“http://localhost:8080/ch14b/goods/input”测试 ch14b 应用。

14.4　本章小结

本章重点讲解了 Spring 验证的编写和 JSR 303 验证的使用方法，不管使用哪种验证方式，用户都需要注意验证流程。

习题 14

1. 如何创建 Spring 验证器类？
2. 举例说明 JSR 303 验证的标注类型的使用方法。

第15章 国际化

学习目的与要求

本章重点讲解了 Spring MVC 国际化的实现方法。通过本章的学习，读者应理解 Spring MVC 国际化的设计思想，掌握 Spring MVC 国际化的实现方法。

本章主要内容

- Java 国际化的思想;
- Spring MVC 的国际化;
- 用户自定义切换语言。

国际化是商业软件系统的一个基本要求，因为当今的软件系统需要面对全球的浏览者。国际化的目的就是根据用户的语言环境的不同向用户输出与之相应的页面，以示友好。

Spring MVC 的国际化主要有 JSP 页面信息国际化和错误消息国际化，错误消息在 14.2.3 节中已经讲解，本章主要介绍如何在 JSP 页面输出国际化消息。最后，本章将示范一个让用户自行选择语言的示例。

15.1 程序国际化概述

程序国际化已成为 Web 应用的基本要求。随着网络的发展，大部分 Web 站点面对的已经不再是本地或者本国的浏览者，而是来自全世界各国、各地区的浏览者，因此国际化成为了 Web 应用不可或缺的一部分。

视频讲解

15.1.1 Java 国际化的思想

Java 国际化的思想是将程序中的信息放在资源文件中，程序根据支持的国家及语言环境读取相应的资源文件。资源文件是 key-value 对，每个资源文件中的 key 是不变的，但 value 随不同国家/语言变化。

Java 程序的国际化主要通过两个类来完成。

- java.util.Locale：用于提供本地信息，通常称它为语言环境。不同的语言、不同的国家和地区采用不同的 Locale 对象来表示。
- java.util.ResourceBundle：该类称为资源包，包含了特定于语言环境的资源对象。当程序需要一个特定于语言环境的资源时（例如字符串资源），程序可以从适合当前用户语言环境的资源包中加载它。采用这种方式可以编写独立于用户语言环境的程序代码，而与特定语言环境相关的信息则通过资源包来提供。

为了实现 Java 程序的国际化，必须事先提供程序所需要的资源文件。资源文件的内容由很多 key-value 对组成，其中 key 是程序使用的部分，而 value 是程序界面的显示。

资源文件的命名可以有如下 3 种形式：

- baseName.properties；
- baseName_language.properties；
- baseName_language_country.properties。

baseName 是资源文件的基本名称，由用户自由定义，但是 language 和 country 必须为 Java 所支持的语言和国家/地区代码。例如：

- 中国大陆：baseName_zh_CN.properties；
- 美国：baseName_en_US.properties。

Java 中的资源文件只支持 ISO-8859-1 编码格式字符，直接编写中文会出现乱码。用户可以使用 Java 命令 native2ascii.exe 解决资源文件的中文乱码问题，使用 Eclipse 编写资源属性文件，在保存资源文件时 Eclipse 自动执行 native2ascii.exe 命令，因此在 Eclipse 中资源文件不会出现中文乱码问题。

15.1.2　Java 支持的语言和国家

java.util.Locale 类的常用构造方法如下：

- public Locale(String language)；
- public Locale(String language,String country)。

其中，language 表示语言，它的取值是由小写的两个字母组成的语言代码；country 表示国家或地区，它的取值是由大写的两个字母组成的国家或地区代码。

实际上，Java 并不能支持所有国家和语言，如果需要获取 Java 所支持的语言和国家，开发者可以通过调用 Locale 类的 getAvailableLocales 方法获取，该方法返回一个 Locale 数组，该数组中包含了 Java 所支持的语言和国家。

下面的 Java 程序简单示范了如何获取 Java 所支持的国家和语言：

```
import java.util.Locale;
public class Test {
    public static void main(String[] args) {
        //返回 Java 所支持的语言和国家的数组
        Locale locales[]=Locale.getAvailableLocales();
        //遍历数组元素，依次获取所支持的国家和语言
```

```
        for (int i=0; i<locales.length; i++) {
            //打印出所支持的国家和语言
            System.out.println(locales[i].getDisplayCountry()+"="
                    +locales[i].getCountry()+" "
                    +locales[i].getDisplayLanguage()+"="
                    +locales[i].getLanguage());
        }
    }
}
```

15.1.3　Java 程序的国际化

假设有如下简单 Java 程序：

```
public class TestI18N {
    public static void main(String[] args) {
        System.out.println("我要向不同国家的人民问好：您好！");
    }
}
```

为了让该程序支持国际化，需要将“我要向不同国家的人民问好：您好！”对应不同语言环境的字符串，定义在不同的资源文件中。

在 Web 应用的 src 目录下新建文件 messageResource_zh_CN.properties 和 messageResource_en_US.properties。然后给资源文件 messageResource_zh_CN.properties 添加“hello=我要向不同国家的人民问好：您好！”内容，保存后可看到如图 15.1 所示的效果。

```
Test.java    TestI18N.java    messageResource_zh_CN.properties
1 hello=\u6211\u8981\u5411\u4E0D\u540C\u56FD\u5BB6\u7684\u4EBA\u6C11\
```

图 15.1　Unicode 编码资源文件

图 15.1 显示的内容看似是很多乱码，实际上是 Unicode 编码文件内容。至此，资源文件 messageResource_zh_CN.properties 创建完成。

最后给资源文件 messageResource_en_US.properties 添加“hello=I want to say hello to all world!”内容。

现在将 TestI18N.java 程序修改成如下形式：

```
import java.util.Locale;
import java.util.ResourceBundle;
public class TestI18N {
    public static void main(String[] args) {
        //取得系统默认的国家语言环境
        Locale lc=Locale.getDefault();
        //根据国家语言环境加载资源文件
```

```
        ResourceBundle rb=ResourceBundle.getBundle("messageResource", lc);
        //打印出从资源文件中取得的信息
        System.out.println(rb.getString("hello"));
    }
}
```

上面程序中的打印语句打印的内容是从资源文件中读取的信息。如果在中文环境下运行程序，将打印“我要向不同国家的人民问好：您好！”；如果在“控制面板”中将计算机的语言环境设置成美国，然后再次运行该程序，将打印“I want to say hello to all world!”。

需要注意的是，如果程序找不到对应国家/语言的资源文件，系统该怎么办？假设以简体中文环境为例，先搜索如下文件：

```
messageResource_zh_CN.properties
```

如果没有找到国家/语言都匹配的资源文件，再搜索语言匹配文件，即搜索如下文件：

```
messageResource_zh.properties
```

如果上面的文件还没有搜索到，则搜索 baseName 匹配的文件，即搜索如下文件：

```
messageResource.properties
```

如果上面 3 个文件都找不到，则系统将出现异常。

15.1.4　带占位符的国际化信息

在资源文件中消息文本可以带有参数，例如：

```
welcome={0}，欢迎学习 Spring MVC。
```

花括号中的数字是一个占位符，可以被动态的数据替换。在消息文本中占位符可以使用 0～9 的数字，也就是说消息文本的参数最多可以有 10 个。例如：

```
welcome={0}，欢迎学习 Spring MVC，今天是星期{1}。
```

如果要替换消息文本中的占位符，可以使用 java.text.MessageFormat 类，该类提供了一个静态方法 format，用来格式化带参数的文本。format 方法的定义如下：

```
public static String format(String pattern,Object …arguments)
```

其中，pattern 字符串就是一个带占位符的字符串，消息文本中的数字占位符将按照方法参数的顺序（从第二个参数开始）被替换。

替换占位符的示例代码如下：

```
import java.text.MessageFormat;
import java.util.Locale;
import java.util.ResourceBundle;
public class TestFormat {
    public static void main(String[] args) {
        //取得系统默认的国家语言环境
```

```
        Locale lc=Locale.getDefault();
        //根据国家语言环境加载资源文件
        ResourceBundle rb=ResourceBundle.getBundle("messageResource", lc);
        //从资源文件中取得的信息
        String msg=rb.getString("welcome");
        //替换消息文本中的占位符，消息文本中的数字占位符将按照参数的顺序
        //（从第二个参数开始）被替换，即“我”替换{0}、“5”替换{1}
        String msgFor=MessageFormat.format(msg, "我","5");
        System.out.println(msgFor);
    }
}
```

15.2 Spring MVC 的国际化

Spring MVC 的国际化是建立在 Java 国际化的基础之上的，Spring MVC 框架的底层国际化与 Java 国际化是一致的，作为一个良好的 MVC 框架，Spring MVC 将 Java 国际化的功能进行了封装和简化，开发者使用起来会更加简单、快捷。

视频讲解

由 15.1 节可知国际化和本地化应用程序时需要具备以下两个条件：

（1）将文本信息放到资源属性文件中。

（2）选择和读取正确位置的资源属性文件。

下面讲解第二个条件的实现。

15.2.1 Spring MVC 加载资源属性文件

在 Spring MVC 中不能直接使用 ResourceBundle 加载资源属性文件，而是利用 bean（messageSource）告知 Spring MVC 框架要将资源属性文件放到哪里。示例代码如下：

```
<bean id="messageSource"
    class="org.springframework.context.support.ReloadableResourceBundleM
    essageSource">
    <!-- <property name="basename" value="classpath:messages" /> -->
    <property name="basename" value="/WEB-INF/resource/messages" />
</bean>
```

上述 bean 配置的是国际化资源文件的路径，classpath:messages 指的是 classpath 路径下的 messages_zh_CN.properties 文件和 messages_en_US.properties 文件。当然也可以将国际化资源文件放在其他的路径下，例如/WEB-INF/resource/messages。

另外，“messageSource” bean 是由 ReloadableResourceBundleMessageSource 类实现的，它是不能重新加载的，如果修改了国际化资源文件，需要重启 JVM。

最后还需要注意，如果有一组属性文件，则用 basenames 替换 basename，示例代码如下：

```
<bean id="messageSource"
    class="org.springframework.context.support.ReloadableResourceBundleM
    essageSource">
    <property name="basenames">
        <list>
            <value>/WEB-INF/resource/messages</value>
            <value>/WEB-INF/resource/labels</value>
        </list>
    </property>
</bean>
```

15.2.2　语言区域的选择

在 Spring MVC 中可以使用语言区域解析器 bean 选择语言区域，该 bean 有 3 个常见实现，即 AcceptHeaderLocaleResolver、SessionLocaleResolver 和 CookieLocaleResolver。

❶ AcceptHeaderLocaleResolver

根据浏览器 Http Header 中的 accept-language 域设定（accept-language 域中一般包含了当前操作系统的语言设定，可通过 HttpServletRequest.getLocale 方法获得此域的内容）。改变 Locale 是不支持的，即不能调用 LocaleResolver 接口的 setLocale(HttpServletRequest request, HttpServletResponse response, Locale locale)方法设置 Locale。

❷ SessionLocaleResolver

根据用户本次会话过程中的语言设定决定语言区域（例如用户进入首页时选择语言种类，则此次会话周期内统一使用该语言设定）。

❸ CookieLocaleResolver

根据 Cookie 判定用户的语言设定（Cookie 中保存着用户前一次的语言设定参数）。

由上述分析可知，SessionLocaleResolver 实现比较方便用户选择喜欢的语言种类，本章使用该方法进行国际化实现。

下面是使用 SessionLocaleResolver 实现的 bean 定义：

```
<bean  id="localeResolver"  class="org.springframework.web.servlet.i18n.
SessionLocaleResolver">
    <property name="defaultLocale" value="zh_CN"></property>
 </bean>
```

如果采用基于 SessionLocaleResolver 和 CookieLocaleResolver 的国际化实现，必须配置 LocaleChangeInterceptor 拦截器，示例代码如下：

```
<mvc:interceptors>
    <bean class="org.springframework.web.servlet.i18n.LocaleChangeInterceptor"/>
</mvc:interceptors>
```

15.2.3　使用 message 标签显示国际化信息

在 Spring MVC 框架中可以使用 Spring 的 message 标签在 JSP 页面中显示国际化消息。

在使用 message 标签时需要在 JSP 页面的最前面使用 taglib 指令声明 spring 标签，代码如下：

```
<%@taglib prefix="spring" uri="http://www.springframework.org/tags" %>
```

message 标签有以下常用属性。

- code：获得国际化消息的 key。
- arguments：代表该标签的参数。如果替换消息中的占位符，示例代码为“<spring:message code="third" arguments="888,999" />”，third 对应的消息有两个占位符{0}和{1}。
- argumentSeparator：用来分隔该标签参数的字符，默认为逗号。
- text：code 属性不存在，或指定的 key 无法获取消息时所显示的默认文本信息。

15.3 用户自定义切换语言示例

在许多成熟的商业软件系统中可以让用户自由切换语言，而不是修改浏览器的语言设置。一旦用户选择了自己需要使用的语言环境，整个系统的语言环境将一直是这种语言环境。Spring MVC 也可以允许用户自行选择程序语言。本章通过 Web 应用 ch15 演示用户自定义切换语言，在该应用中使用 SessionLocaleResolver 实现国际化，具体步骤如下：

视频讲解

❶ 创建应用

创建应用 ch15，并导入 Spring MVC 相关的 JAR 包。

❷ 创建国际化资源文件

在 WEB-INF/resource 目录下创建中英文资源文件 messages_en_US.properties 和 messages_zh_CN.properties。

messages_en_US.properties 的内容如下：

```
first=first
second=second
third={0} third {1}
language.en=English
language.cn=Chinese
```

messages_zh_CN.properties 的内容如下：

```
first=\u7B2C\u4E00\u9875
second=\u7B2C\u4E8C\u9875
third={0} \u7B2C\u4E09\u9875 {1}
language.cn=\u4E2D\u6587
language.en=\u82F1\u6587
```

❸ 创建视图 JSP 文件

在 WEB-INF/jsp 目录下创建 3 个 JSP 文件，即 first.jsp、second.jsp 和 third.jsp。

first.jsp 的代码如下：

```
<%@ page language="java" contentType="text/html; charset=UTF-8"
pageEncoding="UTF-8"%>
<%@taglib prefix="spring" uri="http://www.springframework.org/tags" %>
<!DOCTYPE html PUBLIC "-//W3C//DTD HTML 4.01 Transitional//EN"
"http://www.w3.org/TR/html4/loose.dtd">
<html>
<head>
<meta http-equiv="Content-Type" content="text/html; charset=UTF-8">
<title>Insert title here</title>
</head>
<body>
    <a href="${pageContext.request.contextPath }/i18nTest?locale=zh_CN">
    <spring:message code="language.cn" /></a> --
   <a href="${pageContext.request.contextPath }/i18nTest?locale=en_US">
   <spring:message code="language.en" /></a>
   <br><br>
    <spring:message code="first"/><br><br>
    <a href="${pageContext.request.contextPath }/my/second"><spring:
    message code="second"/></a>
</body>
</html>
```

second.jsp 的代码如下：

```
<%@ page language="java" contentType="text/html; charset=UTF-8"
pageEncoding="UTF-8"%>
<%@taglib prefix="spring" uri="http://www.springframework.org/tags" %>
<!DOCTYPE html PUBLIC "-//W3C//DTD HTML 4.01 Transitional//EN"
"http://www.w3.org/TR/html4/loose.dtd">
<html>
<head>
<meta http-equiv="Content-Type" content="text/html; charset=UTF-8">
<title>Insert title here</title>
</head>
<body>
    <spring:message code="second"/><br><br>
    <a href="${pageContext.request.contextPath }/my/third"><spring:
    message code="third" arguments="888,999"  /></a>
</body>
</html>
```

third.jsp 的代码如下：

```
<%@ page language="java" contentType="text/html; charset=UTF-8"
pageEncoding="UTF-8"%>
<%@taglib prefix="spring" uri="http://www.springframework.org/tags" %>
<!DOCTYPE html PUBLIC "-//W3C//DTD HTML 4.01 Transitional//EN"
"http://www.w3.org/TR/html4/loose.dtd">
<html>
```

```
<head>
<meta http-equiv="Content-Type" content="text/html; charset=UTF-8">
<title>Insert title here</title>
</head>
<body>
    <spring:message code="third" arguments="888,999" /><br><br>
    <a href="${pageContext.request.contextPath }/my/first"><spring:
    message code="first"/></a>
</body>
</html>
```

❹ 创建控制器类

该应用有两个控制器类，一个是 I18NTestController 处理语言种类选择请求，一个是 MyController 进行页面导航。在 src 目录中创建一个名为 controller 的包，并在该包中创建这两个控制器类。

I18NTestController.java 的代码如下：

```
package controller;
import java.util.Locale;
import org.springframework.stereotype.Controller;
import org.springframework.web.bind.annotation.RequestMapping;
@Controller
public class I18NTestController {
    @RequestMapping("/i18nTest")
    /**
     * locale 接收请求参数 locale 值，并存储到 session 中
     */
    public String first(Locale locale){
        return "first";
    }
}
```

MyController 的代码如下：

```
package controller;
import org.springframework.stereotype.Controller;
import org.springframework.web.bind.annotation.RequestMapping;
@Controller
@RequestMapping("/my")
public class MyController {
    @RequestMapping("/first")
    public String first(){
        return "first";
    }
    @RequestMapping("/second")
    public String second(){
```

```
        return "second";
    }
    @RequestMapping("/third")
    public String third(){
        return "third";
    }
}
```

❺ 创建配置文件

在 WEB-INF 目录下创建配置文件 springmvc-servlet.xml 和 web.xml。web.xml 的代码与 Spring MVC 简单应用的相同，这里不再赘述。springmvc-servlet.xml 的代码如下：

```
<?xml version="1.0" encoding="UTF-8"?>
<beans xmlns="http://www.springframework.org/schema/beans"
    xmlns:xsi="http://www.w3.org/2001/XMLSchema-instance"
    xmlns:context="http://www.springframework.org/schema/context"
    xmlns:mvc="http://www.springframework.org/schema/mvc"
    xsi:schemaLocation="
    http://www.springframework.org/schema/beans
    http://www.springframework.org/schema/beans/spring-beans.xsd
    http://www.springframework.org/schema/context
    http://www.springframework.org/schema/context/spring-context.xsd
    http://www.springframework.org/schema/mvc
    http://www.springframework.org/schema/mvc/spring-mvc.xsd">
    <!-- 使用扫描机制扫描包 -->
    <context:component-scan base-package="controller" />
    <!-- 配置视图解析器 -->
    <bean
        class="org.springframework.web.servlet.view.InternalResourceView
        Resolver"
        id="internalResourceViewResolver">
        <!-- 前缀 -->
        <property name="prefix" value="/WEB-INF/jsp/" />
        <!-- 后缀 -->
        <property name="suffix" value=".jsp" />
    </bean>
    <!-- 国际化操作拦截器，如果采用基于 Session/Cookie 则必须配置 -->
    <mvc:interceptors>
       <bean class="org.springframework.web.servlet.i18n.LocaleChange
        Interceptor"/>
    </mvc:interceptors>
    <!-- 存储区域设置信息 -->
    <bean id="localeResolver"
       class="org.springframework.web.servlet.i18n.SessionLocaleResolver" >
       <property name="defaultLocale" value="zh_CN"></property>
     </bean>
```

```
    <!-- 加载国际化资源文件 -->
    <bean id="messageSource" class="org.springframework.context.support.
    ReloadableResourceBundleMessageSource">
        <!-- <property name="basename" value="classpath:messages" /> -->
        <property name="basename" value="/WEB-INF/resource/messages" />
    </bean>
</beans>
```

❻ 发布应用并测试

首先将 ch15 应用发布到 Tomcat 服务器并启动 Tomcat 服务器，然后通过地址“http://localhost:8080/ch15/my/first”测试第一个页面，运行结果如图 15.2 所示。

http://localhost:8080/ch15/my/first

中文 -- 英文

第一页

第二页

图 15.2 中文环境下 first.jsp 的运行结果

单击图 15.2 中的“第二页”超链接，打开 second.jsp 页面，运行结果如图 15.3 所示。

http://localhost:8080/ch15/my/second

第二页

888 第三页 999

图 15.3 中文环境下 second.jsp 的运行结果

单击图 15.3 中的“第三页”超链接，打开 third.jsp 页面，运行结果如图 15.4 所示。

http://localhost:8080/ch15/my/third

888 第三页 999

第一页

图 15.4 中文环境下 third.jsp 的运行结果

单击图 15.2 中的“英文”超链接，打开英文环境下的 first.jsp 页面，运行结果如图 15.5 所示。

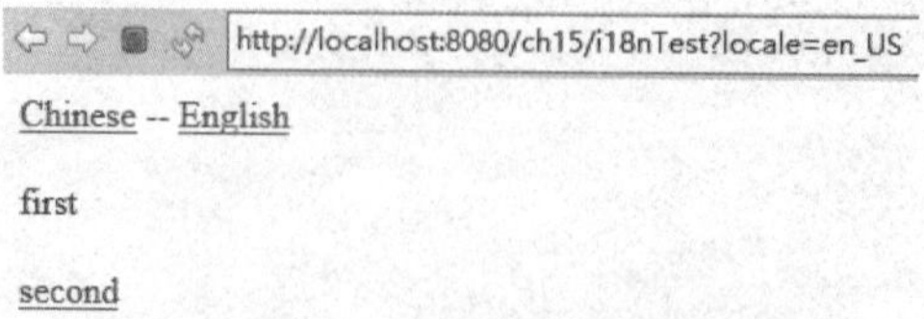

http://localhost:8080/ch15/i18nTest?locale=en_US

Chinese -- English

first

second

图 15.5 英文环境下 first.jsp 的运行结果

单击图 15.5 中的 second 超链接，打开英文环境下的 second.jsp 页面，运行结果如图 15.6 所示。

http://localhost:8080/ch15/my/second

second

888 third 999

图 15.6　英文环境下 second.jsp 的运行结果

单击图 15.6 中的 third 超链接，打开英文环境下的 third.jsp 页面，运行结果如图 15.7 所示。

图 15.7　英文环境下 third.jsp 的运行结果

15.4　本章小结

本章主要讲解了 Spring MVC 的国际化知识，详细讲述了国际化资源文件的加载方式、语言区域选择、国际化信息显示，最后给出了一个让用户自行选择语言的示例。

习题 15

1．在 JSP 页面中可以通过 Spring 提供的（　　）标签来输出国际化信息。

A．input　　B．message　　C．submit　　D．text

2．资源文件的扩展名为（　　）。

A．txt　　B．doc　　C．property　　D．properties

3．什么是国际化？国际化资源文件的命名格式是什么？

第 16 章

统一异常处理

视频讲解

学习目的与要求

本章重点讲解如何使用 Spring MVC 框架进行异常的统一处理。通过本章的学习，读者能够掌握 Spring MVC 框架统一异常处理的使用方法。

本章主要内容

- 简单异常处理 SimpleMappingExceptionResolver;
- 实现 HandlerExceptionResolver 接口自定义异常;
- 使用@ExceptionHandler 注解实现异常处理。

在 Spring MVC 应用的开发中，不管是对底层数据库操作，还是业务层或控制层操作，都会不可避免地遇到各种可预知的、不可预知的异常需要处理。如果每个过程都单独处理异常，那么系统的代码耦合度高，工作量大且不好统一，以后维护的工作量也很大。

如果能将所有类型的异常处理从各层中解耦出来，这样既保证了相关处理过程的功能单一，又实现了异常信息的统一处理和维护。幸运的是，Spring MVC 框架支持这样的实现。本章将从使用 Spring MVC 提供的简单异常处理器 SimpleMappingExceptionResolver、实现 Spring 的异常处理接口 HandlerExceptionResolver 自定义自己的异常处理器、使用@ExceptionHandler 注解实现异常处理 3 种处理方式讲解 Spring MVC 应用的异常统一处理。

16.1 示例介绍

为了验证 Spring MVC 框架的 3 种异常处理方式的实际效果，需要开发一个测试应用 ch16，从 Dao 层、Service 层、Controller 层分别抛出不同的异常（SQLException、自定义异常和未知异常），然后分别集成 3 种方式进行异常处理，进而比较其优缺点。ch16 应用的结构如图 16.1 所示。

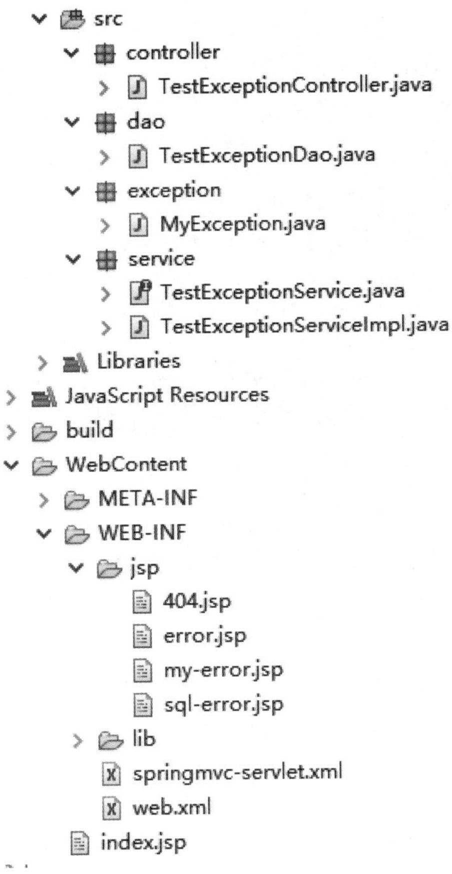

```
src
  controller
    TestExceptionController.java
  dao
    TestExceptionDao.java
  exception
    MyException.java
  service
    TestExceptionService.java
    TestExceptionServiceImpl.java
Libraries
JavaScript Resources
build
WebContent
  META-INF
  WEB-INF
    jsp
      404.jsp
      error.jsp
      my-error.jsp
      sql-error.jsp
    lib
    springmvc-servlet.xml
    web.xml
  index.jsp
```

图 16.1　ch16 应用的结构

3 种异常处理方式的相似部分有 Dao 层、Service 层、View 层、MyException、TestException Controller 以及 web.xml，下面分别介绍这些相似部分。

❶ 创建应用 ch16

创建应用 ch16，并导入 Spring MVC 相关的 JAR 包。

❷ 创建自定义异常类

在 src 目录下创建 exception 包，并在该包中创建自定义异常类 MyException。具体代码如下：

```
package exception;
public class MyException extends Exception {
    private static final long serialVersionUID = 1L;
    public MyException() {
        super();
    }
    public MyException(String message) {
        super(message);
    }
}
```

❸ **创建 Dao 层**

在 src 目录下创建 dao 包，并在该包中创建 TestExceptionDao 类，在该类中定义 3 个方法，分别抛出“数据库异常”“自定义异常”和“未知异常”。具体代码如下：

```
package dao;
import java.sql.SQLException;
import org.springframework.stereotype.Repository;
import exception.MyException;
@Repository("testExceptionDao")
public class TestExceptionDao {
    public void daodb() throws Exception {
        throw new SQLException("Dao 中数据库异常");
    }
    public void daomy() throws Exception {
        throw new MyException("Dao 中自定义异常");
    }
    public void daono() throws Exception {
        throw new Exception("Dao 中未知异常");
    }
}
```

❹ **创建 Service 层**

在 src 目录下创建 service 包，并在该包中创建 TestExceptionService 接口和 TestExceptionServiceImpl 实现类，在该接口中定义 6 个方法，其中有 3 个方法调用 Dao 层中的方法，有 3 个是 Service 层的方法。Service 层的方法是为演示 Service 层的“数据库异常”“自定义异常”和“未知异常”而定义的。

TestExceptionService 接口的代码如下：

```
package service;
public interface TestExceptionService {
   public void servicemy() throws Exception;
   public void servicedb() throws Exception;
   public void daomy() throws Exception;
   public void daodb() throws Exception;
   public void serviceno() throws Exception;
   public void daono() throws Exception;
}
```

TestExceptionServiceImpl 实现类的代码如下：

```
package service;
import java.sql.SQLException;
import org.springframework.beans.factory.annotation.Autowired;
import org.springframework.stereotype.Service;
import dao.TestExceptionDao;
import exception.MyException;
@Service("testExceptionService")
```

```
public class TestExceptionServiceImpl implements TestExceptionService{
    @Autowired
    private TestExceptionDao testExceptionDao;
    @Override
    public void servicemy() throws Exception {
        throw new MyException("Service 中自定义异常");
    }
    @Override
    public void servicedb() throws Exception {
        throw new SQLException("Service 中数据库异常");
    }
    @Override
    public void serviceno() throws Exception {
        throw new Exception("Service 中未知异常");
    }
    @Override
    public void daomy() throws Exception {
        testExceptionDao.daomy();
    }
    @Override
    public void daodb() throws Exception {
        testExceptionDao.daodb();
    }
     public void daono() throws Exception{
         testExceptionDao.daono();
     }
}
```

❺ 创建控制器类

在 src 目录下创建 controller 包，并在该包中创建 TestExceptionController 控制器类，代码如下：

```
package controller;
import java.sql.SQLException;
import org.springframework.beans.factory.annotation.Autowired;
import org.springframework.stereotype.Controller;
import org.springframework.web.bind.annotation.RequestMapping;
import exception.MyException;
import service.TestExceptionService;
@Controller
public class TestExceptionController{
    @Autowired
    private TestExceptionService testExceptionService;
    @RequestMapping("/db")
    public void db() throws Exception {
        throw new SQLException("控制器中数据库异常");
```

```
    }
    @RequestMapping("/my")
    public void my() throws Exception {
        throw new MyException("控制器中自定义异常");
    }
    @RequestMapping("/no")
    public void no() throws Exception {
        throw new Exception("控制器中未知异常");
    }
    @RequestMapping("/servicedb")
    public void servicedb() throws Exception {
         testExceptionService.servicedb();;
    }
    @RequestMapping("/servicemy")
    public void servicemy() throws Exception {
         testExceptionService.servicemy();
    }
    @RequestMapping("/serviceno")
    public void serviceno() throws Exception {
         testExceptionService.serviceno();
    }
    @RequestMapping("/daodb")
    public void daodb() throws Exception {
         testExceptionService.daodb();
    }
    @RequestMapping("/daomy")
    public void daomy() throws Exception {
         testExceptionService.daomy();
    }
    @RequestMapping("/daono")
    public void daono() throws Exception {
         testExceptionService.daono();
    }
}
```

❻ 创建 View 层

View 层中共有 5 个 JSP 页面，下面分别介绍。

测试应用首页面 index.jsp 的代码如下：

```
<%@ page language="java" contentType="text/html; charset=UTF-8"
pageEncoding="UTF-8"%>
<!DOCTYPE html PUBLIC "-//W3C//DTD HTML 4.01 Transitional//EN"
"http://www.w3.org/TR/html4/loose.dtd">
<html>
<head>
<meta http-equiv="Content-Type" content="text/html; charset=UTF-8">
```

```
<title>Insert title here</title>
</head>
<body>
<h1>所有的演示例子</h1>
<h3><a href="${pageContext.request.contextPath }/daodb">处理 dao 中数据库异常</a></h3>
<h3><a href="${pageContext.request.contextPath }/daomy">处理 dao 中自定义异常</a></h3>
<h3><a href="${pageContext.request.contextPath }/daono">处理 dao 未知错误</a></h3>
<hr>
<h3><a href="${pageContext.request.contextPath }/servicedb">处理 service 中数据库异常</a></h3>
<h3><a href="${pageContext.request.contextPath }/servicemy">处理 service 中自定义异常</a></h3>
<h3><a href="${pageContext.request.contextPath }/serviceno">处理 service 未知错误</a></h3>
<hr>
<h3><a href="${pageContext.request.contextPath }/db">处理 controller 中数据库异常</a></h3>
<h3><a href="${pageContext.request.contextPath }/my">处理 controller 中自定义异常</a></h3>
<h3><a href="${pageContext.request.contextPath }/no">处理 controller 未知错误</a></h3>
<hr>
<!-- 在 web.xml 中配置 404 -->
<h3><a href="${pageContext.request.contextPath }/404">404 错误</a></h3>
</body>
</html>
```

404 错误对应页面 404.jsp 的代码如下：

```
<%@ page language="java" contentType="text/html; charset=UTF-8"
pageEncoding="UTF-8" isErrorPage="true"%>
<!DOCTYPE html PUBLIC "-//W3C//DTD HTML 4.01 Transitional//EN"
"http://www.w3.org/TR/html4/loose.dtd">
<html>
<head>
<meta http-equiv="Content-Type" content="text/html; charset=UTF-8">
<title>Insert title here</title>
</head>
<body>
资源已不在。
</body>
</html>
```

未知异常对应页面 error.jsp 的代码如下：

```
<%@ page language="java" contentType="text/html; charset=UTF-8"
pageEncoding="UTF-8" isErrorPage="true"%>
<!DOCTYPE html PUBLIC "-//W3C//DTD HTML 4.01 Transitional//EN"
"http://www.w3.org/TR/html4/loose.dtd">
<html>
<head>
<meta http-equiv="Content-Type" content="text/html; charset=UTF-8">
<title>Insert title here</title>
</head>
<body>
<H1>未知错误：</H1><%=exception%>
<H2>错误内容：</H2>
<%
    exception.printStackTrace(response.getWriter());
%>
</body>
</html>
```

自定义异常对应页面 my-error.jsp 的代码如下：

```
<%@ page language="java" contentType="text/html; charset=UTF-8"
pageEncoding="UTF-8" isErrorPage="true"%>
<!DOCTYPE html PUBLIC "-//W3C//DTD HTML 4.01 Transitional//EN"
"http://www.w3.org/TR/html4/loose.dtd">
<html>
<head>
<meta http-equiv="Content-Type" content="text/html; charset=UTF-8">
<title>Insert title here</title>
</head>
<body>
<H1>自定义异常错误：</H1><%=exception%>
<H2>错误内容：</H2>
<%
exception.printStackTrace(response.getWriter());
%>
</body>
</html>
```

SQL 异常对应页面 sql-error.jsp 的代码如下：

```
<%@ page language="java" contentType="text/html; charset=UTF-8"
pageEncoding="UTF-8" isErrorPage="true"%>
<!DOCTYPE html PUBLIC "-//W3C//DTD HTML 4.01 Transitional//EN"
"http://www.w3.org/TR/html4/loose.dtd">
<html>
```

```
<head>
<meta http-equiv="Content-Type" content="text/html; charset=UTF-8">
<title>Insert title here</title>
</head>
<body>
<H1>数据库异常错误：</H1><%=exception%>
<H2>错误内容：</H2>
<%
exception.printStackTrace(response.getWriter());
%>
</body>
</html>
```

❼ **web.xml**

对于 Unchecked Exception 而言，由于代码不强制捕获，往往被忽略，如果运行期产生了 Unchecked Exception，而代码中又没有进行相应的捕获和处理，则可能不得不面对 404、500 等服务器内部错误提示页面，所以在 web.xml 文件中添加了全局异常 404 处理。具体代码如下：

```
<error-page>
  <error-code>404</error-code>
  <location>/WEB-INF/jsp/404.jsp</location>
</error-page>
```

从上述 Dao 层、Service 层以及 Controller 层的代码中可以看出，它们只管通过 throw 和 throws 语句抛出异常，并不处理。下面分别从 3 种方式统一处理这些异常。

16.2 SimpleMappingExceptionResolver 类

使用 org.springframework.web.servlet.handler.SimpleMappingExceptionResolver 类统一处理异常时需要在配置文件中提前配置异常类和 View 的对应关系。配置文件 springmvc-servlet.xml 的具体代码如下：

```
<?xml version="1.0" encoding="UTF-8"?>
<beans xmlns="http://www.springframework.org/schema/beans"
    xmlns:xsi="http://www.w3.org/2001/XMLSchema-instance"
    xmlns:context="http://www.springframework.org/schema/context"
    xsi:schemaLocation="
    http://www.springframework.org/schema/beans
    http://www.springframework.org/schema/beans/spring-beans.xsd
    http://www.springframework.org/schema/context
    http://www.springframework.org/schema/context/spring-context.xsd">
    <!-- 使用扫描机制扫描包 -->
    <context:component-scan base-package="controller" />
```

```
    <context:component-scan base-package="service" />
    <context:component-scan base-package="dao" />
    <!-- 配置视图解析器 -->
    <bean
        class="org.springframework.web.servlet.view.InternalResourceView
        Resolver"
        id="internalResourceViewResolver">
        <!-- 前缀 -->
        <property name="prefix" value="/WEB-INF/jsp/" />
        <!-- 后缀 -->
        <property name="suffix" value=".jsp" />
    </bean>
    <!-- 配置 SimpleMappingExceptionResolver（异常类与 View 的对应关系）-->
    <bean   class="org.springframework.web.servlet.handler.SimpleMapping
    ExceptionResolver">
        <!-- 定义默认的异常处理页面，当该异常类型注册时使用 -->
        <property name="defaultErrorView" value="error"></property>
        <!-- 定义异常处理页面用来获取异常信息的变量名，默认名为 exception -->
        <property name="exceptionAttribute" value="ex"></property>
        <!-- 定义需要特殊处理的异常，用类名或完全路径名作为 key，异常页名作为值 -->
        <property name="exceptionMappings">
            <props>
                <prop key="exception.MyException">my-error</prop>
                <prop key="java.sql.SQLException">sql-error</prop>
                <!-- 在这里还可以继续扩展对不同异常类型的处理 -->
            </props>
        </property>
    </bean>
</beans>
```

在配置完成后就可以通过 SimpleMappingExceptionResolver 异常处理器统一处理 16.1 节中的异常。

发布 ch16 应用到 Tomcat 服务器并启动服务器，然后即可通过地址“http://localhost:8080/ch16/”测试应用。

16.3 HandlerExceptionResolver 接口

org.springframework.web.servlet.HandlerExceptionResolver 接口用于解析请求处理过程中所产生的异常。开发者可以开发该接口的实现类进行 Spring MVC 应用的异常统一处理。在 ch16 应用的 exception 包中创建一个 HandlerExceptionResolver 接口的实现类 MyExceptionHandler，具体代码如下：

```
package exception;
```

```
import java.sql.SQLException;
import java.util.HashMap;
import java.util.Map;
import javax.servlet.http.HttpServletRequest;
import javax.servlet.http.HttpServletResponse;
import org.springframework.web.servlet.HandlerExceptionResolver;
import org.springframework.web.servlet.ModelAndView;
public class MyExceptionHandler implements HandlerExceptionResolver {
    @Override
public ModelAndView resolveException(HttpServletRequest arg0, HttpServlet
Response arg1, Object arg2,
            Exception arg3) {
            Map<String, Object> model=new HashMap<String, Object>();
            model.put("ex", arg3);
           //根据不同错误转向不同页面（统一处理），即异常与 View 的对应关系
           if(arg3 instanceof MyException) {
              return new ModelAndView("my-error", model);
           }else if(arg3 instanceof SQLException) {
              return new ModelAndView("sql-error", model);
           } else {
              return new ModelAndView("error", model);
           }
    }
}
```

需要将实现类 MyExceptionHandler 在配置文件中托管给 Spring MVC 框架才能进行异常的统一处理，配置代码为<bean class="exception.MyExceptionHandler"/>。

在实现 HandlerExceptionResolver 接口统一处理异常时将配置文件的代码修改如下：

```
<?xml version="1.0" encoding="UTF-8"?>
<beans xmlns="http://www.springframework.org/schema/beans"
    xmlns:xsi="http://www.w3.org/2001/XMLSchema-instance"
    xmlns:context="http://www.springframework.org/schema/context"
    xsi:schemaLocation="
   http://www.springframework.org/schema/beans
   http://www.springframework.org/schema/beans/spring-beans.xsd
   http://www.springframework.org/schema/context
   http://www.springframework.org/schema/context/spring-context.xsd">
    <!-- 使用扫描机制扫描包 -->
    <context:component-scan base-package="controller" />
    <context:component-scan base-package="service" />
    <context:component-scan base-package="dao" />
    <!-- 配置视图解析器 -->
    <bean
        class="org.springframework.web.servlet.view.InternalResource
        ViewResolver"
```

```
        id="internalResourceViewResolver">
        <!-- 前缀 -->
        <property name="prefix" value="/WEB-INF/jsp/" />
        <!-- 后缀 -->
        <property name="suffix" value=".jsp" />
    </bean>
    <!-- 托管 MyExceptionHandler -->
    <bean class="exception.MyExceptionHandler"/>
</beans>
```

发布 ch16 应用到 Tomcat 服务器并启动服务器，然后即可通过地址“http://localhost:8080/ch16/”测试应用。

16.4 @ExceptionHandler 注解

创建 BaseController 类，并在该类中使用@ExceptionHandler 注解声明异常处理方法，具体代码如下：

```
package controller;
import java.sql.SQLException;
import javax.servlet.http.HttpServletRequest;
import org.springframework.web.bind.annotation.ExceptionHandler;
import exception.MyException;
public abstract class BaseController {
     /** 基于@ExceptionHandler 异常处理 */
    @ExceptionHandler
    public String exception(HttpServletRequest request, Exception ex)  {
        request.setAttribute("ex", ex);
        //根据不同错误转向不同页面，即异常与 View 的对应关系
        if(ex instanceof SQLException) {
        return "sql-error";
        }else if(ex instanceof MyException) {
        return "my-error";
        } else {
        return "error";
        }
    }
}
```

将所有需要异常处理的 Controller 都继承 BaseController 类，示例代码如下：

```
@Controller
public class TestExceptionController extends BaseController{
    ...
}
```

在使用@ExceptionHandler 注解声明统一处理异常时不需要配置任何信息，此时将配置文件的代码修改如下：

```
<?xml version="1.0" encoding="UTF-8"?>
<beans xmlns="http://www.springframework.org/schema/beans"
    xmlns:xsi="http://www.w3.org/2001/XMLSchema-instance"
    xmlns:context="http://www.springframework.org/schema/context"
    xsi:schemaLocation="
   http://www.springframework.org/schema/beans
   http://www.springframework.org/schema/beans/spring-beans.xsd
   http://www.springframework.org/schema/context
   http://www.springframework.org/schema/context/spring-context.xsd">
    <!-- 使用扫描机制扫描包 -->
    <context:component-scan base-package="controller" />
    <context:component-scan base-package="service" />
    <context:component-scan base-package="dao" />
    <!-- 配置视图解析器 -->
    <bean
        class="org.springframework.web.servlet.view.InternalResource
        ViewResolver"
        id="internalResourceViewResolver">
        <!-- 前缀 -->
        <property name="prefix" value="/WEB-INF/jsp/" />
        <!-- 后缀 -->
        <property name="suffix" value=".jsp" />
    </bean>
</beans>
```

发布 ch16 应用到 Tomcat 服务器并启动服务器，然后即可通过地址“http://localhost:8080/ch16/”测试应用。

16.5　本章小结

本章重点介绍了 Spring MVC 框架应用程序的统一异常处理的 3 种方法。从上面的处理过程可知，使用@ExceptionHandler 注解实现异常处理具有集成简单、可扩展性好（只需要将要异常处理的 Controller 类继承于 BaseController 即可）、不需要附加 Spring 配置等优点，但该方法对已有代码存在入侵性（需要修改已有代码，使相关类继承于 BaseController）。

习题 16

1. 简述 Spring MVC 框架中统一异常处理的常用方式。
2. 如何使用@ExceptionHandler 注解进行统一异常处理？

第17章 文件的上传和下载

学习目的与要求

本章重点讲解如何使用 Spring MVC 框架进行文件的上传与下载。通过本章的学习，读者能够掌握 Spring MVC 框架单文件上传、多文件上传以及文件下载。

本章主要内容

- 单文件上传;
- 多文件上传;
- 文件下载。

文件上传是 Web 应用经常需要面对的问题。对于 Java 应用而言上传文件有多种方式，包括使用文件流手工编程上传、基于 commons-fileupload 组件的文件上传、基于 Servlet 3 及以上版本的文件上传等方式。后两种方式在编者的另一本教材（《基于 Eclipse 平台的 JSP 应用教程》）中已经阐述，本章将重点介绍如何使用 Spring MVC 框架进行文件上传。

17.1 文件上传

Spring MVC 框架的文件上传是基于 commons-fileupload 组件的文件上传，只不过 Spring MVC 框架在原有文件上传组件上做了进一步封装，简化了文件上传的代码实现，取消了不同上传组件上的编程差异。

视频讲解

17.1.1 commons-fileupload 组件

由于 Spring MVC 框架的文件上传是基于 commons-fileupload 组件的文件上传，因此需要将 commons-fileupload 组件相关的 JAR（commons-fileupload-1.3.1.jar 和 commons-io-2.4.jar）复制到 Spring MVC 应用的 WEB-INF/lib 目录下。下面讲解如何下载相关 JAR 包。

Commons 是 Apache 开放源代码组织中的一个 Java 子项目，该项目包括文件上传、命令行处理、数据库连接池、XML 配置文件处理等模块。fileupload 就是其中用来处理基于表单的文件上传的子项目，commons-fileupload 组件性能优良，并支持任意大小文件的上传。

commons-fileupload 组件可以从“http://commons.apache.org/proper/commons-fileupload/”下载，本书采用的版本是 1.3.1。下载它的 Binaries 压缩包（commons-fileupload-1.3.1-bin.zip），解压缩后的目录中有两个子目录，分别是 lib 和 site。在 lib 目录下有一个 JAR 文件——commons-fileupload-1.3.1.jar，该文件是 commons-fileupload 组件的类库。在 site 目录中是 commons-fileupload 组件的文档，也包括 API 文档。

commons-fileupload 组件依赖于 Apache 的另外一个项目——commons-io，该组件可以从“http://commons.apache.org/proper/commons-io/”下载，本书采用的版本是 2.4。下载它的 Binaries 压缩包（commons-io-2.4-bin.zip），解压缩后的目录中有 4 个 JAR 文件，其中有一个 commons-io-2.4.jar 文件，该文件是 commons-io 的类库。

17.1.2　基于表单的文件上传

标签<input type="file"/>会在浏览器中显示一个输入框和一个按钮，输入框可供用户填写本地文件的文件名和路径名，按钮可以让浏览器打开一个文件选择框供用户选择文件。

文件上传的表单例子如下：

```
<form action="upload" method="post" enctype="multipart/form-data">
    <input type="file" name="myfile"/>
    …
</form>
```

对于基于表单的文件上传，不要忘记使用 enctype 属性，并将它的值设置为 multipart/form-data，同时将表单的提交方式设置为 post。为什么要这样呢？下面从 enctype 属性说起。

表单的 enctype 属性指定的是表单数据的编码方式，该属性有以下 3 个值。

- application/x-www-form-urlencoded：这是默认的编码方式，它只处理表单域里的 value 属性值。
- multipart/form-data：该编码方式以二进制流的方式来处理表单数据，并将文件域指定文件的内容封装到请求参数里。
- text/plain：该编码方式只有当表单的 action 属性为 mailto:URL 的形式时才使用，主要适用于直接通过表单发送邮件的方式。

由上面 3 个属性的解释可知，在基于表单上传文件时 enctype 的属性值应为 multipart/form-data。

17.1.3　MultipartFile 接口

在 Spring MVC 框架中上传文件时将文件相关信息及操作封装到 MultipartFile 对象中，

因此开发者只需要使用 MultipartFile 类型声明模型类的一个属性即可对被上传文件进行操作。该接口具有如下方法。

- byte[] getBytes()：以字节数组的形式返回文件的内容。
- String getContentType()：返回文件的内容类型。
- InputStream getInputStream()：返回一个 InputStream，从中读取文件的内容。
- String getName()：返回请求参数的名称。
- String getOriginalFilename()：返回客户端提交的原始文件名称。
- long getSize()：返回文件的大小，单位为字节。
- boolean isEmpty()：判断被上传文件是否为空。
- void transferTo(File destination)：将上传文件保存到目标目录下。

在上传文件时需要在配置文件中使用 Spring 的 org.springframework.web.multipart.commons.CommonsMultipartResolver 类配置 MultipartResolver 用于文件上传。下面从单文件上传开始讲解该接口的使用方法。

17.1.4 单文件上传

本节通过一个应用案例 ch17 讲解 Spring MVC 框架如何实现单文件上传，具体步骤如下：

❶ 创建应用并导入 JAR 包

创建应用 ch17，将 Spring MVC 相关的 JAR 包、commons-fileupload 组件相关的 JAR 包以及 JSTL 相关的 JAR 包导入应用的 lib 目录中，如图 17.1 所示。

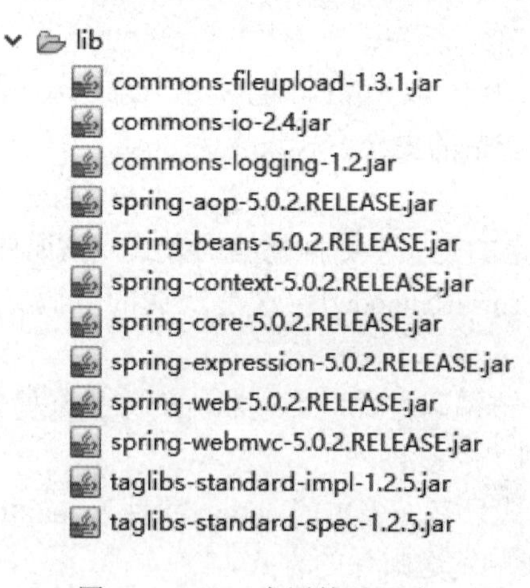

图 17.1　ch17 应用的 JAR 包

❷ 创建 web.xml 文件

在 WEB-INF 目录下创建 web.xml 文件。为防止中文乱码，需要在 web.xml 文件中添加字符编码过滤器，这里不再赘述。

❸ 创建文件选择页面

在 WebContent 目录下创建 JSP 页面 oneFile.jsp，在该页面中使用表单上传单个文件，具体代码如下：

```
<%@ page language="java" contentType="text/html; charset=UTF-8"
pageEncoding="UTF-8"%>
<!DOCTYPE html PUBLIC "-//W3C//DTD HTML 4.01 Transitional//EN" "http://www.
w3.org/TR/html4/loose.dtd">
<html>
<head>
<meta http-equiv="Content-Type" content="text/html; charset=UTF-8">
<title>Insert title here</title>
</head>
<body>
<form action="${pageContext.request.contextPath }/onefile" method="post"
enctype="multipart/form-data">
    选择文件:<input type="file" name="myfile">  <br>
    文件描述:<input type="text" name="description"> <br>
 <input type="submit" value="提交">
</form>
</body>
</html>
```

❹ 创建 POJO 类

在 src 目录下创建 pojo 包，在该包中创建 POJO 类 FileDomain。然后在该 POJO 类中声明一个 MultipartFile 类型的属性封装被上传的文件信息，属性名与文件选择页面 oneFile.jsp 中的 file 类型的表单参数名 myfile 相同。具体代码如下：

```
package pojo;
import org.springframework.web.multipart.MultipartFile;
public class FileDomain {
    private String description;
    private MultipartFile myfile;
    //省略 setter 和 getter 方法
}
```

❺ 创建控制器类

在 src 目录下创建 controller 包，并在该包中创建 FileUploadController 控制器类。具体代码如下：

```
package controller;
import java.io.File;
import javax.servlet.http.HttpServletRequest;
import org.apache.commons.logging.Log;
import org.apache.commons.logging.LogFactory;
import org.springframework.stereotype.Controller;
import org.springframework.web.bind.annotation.ModelAttribute;
import org.springframework.web.bind.annotation.RequestMapping;
import pojo.FileDomain;
```

```
@Controller
public class FileUploadController {
    //得到一个用来记录日志的对象，这样在打印信息时能够标记打印的是哪个类的信息
    private static final Log logger=LogFactory.getLog
    (FileUploadController.class);
    /**
     * 单文件上传
     */
    @RequestMapping("/onefile")
    public String oneFileUpload(@ModelAttribute FileDomain fileDomain,
    HttpServletRequest request){
        /*文件上传到服务器的位置"/uploadfiles"，该位置是指
        workspace\.metadata\.plugins\org.eclipse.wst.server.core\tmp0\
        wtpwebapps,
        发布后使用*/
        String realpath=request.getServletContext().getRealPath
        ("uploadfiles");
        String fileName=fileDomain.getMyfile().getOriginalFilename();
        File targetFile=new File(realpath, fileName);
        if(!targetFile.exists()){
           targetFile.mkdirs();
       }
        //上传
       try {
        fileDomain.getMyfile().transferTo(targetFile);
        logger.info("成功");
       } catch (Exception e) {
          e.printStackTrace();
       }
        return "showOne";
    }
}
```

❻ 创建 Spring MVC 的配置文件

在上传文件时需要在配置文件中使用 Spring 的 CommonsMultipartResolver 类配置 MultipartResolver 用于文件上传，应用的配置文件 springmvc-servlet.xml 的代码如下：

```
<?xml version="1.0" encoding="UTF-8"?>
<beans xmlns="http://www.springframework.org/schema/beans"
    xmlns:xsi="http://www.w3.org/2001/XMLSchema-instance"
    xmlns:p="http://www.springframework.org/schema/p"
    xmlns:context="http://www.springframework.org/schema/context"
    xsi:schemaLocation="
    http://www.springframework.org/schema/beans
    http://www.springframework.org/schema/beans/spring-beans.xsd
    http://www.springframework.org/schema/context
```

```
    http://www.springframework.org/schema/context/spring-context.xsd">
    <!-- 使用扫描机制扫描包 -->
    <context:component-scan base-package="controller" />
    <!-- 配置视图解析器 -->
    <bean
        class="org.springframework.web.servlet.view.
        InternalResourceViewResolver"
        id="internalResourceViewResolver">
        <!-- 前缀 -->
        <property name="prefix" value="/WEB-INF/jsp/" />
        <!-- 后缀 -->
        <property name="suffix" value=".jsp" />
    </bean>
    <!-- 使用 Spring 的 CommosMultipartResolver 配置 MultipartResolver 用于文
    件上传 -->
    <bean id="multipartResolver"
       class="org.springframework.web.multipart.commons.
       CommonsMultipartResolver"
       p:defaultEncoding="UTF-8"
       p:maxUploadSize="5400000"
       p:uploadTempDir="fileUpload/temp"
     >
      <!-- D:\spring mvc workspace\.metadata\.plugins\org.eclipse.wst.
      server.core\tmp0\wtpwebapps\fileUpload -->
    </bean>
    <!-- defaultEncoding="UTF-8" 是请求的编码格式，默认为 iso-8859-1;
    maxUploadSize="5400000" 是允许上传文件的最大值，单位为字节;
    uploadTempDir="fileUpload/temp" 为上传文件的临时路径 -->
</beans>
```

❼ 创建成功显示页面

在 WEB-INF 目录下创建 JSP 文件夹，并在该文件夹中创建单文件上传成功显示页面 showOne.jsp。具体代码如下：

```
<%@ page language="java" contentType="text/html; charset=UTF-8"
pageEncoding="UTF-8"%>
<!DOCTYPE html PUBLIC "-//W3C//DTD HTML 4.01 Transitional//EN"
"http://www.w3.org/TR/html4/loose.dtd">
<html>
<head>
<meta http-equiv="Content-Type" content="text/html; charset=UTF-8">
<title>Insert title here</title>
</head>
<body>
    ${fileDomain.description }<br>
    <!-- fileDomain.getMyfile().getOriginalFilename() -->
```

```
    ${fileDomain.myfile.originalFilename }
</body>
</html>
```

❽ 测试文件上传

发布 ch17 应用到 Tomcat 服务器并启动 Tomcat 服务器，然后通过地址“http://localhost:8080/ch17/oneFile.jsp”运行文件选择页面，运行结果如图 17.2 所示。

在图 17.2 中选择文件并输入文件描述，然后单击“提交”按钮上传文件，若成功则显示如图 17.3 所示的结果。

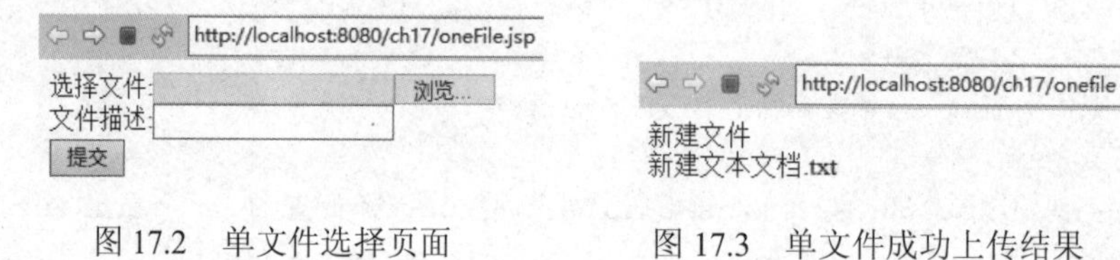

图 17.2　单文件选择页面　　　　图 17.3　单文件成功上传结果

17.1.5　多文件上传

本小节继续通过 ch17 应用案例讲解 Spring MVC 框架如何实现多文件上传，具体步骤如下：

❶ 创建多文件选择页面

在 WebContent 目录下创建 JSP 页面 multiFiles.jsp，在该页面中使用表单上传多个文件，具体代码如下：

```
<%@ page language="java" contentType="text/html; charset=UTF-8"
pageEncoding="UTF-8"%>
<!DOCTYPE html PUBLIC "-//W3C//DTD HTML 4.01 Transitional//EN"
"http://www.w3.org/TR/html4/loose.dtd">
<html>
<head>
<meta http-equiv="Content-Type" content="text/html; charset=UTF-8">
<title>Insert title here</title>
</head>
<body>
<form action="${pageContext.request.contextPath }/multifile" method=
"post" enctype="multipart/form-data">
    选择文件 1:<input type="file" name="myfile">  <br>
    文件描述 1:<input type="text" name="description"> <br>
    选择文件 2:<input type="file" name="myfile">  <br>
    文件描述 2:<input type="text" name="description"> <br>
    选择文件 3:<input type="file" name="myfile">  <br>
    文件描述 3:<input type="text" name="description"> <br>
 <input type="submit" value="提交">
</form>
```

```
</body>
</html>
```

❷ 创建 POJO 类

在上传多文件时需要 POJO 类 MultiFileDomain 封装文件信息，MultiFileDomain 类的具体代码如下：

```
package pojo;
import java.util.List;
import org.springframework.web.multipart.MultipartFile;
public class MultiFileDomain {
    private List<String> description;
    private List<MultipartFile> myfile;
    //省略 setter 和 getter 方法
}
```

❸ 添加多文件上传处理方法

在控制器类 FileUploadController 中添加多文件上传的处理方法 multiFileUpload，具体代码如下：

```
/**
 * 多文件上传
 */
@RequestMapping("/multifile")
public String multiFileUpload(@ModelAttribute MultiFileDomain
multiFileDomain, HttpServletRequest request){
    String realpath=request.getServletContext().getRealPath
    ("uploadfiles");
    //String realpath="D:/spring mvc workspace/ch17/WebContent/
    uploadfiles";
    File targetDir=new File(realpath);
    if(!targetDir.exists()){
        targetDir.mkdirs();
    }
    List<MultipartFile> files=multiFileDomain.getMyfile();
    for (int i=0; i < files.size(); i++) {
        MultipartFile file=files.get(i);
        String fileName=file.getOriginalFilename();
        File targetFile=new File(realpath,fileName);
        //上传
        try {
          file.transferTo(targetFile);
        } catch (Exception e) {
            e.printStackTrace();
        }
    }
    logger.info("成功");
    return "showMulti";
}
```

❹ 创建成功显示页面

在 JSP 文件夹中创建多文件上传成功显示页面 showMulti.jsp，具体代码如下：

```
<%@ page language="java" contentType="text/html; charset=UTF-8"
pageEncoding="UTF-8"%>
<%@ taglib uri="http://java.sun.com/jsp/jstl/core" prefix="c" %>
<!DOCTYPE html PUBLIC "-//W3C//DTD HTML 4.01 Transitional//EN"
"http://www.w3.org/TR/html4/loose.dtd">
<html>
<head>
<meta http-equiv="Content-Type" content="text/html; charset=UTF-8">
<title>Insert title here</title>
</head>
<body>
    <table>
        <tr>
            <td>详情</td><td>文件名</td>
        </tr>
        <!-- 同时取两个数组的元素 -->
        <c:forEach items="${multiFileDomain.description}" var=
        "description" varStatus="loop">
            <tr>
                <td>${description}</td>
                <td>${multiFileDomain.myfile[loop.count-1].
                originalFilename}</td>
            </tr>
        </c:forEach>
        <!-- fileDomain.getMyfile().getOriginalFilename() -->
    </table>
</body>
</html>
```

❺ 测试文件上传

发布 ch17 应用到 Tomcat 服务器并启动 Tomcat 服务器，然后通过地址"http://localhost:8080/ch17/multiFiles.jsp"运行多文件选择页面，运行结果如图 17.4 所示。

在图 17.4 中选择文件并输入文件描述，然后单击"提交"按钮上传多个文件，若成功则显示如图 17.5 所示的结果。

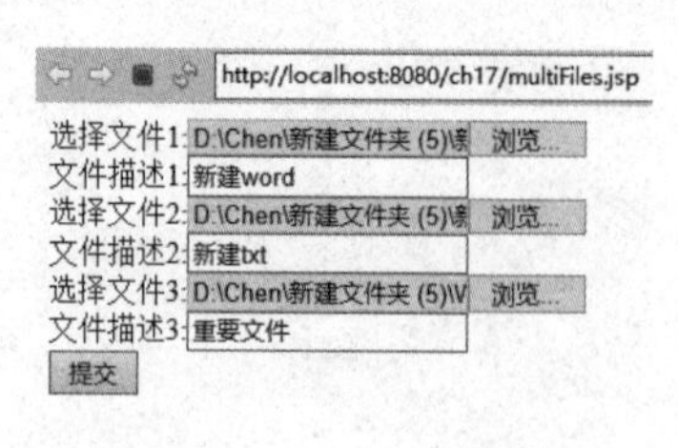

图 17.4　多文件选择页面

http://localhost:8080/ch17/multifile

详情　文件名
新建word 新建 Microsoft Word 文档.docx
新建txt　新建文本文档.txt
重要文件 W020170628400240077078.pdf

图 17.5　多文件成功上传结果

17.2　文件下载

视频讲解

17.2.1　文件下载的实现方法

实现文件下载有两种方法：一种是通过超链接实现下载，另一种是利用程序编码实现下载。通过超链接实现下载固然简单，但暴露了下载文件的真实位置，并且只能下载存放在 Web 应用程序所在的目录下的文件。利用程序编码实现下载可以增加安全访问控制，还可以从任意位置提供下载的数据，可以将文件存放到 Web 应用程序以外的目录中，也可以将文件保存到数据库中。

利用程序实现下载需要设置两个报头：

（1）Web 服务器需要告诉浏览器其所输出内容的类型不是普通文本文件或 HTML 文件，而是一个要保存到本地的下载文件，这需要设置 Content-Type 的值为 application/x-msdownload。

（2）Web 服务器希望浏览器不直接处理相应的实体内容，而是由用户选择将相应的实体内容保存到一个文件中，这需要设置 Content-Disposition 报头。该报头指定了接收程序处理数据内容的方式，在 HTTP 应用中只有 attachment 是标准方式，attachment 表示要求用户干预。在 attachment 后面还可以指定 filename 参数，该参数是服务器建议浏览器将实体内容保存到文件中的文件名称。

设置报头的示例如下：

```
response.setHeader("Content-Type", "application/x-msdownload" );
response.setHeader("Content-Disposition", "attachment; filename=" +
filename);
```

17.2.2　文件下载的过程

下面继续通过 ch17 应用讲述利用程序实现下载的过程，要求从 17.1 节上传文件的目录（workspace\.metadata\.plugins\org.eclipse.wst.server.core\tmp0\wtpwebapps\ch17\uploadfiles）中下载文件，具体开发步骤如下：

❶ 编写控制器类

首先编写控制器类 FileDownController，在该类中有 3 个方法，即 show、down 和 toUTF8String。其中，show 方法获取被下载的文件名称；down 方法执行下载功能；toUTF8String 方法是下载保存时中文文件名的字符编码转换方法。FileDownController 类的代码如下：

```
package controller;
import java.io.File;
import java.io.FileInputStream;
import java.io.UnsupportedEncodingException;
```

```
import java.util.ArrayList;
import javax.servlet.ServletOutputStream;
import javax.servlet.http.HttpServletRequest;
import javax.servlet.http.HttpServletResponse;
import org.apache.commons.logging.Log;
import org.apache.commons.logging.LogFactory;
import org.springframework.stereotype.Controller;
import org.springframework.ui.Model;
import org.springframework.web.bind.annotation.RequestMapping;
import org.springframework.web.bind.annotation.RequestParam;
@Controller
public class FileDownController {
    //得到一个用来记录日志的对象，在打印时标记打印的是哪个类的信息
    private static final Log logger=LogFactory.getLog
    (FileDownController.class);
    /**
     * 显示要下载的文件
     */
    @RequestMapping("showDownFiles")
    public String show(HttpServletRequest request, Model model){
        //从 workspace\.metadata\.plugins\org.eclipse.wst.server.core\
        //tmp0\wtpwebapps\ch17\下载
        String realpath=request.getServletContext().getRealPath
        ("uploadfiles");
        File dir=new File(realpath);
        File files[]=dir.listFiles();
        //获取该目录下的所有文件名
        ArrayList<String> fileName=new ArrayList<String>();
        for (int i=0; i < files.length; i++) {
            fileName.add(files[i].getName());
        }
        model.addAttribute("files", fileName);
        return "showDownFiles";
    }
    /**
     * 执行下载
     */
    @RequestMapping("down")
    public String down(@RequestParam String filename, HttpServletRequest
    request, HttpServletResponse response){
        String aFilePath=null;          //要下载的文件路径
        FileInputStream in=null;        //输入流
        ServletOutputStream out=null;   //输出流
        try {
            //从 workspace\.metadata\.plugins\org.eclipse.wst.server.core\
```

```
            //tmp0\wtpwebapps 下载
            aFilePath=request.getServletContext().getRealPath
            ("uploadfiles");
            //设置下载文件使用的报头
            response.setHeader("Content-Type", "application/x-msdownload" );
            response.setHeader("Content-Disposition", "attachment; filename="
                    + toUTF8String(filename));
            //读入文件
            in=new FileInputStream(aFilePath + "\\"+ filename);
            //得到响应对象的输出流，用于向客户端输出二进制数据
            out=response.getOutputStream();
            out.flush();
            int aRead=0;
            byte b[]=new byte[1024];
            while ((aRead=in.read(b))!=-1 & in!=null) {
                out.write(b,0,aRead);
            }
            out.flush();
            in.close();
            out.close();
        } catch (Throwable e) {
            e.printStackTrace();
        }
        logger.info("下载成功");
        return null;
    }
    /**
     * 下载保存时中文文件名的字符编码转换方法
     */
    public String toUTF8String(String str){
        StringBuffer sb=new StringBuffer();
        int len=str.length();
        for(int i=0; i<len; i++){
            //取出字符中的每个字符
            char c=str.charAt(i);
            //Unicode 码值为 0~255 时，不做处理
            if(c>=0 && c<=255){
                sb.append(c);
            }else{  //转换 UTF-8 编码
                byte b[];
                try {
                    b=Character.toString(c).getBytes("UTF-8");
                } catch (UnsupportedEncodingException e) {
                    e.printStackTrace();
                    b=null;
```

```
                }
                //转换为%HH 的字符串形式
                for(int j=0; j<b.length; j++){
                    int k=b[j];
                    if(k<0){
                        k&=255;
                    }
                    sb.append("%" + Integer.toHexString(k).toUpperCase());
                }
            }
        }
        return sb.toString();
    }
}
```

❷ 创建文件列表页面

下载文件示例需要一个显示被下载文件的 JSP 页面 showDownFiles.jsp，代码如下：

```
<%@ page language="java" contentType="text/html; charset=UTF-8"
pageEncoding="UTF-8"%>
<%@ taglib uri="http://java.sun.com/jsp/jstl/core" prefix="c" %>
<!DOCTYPE html PUBLIC "-//W3C//DTD HTML 4.01 Transitional//EN"
"http://www.w3.org/TR/html4/loose.dtd">
<html>
<head>
<meta http-equiv="Content-Type" content="text/html; charset=UTF-8">
<title>Insert title here</title>
</head>
<body>
    <table>
        <tr>
            <td>被下载的文件名</td>
        </tr>
        <!-- 遍历 model 中的 files -->
        <c:forEach items="${files}" var="filename">
            <tr>
                <td><a href="${pageContext.request.contextPath }/down?
                filename=${filename}">${filename}</a></td>
            </tr>
        </c:forEach>
    </table>
</body>
</html>
```

❸ 测试下载功能

发布 ch17 应用到 Tomcat 服务器并启动 Tomcat 服务器，然后通过地址“http://localhost:

8080/ch17/showDownFiles”测试下载示例，运行结果如图 17.6 所示。

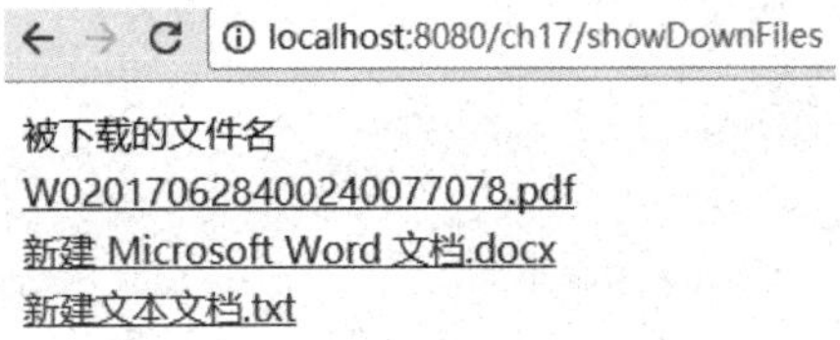

localhost:8080/ch17/showDownFiles

被下载的文件名
W020170628400240077078.pdf
新建 Microsoft Word 文档.docx
新建文本文档.txt

图 17.6　被下载文件列表页面

单击图 17.6 中的超链接下载文件，需要注意的是，使用浏览器演示该案例，不能在 Eclipse 中演示下载案例。

17.3　本章小结

本章重点介绍了 Spring MVC 的文件上传，主要包括如何使用 MultipartFile 接口封装文件信息，最后介绍了如何利用文件流进行文件下载。

习题 17

1．基于表单的文件上传应将表单的 enctype 属性值设置为（　　）。

A．multipart/form-data

B．application/x-www-form-urlencoded

C．text/plain

D．html/text

2．在 Spring MVC 框架中如何限定上传文件的大小？

3．单文件上传与多文件上传有什么区别？

第18章 EL与JSTL

学习目的与要求

本章主要介绍了表达式语言（Expression Language，EL）和 JSP 标准标签库（Java Server Pages Standard Tag Library，JSTL）的基本用法。通过本章的学习，读者能够掌握表达式语言，掌握 EL 隐含对象，了解什么是 JSTL，掌握 JSTL 的核心标签库。

本章主要内容

- EL;
- JSTL。

在 JSP 页面中可以使用 Java 代码来实现页面显示逻辑，但网页中夹杂着 HTML 与 Java 代码，给网页的设计与维护带来困难。用户可以使用 EL 来访问和处理应用程序的数据，也可以使用 JSTL 来替换网页中实现页面显示逻辑的 Java 代码，这样 JSP 页面就尽量减少了 Java 代码的使用，为以后的维护提供了方便。

18.1 表达式语言

EL 是 JSP 2.0 规范中新增加的，它的基本语法如下：

视频讲解

```
${表达式}
```

EL 表达式类似于 JSP 表达式<%=表达式%>，EL 语句中的表达式值会被直接送到浏览器显示，通过 page 指令的 isELIgnored 属性来说明是否支持 EL 表达式。当 isELIgnored 属性值为 false 时，JSP 页面可以使用 EL 表达式；当 isELIgnored 属性值为 true 时，JSP 页面不能使用 EL 表达式。isELIgnored 属性值默认为 false。

18.1.1 基本语法

EL 的语法简单、使用方便，它以“${”开始、以“}”结束。

❶ “[ ]”与“.”运算符

EL 使用“[ ]”和“.”运算符来访问数据，主要使用 EL 获取对象的属性，包括获取 JavaBean 的属性值、获取数组中的元素以及获取集合对象中的元素。对于 null 值直接以空字符串显示，而不是 null，在运算时也不会发生错误或空指针异常，所以在使用 EL 访问对象的属性时不需要判断对象是否为 null 对象，这样就为编写程序提供了方便。

1）获取 JavaBean 的属性值

假设在 JSP 页面中有这样一句话：

```
<%=user.getAge ()%>
```

那么可以使用 EL 获取 user 对象的属性 age，代码如下：

```
${user.age}
```

或

```
${user["age"]}
```

其中，点运算符前面为 JavaBean 的对象 user，后面为该对象的属性 age，表示利用 user 对象的 getAge 方法取值并显示在网页上。

2）获取数组中的元素

假设在 Controller 或 Servlet 中有这样一段话：

```
String dogs[]={"lili","huahua","guoguo"};
request.setAttribute("array", dogs);
```

那么在对应视图 JSP 中可以使用 EL 取出数组中的元素（也可以使用 18.2 节中的 JSTL 遍历数组），代码如下：

```
${array[0]}
${array[1]}
${array[2]}
```

3）获取集合对象中的元素

假设在 Controller 或 Servlet 中有这样一段话：

```
ArrayList<UserBean> users=new ArrayList<UserBean>();
UserBean ub1=new UserBean("zhang",20);
UserBean ub2=new UserBean("zhao",50);
users.add(ub1);
users.add(ub2);
request.setAttribute("array", users);
```

其中，UserBean 有两个属性 name 和 age，那么在对应视图 JSP 页面中可以使用 EL 取出 UserBean 中的属性（也可以使用 18.2 节中的 JSTL 遍历数组），代码如下：

```
${array[0].name}  ${array[0].age}
${array[1].name}  ${array[1].age}
```

❷ **算术运算符**

在 EL 表达式中有 5 个算术运算符，如表 18.1 所示。

表 18.1　EL 的算术运算符

算术运算符	说　明	示　例	结　果
+	加	${13+2}	15
-	减	${13-2}	11
*	乘	${13*2}	26
/（或 div）	除	${13/2}或${13 div 2}	6.5
%（或 mod）	取模（求余）	${13%2}或${13 mod 2}	1

❸ **关系运算符**

在 EL 表达式中有 6 个关系运算符，如表 18.2 所示。

表 18.2　EL 的关系运算符

关系运算符	说　明	示　例	结　果
= =（或 eq）	等于	${13== 2}或${13 eq 2}	false
!=（或 ne）	不等于	${13 != 2}或${13 ne 2}	true
<（或 lt）	小于	${13 < 2}或${13 lt 2}	false
>（或 gt）	大于	${13 > 2}或${13 gt 2}	true
<=（或 le）	小于等于	${13 <= 2}或${13 le 2}	false
>=（或 ge）	大于等于	${13 >= 2}或${13 ge 2}	true

❹ **逻辑运算符**

在 EL 表达式中有 3 个逻辑运算符，如表 18.3 所示。

表 18.3　EL 的逻辑运算符

逻辑运算符	说　明	示　例	结　果
&&（或 and）	逻辑与	如果 A 为 true，B 为 false，则 A && B（或 A and B）	false
||（或 or）	逻辑或	如果 A 为 true，B 为 false，则 A || B（或 A or B）	true
!（或 not）	逻辑非	如果 A 为 true，则!A（或 not A）	false

❺ **empty 运算符**

empty 运算符用于检测一个值是否为 null，例如变量 A 不存在，则${empty A}返回的结果为 true。

❻ **条件运算符**

EL 中的条件运算符是“? :”，例如${A ? B:C}，如果 A 为 true，计算 B 并返回其结果，如果 A 为 false，计算 C 并返回其结果。

18.1.2　EL 隐含对象

EL 隐含对象共有 11 个，在本书中只介绍几个常用的 EL 隐含对象，即 pageScope、requestScope、sessionScope、applicationScope、param 以及 paramValues。

❶ **与作用范围相关的隐含对象**

与作用范围相关的 EL 隐含对象有 pageScope、requestScope、sessionScope 和

applicationScope，分别可以获取 JSP 隐含对象 pageContext、request、session 和 application 中的数据。如果在 EL 中没有使用隐含对象指定作用范围，则会依次从 page、request、session、application 范围查找，若找到就直接返回，不再继续找下去；如果所有范围都没有找到，就返回空字符串。获取数据的格式如下：

```
${EL 隐含对象.关键字对象.属性}
```

或

```
${EL 隐含对象.关键字对象}
```

例如：

```
<jsp:useBean id="user" class="bean.UserBean" scope="page"/><!-- bean 标签 -->
<jsp:setProperty name="user" property="name" value="EL 隐含对象" />
name: ${pageScope.user.name}
```

再如，在 Controller 或 Servlet 中有这样一段话：

```
ArrayList<UserBean> users=new ArrayList<UserBean>();
UserBean ub1=new UserBean("zhang",20);
UserBean ub2=new UserBean("zhao",50);
users.add(ub1);
users.add(ub2);
request.setAttribute("array", users);
```

其中，UserBean 有两个属性 name 和 age，那么在对应视图 JSP 中 request 有效的范围内可以使用 EL 取出 UserBean 的属性（也可以使用 18.2 节中的 JSTL 遍历数组），代码如下：

```
${requestScope.array[0].name}  ${requestScope.array[0].age}
${requestScope.array[1].name}  ${requestScope.array[1].age}
```

❷ 与请求参数相关的隐含对象

与请求参数相关的 EL 隐含对象有 param 和 paramValues。获取数据的格式如下：

```
${EL 隐含对象.参数名}
```

例如，input.jsp 的代码如下：

```
<form method="post" action="param.jsp">
    <p>姓名：<input type="text" name="username" size="15" /></p>
    <p>兴趣：
    <input type="checkbox" name="habit" value="看书"/>看书
    <input type="checkbox" name="habit" value="玩游戏"/>玩游戏
    <input type="checkbox" name="habit" value="旅游"/>旅游
    <p>
    <input type="submit" value="提交"/>
</form>
```

那么在 param.jsp 页面中可以使用 EL 获取参数值，代码如下：

```
<%request.setCharacterEncoding("GBK");%>
<body>
<h2>EL 隐含对象 param、paramValues</h2>
姓名： ${param.username}</br>
兴趣：
${paramValues.habit[0]}
${paramValues.habit[1]}
${paramValues.habit[2]}
```

例 **18-1** 编写一个 Controller，在该控制器类处理方法中使用 request 对象和 Model 对象存储数据，然后从处理方法转发到 show.jsp 页面，在 show.jsp 页面中显示 request 对象的数据。

运行控制器的处理方法，在 IE 地址栏中输入：

```
http://localhost:8080/ch18/input
```

程序运行结果如图 18.1 所示。

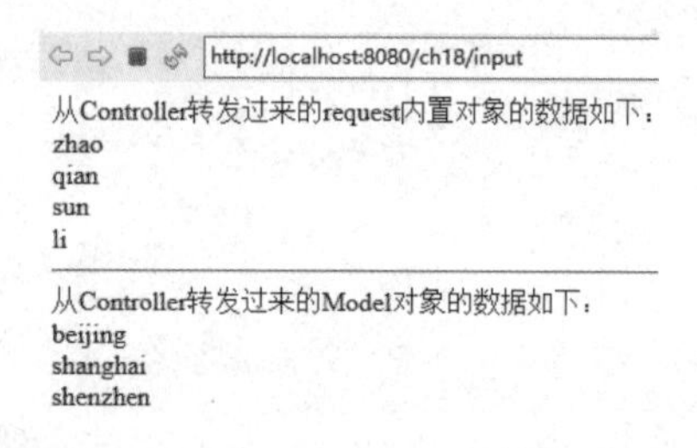

图 18.1 使用 EL 获取数据

实现上述功能的具体步骤如下：

❶ 创建应用并导入相关 JAR 包

创建 ch18 应用，导入 Spring MVC 相关的 JAR 包。

❷ 创建配置文件

在 ch18 应用的 WEB-INF 目录下创建 web.xml 和 springmvc-servlet.xml。这两个文件的代码与普通的 Spring MVC 应用的代码相同，这里不再赘述。

❸ 创建控制器类

在 src 目录下创建一个名为 controller 的包，并在该包下创建一个名为 InputController 的控制器类。代码如下：

```
package controller;
import javax.servlet.http.HttpServletRequest;
import org.springframework.stereotype.Controller;
import org.springframework.ui.Model;
import org.springframework.web.bind.annotation.RequestMapping;
@Controller
public class InputController {
    @RequestMapping("/input")
    public String input(HttpServletRequest request, Model model){
```

```
        String names[]={ "zhao", "qian", "sun", "li" };
        request.setAttribute("name", names);
        String address[]={ "beijing", "shanghai", "shenzhen"};
        model.addAttribute("address", address);
        return "show";
    }
}
```

❹ 创建 show.jsp 页面

在 WEB-INF 目录下创建 jsp 文件夹，在该文件夹中创建 show.jsp 页面。代码如下：

```
<%@ page language="java" contentType="text/html; charset=UTF-8"
pageEncoding="UTF-8"%>
<!DOCTYPE html PUBLIC "-//W3C//DTD HTML 4.01 Transitional//EN"
"http://www.w3.org/TR/html4/loose.dtd">
<html>
<head>
<meta http-equiv="Content-Type" content="text/html; charset=UTF-8">
<title>Insert title here</title>
</head>
<body>
    从 Controller 转发过来的 request 内置对象的数据如下：<br>
    ${requestScope.name[0]}<br>
    ${requestScope.name[1]}<br>
    ${requestScope.name[2]}<br>
    ${requestScope.name[3]}<br>
    <hr>
    从 Controller 转发过来的 Model 对象的数据如下：<br>
    ${address[0]}<br>
    ${address[1]}<br>
    ${address[2]}<br>
</body>
</html>
```

18.2 JSP 标准标签库

JSTL 规范由 Sun 公司制定，由 Apache 的 Jakarta 小组负责实现。JSTL 由 5 个不同功能的标签库组成，包括 Core、I18N、XML、SQL 以及 Functions，本节只简要介绍 JSTL 的 Core 和 Functions 标签库中几个常用的标签。

视频讲解

18.2.1 配置 JSTL

JSTL 现在已经是 Java EE5 的一个组成部分，如果采用支持 Java EE5 或以上版本的集成开发环境开发 Web 应用程序，就不需要再配置 JSTL 了。但本书采用的是 Eclipse 平台，因此需要配置 JSTL。配置 JSTL 的步骤如下：

❶ 复制 JSTL 的标准实现

在 Tomcat 的\webapps\examples\WEB-INF\lib 目录下找到 taglibs-standard-impl-1.2.5.jar 和 taglibs-standard-spec-1.2.5.jar 文件，然后复制到 Web 工程的 WEB-INF\lib 目录下。

❷ 使用 taglib 标记定义前缀与 uri 引用

如果使用 Core 标签库，首先需要在 JSP 页面中使用 taglib 标记定义前缀与 uri 引用，代码如下：

```
<%@ taglib prefix="c" uri="http://java.sun.com/jsp/jstl/core"%>
```

如果使用 Functions 标签库，首先需要在 JSP 页面中使用 taglib 标记定义前缀与 uri 引用，代码如下：

```
<%@ taglib prefix="fn" uri="http://java.sun.com/jsp/jstl/functions"%>
```

18.2.2 核心标签库之通用标签

❶ <c:out>标签

<c:out>标签用来显示数据的内容，与<%=表达式%>或${表达式}类似。格式如下：

```
<c:out value="输出的内容" [default="defaultValue"]/>
```

或

```
<c:out value="输出的内容">
    defaultValue
</c:out>
```

其中，value 值可以是一个 EL 表达式，也可以是一个字符串；default 可有可无，当 value 值不存在时输出 defaultValue。例如：

```
<c:out value="${param.data}" default="没有数据" />
<br>
<c:out value="${param.nothing}" />
<br>
<c:out value="这是一个字符串" />
```

程序输出的结果如图 18.2 所示。

http://localhost:8080/ch18/cout.jsp

没有数据

这是一个字符串

图 18.2 <c:out>标签

❷ <c:set>标签

1）设置作用域变量

用户可以使用<c:set>在 page、request、session、application 等范围内设置一个变量。

格式如下：

```
<c:set value="value" var="varName" [scope="page|request|session|
application"]/>
```

该代码将 value 值赋给变量 varName。

```
<c:set value="zhao" var="userName" scope="session"/>
```

相当于

```
<% session.setAttribute("userName","zhao"); %>
```

2）设置 JavaBean 的属性

在使用<c:set>设置 JavaBean 的属性时必须使用 target 属性进行设置。格式如下：

```
<c:set value="value" target="target" property="propertyName"/>
```

该代码将 value 值赋给 target 对象（JaveBean 对象）的 propertyName 属性。如果 target 为 null 或没有 set 方法则抛出异常。

❸ <c:remove>标签

如果要删除某个变量，可以使用<c:remove>标签。例如：

```
<c:remove var="userName" scope="session"/>
```

相当于

```
<%session.removeAttribute("userName") %>
```

18.2.3　核心标签库之流程控制标签

❶ <c:if>标签

<c:if>标签实现 if 语句的作用，具体语法格式如下：

```
<c:if test="条件表达式">
    主体内容
</c:if>
```

其中，条件表达式可以是 EL 表达式，也可以是 JSP 表达式。如果表达式的值为 true，则会执行<c:if>的主体内容，但是没有相对应的<c:else>标签。如果想在条件成立时执行一块内容，不成立时执行另一块内容，可以使用<c:choose>、<c:when>及<c:otherwise>标签。

❷ <c:choose>、<c:when>及<c:otherwise>标签

<c:choose>、<c:when>及<c:otherwise>标签实现 if/elseif/else 语句的作用，具体语法格式如下：

```
<c:choose>
    <c:when test="条件表达式 1">
        主体内容 1
    </c:when>
```

```
    <c:when test="条件表达式 2">
        主体内容 2
    </c:when>
    <c:otherwise>
        表达式都不正确时执行的主体内容
    </c:otherwise>
</c:choose>
```

例 **18-2**　编写一个 JSP 页面 ifelse.jsp，在该页面中使用<c:set>标签把两个字符串设置为 request 范围内的变量，使用<c:if>标签求出这两个字符串的最大值（按字典顺序比较大小），使用<c:choose>、<c:when>及<c:otherwise>标签求出这两个字符串的最小值。

例 18-2 页面文件 ifelse.jsp 的代码如下：

```
<%@ page language="java" contentType="text/html; charset=UTF-8"
pageEncoding="UTF-8"%>
<%@ taglib uri="http://java.sun.com/jsp/jstl/core" prefix="c" %>
<!DOCTYPE html PUBLIC "-//W3C//DTD HTML 4.01 Transitional//EN"
  "http://www.w3.org/TR/html4/loose.dtd">
<html>
<head>
<meta http-equiv="Content-Type" content="text/html; charset=UTF-8">
<title>Insert title here</title>
</head>
<body>
    <c:set value="if" var="firstNumber" scope="request" />
    <c:set value="else" var="secondNumber" scope="request" />
    <c:if test="${firstNumber>secondNumber}">
        最大值为${firstNumber}
    </c:if>
    <c:if test="${firstNumber<secondNumber}">
        最大值为${secondNumber}
    </c:if>
    <c:choose>
        <c:when test="${firstNumber<secondNumber}">
            最小值为${firstNumber}
        </c:when>
        <c:otherwise>
            最小值为${secondNumber}
        </c:otherwise>
    </c:choose>
</body>
</html>
```

<c:when>及<c:otherwise>必须放在<c:choose>之中。当<c:when>的 test 结果为 true 时会输出<c:when>的主体内容，而不理会<c:otherwise>的内容。在<c:choose>中可以有多个<c:when>，程序会从上到下进行条件判断，如果有一个<c:when>的 test 结果为 true 就输出其主体内容，之后的<c:when>不再执行。如果所有的<c:when>的 test 结果都为 false，则会

输出<c:otherwise>的内容。<c:if>与<c:choose>也可以嵌套使用，例如：

```
<c:set value="fda" var="firstNumber" scope="request"/>
<c:set value="else" var="secondNumber" scope="request"/>
<c:set value="ddd" var="threeNumber" scope="request"/>
<c:if test="${firstNumber>secondNumber}">
    <c:choose>
        <c:when test="${firstNumber<threeNumber}">
            最大值为${threeNumber}
        </c:when>
        <c:otherwise>
            最大值为${firstNumber}
        </c:otherwise>
    </c:choose>
</c:if>
<c:if test="${secondNumber>firstNumber}">
    <c:choose>
        <c:when test="${secondNumber<threeNumber}">
            最大值为${threeNumber}
        </c:when>
        <c:otherwise>
            最大值为${secondNumber}
        </c:otherwise>
    </c:choose>
</c:if>
```

18.2.4 核心标签库之迭代标签

❶ <c:forEach>标签

<c:forEach>标签可以实现程序中的 for 循环。语法格式如下：

```
<c:forEach var="变量名" items="数组或 Collection 对象">
    循环体
</c:forEach>
```

其中，items 属性可以是数组或 Collection 对象，每次循环读取对象中的一个元素，并赋值给 var 属性指定的变量，之后就可以在循环体中使用 var 指定的变量获取对象的元素。例如，在 Controller 或 Servlet 中有这样一段代码：

```
ArrayList<UserBean> users=new ArrayList<UserBean>();
UserBean ub1=new UserBean("zhao",20);
UserBean ub2=new UserBean("qian",40);
UserBean ub3=new UserBean("sun",60);
UserBean ub4=new UserBean("li",80);
users.add(ub1);
users.add(ub2);
users.add(ub3);
users.add(ub4);
```

```
request.setAttribute("usersKey", users);
```

那么在对应 JSP 页面中可以使用<c:forEach>循环遍历出数组中的元素。代码如下：

```
<table>
  <tr>
    <th>姓名</th>
    <th>年龄</th>
  </tr>
<c:forEach var="user" items="${requestScope.usersKey}">
  <tr>
    <td>${user.name}</td>
    <td>${user.age}</td>
  </tr>
</c:forEach>
</table>
```

在有些情况下需要为<c:forEach>标签指定 begin、end、step 和 varStatus 属性。begin 为迭代时的开始位置，默认值为 0；end 为迭代时的结束位置，默认值是最后一个元素；step 为迭代步长，默认值为 1；varStatus 代表迭代变量的状态，包括 count（迭代的次数）、index（当前迭代的索引，第一个索引为 0）、first（是否为第一个迭代对象）和 last（是否为最后一个迭代对象）。例如：

```
<table border=1>
        <tr>
            <th>Value</th>
            <th>Square</th>
            <th>Index</th>
        </tr>
        <c:forEach var="x" varStatus="status" begin="0" end="10" step="2">
            <tr>
                <td>${x}</td>
                <td>${x * x}</td>
                <td>${status.index}</td>
            </tr>
        </c:forEach>
</table>
```

上述程序的运行结果如图 18.3 所示。

http://localhost:8080/ch18/cforEach.jsp

Value	Square	Index
0	0	0
2	4	2
4	16	4
6	36	6
8	64	8
10	100	10

图 18.3　<c:forEach>标签

❷ <c:forTokens>标签

<c:forTokens>用于迭代字符串中由分隔符分隔的各成员，它通过 java.util.StringTokenizer 实例来完成字符串的分隔，属性 items 和 delims 作为构造 StringTokenizer 实例的参数。语法格式如下：

```
<c:forTokens var="变量名" items="要迭代的 String 对象" delims="指定分隔字符串的分隔符">
        循环体
</c:forTokens>
```

例如：

```
<c:forTokens items="chenheng1:chenheng2:chenheng3" delims=":" var="name">
    ${name}<br>
</c:forTokens>
```

上述程序的运行结果如图 18.4 所示。

图 18.4 <c:forTokens>标签

<c:forTokens>标签与<c:forEach>标签一样，也有 begin、end、step 和 varStatus 属性，并且用法相同，这里不再赘述。

18.2.5 函数标签库

在 JSP 页面中调用 JSTL 中的函数时需要使用 EL 表达式，调用语法格式如下：

```
${fn:函数名(参数 1,参数 2,…)}
```

下面介绍几个常用的函数。

❶ contains 函数

该函数的功能是判断一个字符串中是否包含指定的子字符串，如果包含，则返回 true，否则返回 false。其定义如下：

```
contains(string, substring)
```

该函数的调用示例代码如下：

```
${fn:contains("I am studying", "am") }
```

上述 EL 表达式将返回 true。

❷ containsIgnoreCase 函数

该函数与 contains 函数的功能相似，但判断是不区分大小写的。其定义如下：

```
containsIgnoreCase(string, substring)
```

该函数的调用示例代码如下：

```
${fn:containsIgnoreCase("I AM studying", "am") }
```

上述 EL 表达式将返回 true。

❸ **endsWith 函数**

该函数的功能是判断一个字符串是否以指定的后缀结尾。其定义如下：

```
endsWith(string, suffix)
```

该函数的调用示例代码如下：

```
${fn:endsWith("I AM studying", "am") }
```

上述 EL 表达式将返回 false。

❹ **indexOf 函数**

该函数的功能是返回指定子字符串在某个字符串中第一次出现时的索引，找不到时将返回-1。其定义如下：

```
indexOf(string, substring)
```

该函数的调用示例代码如下：

```
${fn:indexOf("I am studying", "am") }
```

上述 EL 表达式将返回 2。

❺ **join 函数**

该函数的功能是将一个 String 数组中的所有元素合并成一个字符串，并用指定的分隔符分开。其定义如下：

```
join(array, separator)
```

例如，假设一个 String 数组 my，它有 3 个元素，即“I”“am”和“studying”，那么下列 EL 表达式将返回“I,am,studying”。

```
${fn:join(my, ",") }
```

❻ **length 函数**

该函数的功能是返回集合中元素的个数或者字符串中字符的个数。其定义如下：

```
length(input)
```

该函数的调用示例代码如下：

```
${fn:length("aaa")}
```

上述 EL 表达式将返回 3。

❼ **replace 函数**

该函数的功能是将字符串中出现的所有 beforestring 用 afterstring 替换，并返回替换后的结果。其定义如下：

```
replace(string, beforestring, afterstring)
```

该函数的调用示例代码如下：

```
${fn:replace("I am am studying", "am", "do") }
```

上述 EL 表达式将返回“I do do studying”。

❽ split 函数

该函数的功能是将一个字符串使用指定的分隔符 separator 分离成一个子字符串数组。其定义如下：

```
split(string, separator)
```

该函数的调用示例代码如下：

```
<c:set var="my" value="${fn:split('I am studying', ' ') }"/>
<c:forEach var="myArrayElement" items="${my }">
    ${myArrayElement}<br>
</c:forEach>
```

上述示例代码的显示结果如图 18.5 所示。

图 18.5 split 示例结果

❾ startsWith 函数

该函数的功能是判断一个字符串是否以指定的前缀开头。其定义如下：

```
startsWith(string, prefix)
```

该函数的调用示例代码如下：

```
${fn:startsWith("I AM studying", "am") }
```

上述 EL 表达式将返回 false。

❿ substring 函数

该函数的功能是返回一个字符串的子字符串。其定义如下：

```
substring(string, begin, end)
```

该函数的调用示例代码如下：

```
${fn:substring("abcdef", 1, 3)}
```

上述 EL 表达式将返回“bc”。

⓫ toLowerCase 函数

该函数的功能是将一个字符串转换成它的小写版本。其定义如下：

```
toLowerCase(string)
```

该函数的调用示例代码如下：

```
${fn:toLowerCase("I AM studying") }
```

上述 EL 表达式将返回“i am studying”。

⑫ toUpperCase 函数

该函数的功能与 toLowerCase 函数的功能相反，这里不再赘述。

⑬ trim 函数

该函数的功能是将一个字符串开头和结尾的空白去掉。其定义如下：

```
trim(string)
```

该函数的调用示例代码如下：

```
${fn:trim("    I AM studying     ") }
```

上述 EL 表达式将返回“I AM studying”。

18.3 本章小结

本章重点介绍了表达式语言、JSTL 核心标签库以及 JSTL 函数标签库的用法，EL 与 JSTL 的应用大大提高了编程效率，并且降低了维护难度。

习题 18

1. 在 Web 应用程序中有以下程序代码段，执行后转发到某个 JSP 页面：

```
ArrayList<String>  dogNames=new ArrayList<String>();
dogNames.add("goodDog");
request.setAttribute("dogs", dogNames);
```

以下选项中，（　　）可以正确地使用 EL 取得数组中的值。

A. ${ dogs .0}　　B. ${ dogs [0]}
C. ${ dogs .[0]}　　D. ${ dogs "0"}

2. JSTL 标签（　　）可以实现 Java 程序中的 if 语句功能。

A. <c:set>　　B. <c:out>
C. <c:forEach>　　D. <c:if>

3. （　　）不是 EL 的隐含对象。

A. request　　B. pageScope
C. sessionScope　　D. applicationScope

4. JSTL 标签（　　）可以实现 Java 程序中的 for 语句功能。

A. <c:set>　　B. <c:out>
C. <c:forEach>　　D. <c:if>

第 4 部分

SSM 框架

第 19 章

SSM 框架整合

视频讲解

学习目的与要求

本章重点讲解 SSM 框架整合环境构建，通过本章的学习，读者应该了解 SSM 框架整合思路，掌握 SSM 框架整合环境构建。

本章主要内容

- SSM 框架整合思路;
- SSM 框架整合环境构建;
- SSM 框架整合应用测试。

通过前面的学习，读者应该掌握了 Spring、MyBatis 以及 Spring MVC 框架的使用，并且在第 6 章中已经掌握了 Spring 与 MyBatis 的整合。在实际开发中通常将 Spring、MyBatis 以及 Spring MVC 三大框架整合在一起使用。本章将对 Spring、Spring MVC 以及 MyBatis（SSM）框架的整合使用进行详细的讲解。

19.1　SSM 框架整合所需 JAR 包

因为 Spring MVC 是 Spring 框架中的一个子模块，所以 Spring 与 Spring MVC 之间不存在整合的问题。实际上，SSM 框架的整合只涉及 Spring 与 MyBatis 的整合以及 Spring MVC 与 MyBatis 的整合。

实现 SSM 框架的整合首先需要准备 3 个框架的 JAR 包以及其他整合所需要的 JAR 包。在 6.5 节已经讲解了 Spring 与 MyBatis 框架整合所需要的 JAR 包，本章只需再加入 Spring MVC 的相关 JAR 包（spring-web-5.0.2.RELEASE.jar 和 spring-webmvc-5.0.2.RELEASE.jar）即可。因此，SSM 框架整合所需的 JAR 包如图 19.1 所示。

ant-1.9.6.jar
ant-launcher-1.9.6.jar
aopalliance-1.0.jar
asm-5.2.jar
aspectjweaver-1.8.13.jar
cglib-3.2.5.jar
commons-dbcp2-2.2.0.jar
commons-logging-1.2.jar
commons-pool2-2.5.0.jar
javassist-3.22.0-CR2.jar
log4j-1.2.17.jar
log4j-api-2.3.jar
log4j-core-2.3.jar
mybatis-3.4.5.jar
mybatis-spring-1.3.1.jar
mysql-connector-java-5.1.45-bin.jar
ognl-3.1.15.jar
slf4j-api-1.7.25.jar
slf4j-log4j12-1.7.25.jar
spring-aop-5.0.2.RELEASE.jar
spring-aspects-5.0.2.RELEASE.jar
spring-beans-5.0.2.RELEASE.jar
spring-context-5.0.2.RELEASE.jar
spring-core-5.0.2.RELEASE.jar
spring-expression-5.0.2.RELEASE.jar
spring-jdbc-5.0.2.RELEASE.jar
spring-tx-5.0.2.RELEASE.jar
spring-web-5.0.2.RELEASE.jar
spring-webmvc-5.0.2.RELEASE.jar

图 19.1　SSM 框架整合所需的 JAR 包

19.2　SSM 框架整合应用测试

本节是一个应用案例（根据用户名模糊查询用户信息，用户表是 5.1.2 小节中的数据表 user），使用 SSM 框架实现该案例的具体步骤如下：

❶ 创建应用并导入相关 JAR 包

创建应用 ch19，将图 19.1 所示的 JAR 包复制到应用的 lib 中。因为在案例中使用 JSTL 标签显示查询结果，所以还需要将 JSTL 标签相关的 JAR 包 taglibs-standard-impl-1.2.5.jar 和 taglibs-standard-spec-1.2.5.jar 复制到应用的 lib 中。

❷ 创建信息输入页面

在 WebContent 目录下创建 input.jsp 页面，具体代码如下：

```
<%@ page language="java" contentType="text/html; charset=UTF-8"
pageEncoding="UTF-8"%>
<!DOCTYPE html PUBLIC "-//W3C//DTD HTML 4.01 Transitional//EN"
"http://www.w3.org/TR/html4/loose.dtd">
<html>
<head>
```

```
<meta http-equiv="Content-Type" content="text/html; charset=UTF-8">
<title>Insert title here</title>
</head>
<body>
    <form action="${pageContext.request.contextPath }/select" method=
    "post">
        输入用户名：<input type="text" name="uname"/><br>
        <input type="submit" value="提交"/>
    </form>
</body>
</html>
```

❸ **创建持久化类**

在 src 目录下创建一个名为 com.po 的包，并在该包中创建一个 PO 类 MyUser。具体代码如下：

```
package com.po;
/**
 *springtest 数据库中 user 表的持久化类
 */
public class MyUser {
    private Integer uid;
    private String uname;
    private String usex;
    //省略 setter 和 getter 方法
}
```

❹ **创建 Dao 层**

在 src 目录下创建一个名为 com.dao 的包，并在该包中创建一个名为 UserDao 的接口，该接口使用@Mapper 注解自动装配为 MyBatis 的映射接口。具体代码如下：

```
package com.dao;
import java.util.List;
import org.apache.ibatis.annotations.Mapper;
import org.springframework.stereotype.Repository;
import com.po.MyUser;
@Repository("userDao")
@Mapper
/*使用 Spring 自动扫描 MyBatis 的接口并装配
（Spring 将指定包中所有被@Mapper 注解标注的接口自动装配为 MyBatis 的映射接口）*/
public interface UserDao {
    /**
     * 接口方法对应 SQL 映射文件 UserMapper.xml 中的 id
     */
    public List<MyUser> selectUserByUname(MyUser user);
}
```

❺ 创建 Service 层

在 src 目录下创建一个名为 com.service 的包，并在该包中创建一个名为 UserService 的接口和该接口的实现类 UserServiceImpl。

UserService 接口的代码如下：

```
package com.service;
import java.util.List;
import com.po.MyUser;
public interface UserService {
    public List<MyUser> selectUserByUname(MyUser user);
}
```

UserServiceImpl 实现类的代码如下：

```
package com.service;
import java.util.List;
import org.springframework.beans.factory.annotation.Autowired;
import org.springframework.stereotype.Service;
import org.springframework.transaction.annotation.Transactional;
import com.dao.UserDao;
import com.po.MyUser;
@Service("userService")
@Transactional
/**加上注解@Transactional 可以指定这个类需要受 Spring 的事务管理，
注意@Transactional 只能针对 public 属性范围内的方法添加，
本案例并不需要处理事务，在这里只是告诉读者如何使用事务*/
public class UserServiceImpl implements UserService{
    @Autowired
    private UserDao userDao;
    @Override
    public List<MyUser> selectUserByUname(MyUser user) {
        return userDao.selectUserByUname(user);
    }
}
```

❻ 创建 Controller 层

在 src 目录下创建一个名为 com.controller 的包，并在该包中创建一个名为 UserController 的控制器类。具体代码如下：

```
package com.controller;
import java.util.List;
import org.springframework.beans.factory.annotation.Autowired;
import org.springframework.stereotype.Controller;
import org.springframework.ui.Model;
import org.springframework.web.bind.annotation.RequestMapping;
import com.dao.UserDao;
import com.po.MyUser;
```

```
@Controller
public class UserController {
    @Autowired
    private UserDao userDao;
    @RequestMapping("/select")
    public String select(MyUser user, Model model) {
        List<MyUser> list=userDao.selectUserByUname(user);
        model.addAttribute("userList", list);
        return "userList";
    }
}
```

❼ 创建用户信息显示页面

在 WEB-INF 目录下创建文件夹 JSP，并在该文件夹下创建用户信息显示页面 userList.jsp。具体代码如下：

```
<%@ page language="java" contentType="text/html; charset=UTF-8"
pageEncoding="UTF-8"%>
<%@ taglib prefix="c" uri="http://java.sun.com/jsp/jstl/core"%>
<!DOCTYPE   html   PUBLIC   "-//W3C//DTD   HTML   4.01   Transitional//EN"
"http://www.w3.org/TR/html4/loose.dtd">
<html>
<head>
<meta http-equiv="Content-Type" content="text/html; charset=UTF-8">
<title>Insert title here</title>
</head>
<body>
    用户信息<br>
    <c:forEach items="${userList}" var="user">
        ${user.uid}    
        ${user.uname}    
        ${user.usex}<br>
    </c:forEach>
</body>
</html>
```

❽ 创建相关配置文件

1）web.xml

在 WEB-INF 目录下创建 web.xml 文件，并在该文件中实例化 ApplicationContext 容器、启动 Spring 容器、配置 DispatcherServlet 以及配置字符编码过滤器。具体代码如下：

```
<?xml version="1.0" encoding="UTF-8"?>
<web-app xmlns:xsi="http://www.w3.org/2001/XMLSchema-instance"
    xmlns="http://xmlns.jcp.org/xml/ns/javaee"
    xsi:schemaLocation="http://xmlns.jcp.org/xml/ns/javaee
http://xmlns.jcp.org/xml/ns/javaee/web-app_3_1.xsd"
```

```
    id="WebApp_ID" version="3.1">
    <!-- 实例化 ApplicationContext 容器 -->
    <context-param>
        <!-- 加载 src 目录下的 applicationContext.xml 文件 -->
        <param-name>contextConfigLocation</param-name>
        <param-value>
            classpath:applicationContext.xml
        </param-value>
    </context-param>
    <!-- 指定以 ContextLoaderListener 方式启动 Spring 容器 -->
    <listener>
        <listener-class>
            org.springframework.web.context.ContextLoaderListener
        </listener-class>
    </listener>
    <!-- 配置 DispatcherServlet -->
    <servlet>
        <servlet-name>springmvc</servlet-name>
        <servlet-class>org.springframework.web.servlet.DispatcherServlet
        </servlet-class>
        <load-on-startup>1</load-on-startup>
    </servlet>
    <servlet-mapping>
        <servlet-name>springmvc</servlet-name>
        <url-pattern>/</url-pattern>
    </servlet-mapping>
    <!-- 避免中文乱码 -->
    <filter>
    <filter-name>characterEncodingFilter</filter-name>
    <filter-class>org.springframework.web.filter.CharacterEncodingFilter
    </filter-class>
    <init-param>
            <param-name>encoding</param-name>
            <param-value>UTF-8</param-value>
    </init-param>
    <init-param>
            <param-name>forceEncoding</param-name>
            <param-value>true</param-value>
    </init-param>
  </filter>
  <filter-mapping>
    <filter-name>characterEncodingFilter</filter-name>
    <url-pattern>/*</url-pattern>
  </filter-mapping>
</web-app>
```

2）springmvc-servlet.xml

在 WEB-INF 目录下创建 Spring MVC 的核心配置文件 springmvc-servlet.xml，在该文件中仅配置控制器扫描包和视图解析器，具体代码如下：

```
<?xml version="1.0" encoding="UTF-8"?>
<beans xmlns="http://www.springframework.org/schema/beans"
    xmlns:xsi="http://www.w3.org/2001/XMLSchema-instance"
    xmlns:context="http://www.springframework.org/schema/context"
    xsi:schemaLocation="
    http://www.springframework.org/schema/beans
    http://www.springframework.org/schema/beans/spring-beans.xsd
    http://www.springframework.org/schema/context
    http://www.springframework.org/schema/context/spring-context.xsd">
    <!-- 使用扫描机制扫描包 -->
    <context:component-scan base-package="com.controller" />
    <!-- 配置视图解析器 -->
    <bean
        class="org.springframework.web.servlet.view.
        InternalResourceViewResolver"
        id="internalResourceViewResolver">
        <!-- 前缀 -->
        <property name="prefix" value="/WEB-INF/jsp/" />
        <!-- 后缀 -->
        <property name="suffix" value=".jsp" />
    </bean>
</beans>
```

3）log4j.properties

在 src 目录下创建 MyBatis 的日志文件 log4j.properties，具体代码如下：

```
# Global logging configuration
log4j.rootLogger=ERROR, stdout
# MyBatis logging configuration...
log4j.logger.com.dao=DEBUG
# Console output...
log4j.appender.stdout=org.apache.log4j.ConsoleAppender
log4j.appender.stdout.layout=org.apache.log4j.PatternLayout
log4j.appender.stdout.layout.ConversionPattern=%5p [%t] - %m%n
```

4）applicationContext.xml

在 src 目录下创建 Spring 的配置文件 applicationContext.xml，在该文件中配置数据源、添加事务支持、开启事务注解、配置 MyBatis 工厂、进行 Mapper 代理开发以及指定扫描包。具体代码如下：

```
<?xml version="1.0" encoding="UTF-8"?>
<beans xmlns="http://www.springframework.org/schema/beans"
```

```
xmlns:xsi="http://www.w3.org/2001/XMLSchema-instance"
xmlns:context="http://www.springframework.org/schema/context"
xmlns:tx="http://www.springframework.org/schema/tx"
xsi:schemaLocation="http://www.springframework.org/schema/beans
   http://www.springframework.org/schema/beans/spring-beans.xsd
   http://www.springframework.org/schema/context
   http://www.springframework.org/schema/context/spring-context.xsd
   http://www.springframework.org/schema/tx
   http://www.springframework.org/schema/tx/spring-tx.xsd">
<!-- 配置数据源 -->
<bean id="dataSource" class="org.apache.commons.dbcp2.BasicDataSource">
        <property name="driverClassName" value="com.mysql.jdbc.
        Driver" />
    <property name="url" value="jdbc:mysql://localhost:3306/
    springtest?characterEncoding=utf8" />
        <property name="username" value="root" />
        <property name="password" value="root" />
        <!-- 最大连接数 -->
        <property name="maxTotal" value="30"/>
        <!-- 最大空闲连接数 -->
        <property name="maxIdle" value="10"/>
        <!-- 初始化连接数 -->
        <property name="initialSize" value="5"/>
</bean>
<!-- 添加事务支持 -->
<bean id="txManager"
   class="org.springframework.jdbc.datasource.
   DataSourceTransactionManager">
   <property name="dataSource" ref="dataSource" />
</bean>
<!-- 开启事务注解 -->
<tx:annotation-driven transaction-manager="txManager" />
<!-- 配置 MyBatis 工厂，同时指定数据源，并与 MyBatis 完美整合 -->
<bean id="sqlSessionFactory" class="org.mybatis.spring.
SqlSessionFactoryBean">
   <property name="dataSource" ref="dataSource" />
   <!-- configLocation 的属性值为 MyBatis 的核心配置文件 -->
   <property name="configLocation" value="classpath:com/mybatis/
   mybatis-config.xml"/>
</bean>
<!-- Mapper 代理开发，使用 Spring 自动扫描 MyBatis 的接口并装配
（Spring 将指定包中所有接口自动装配为 MyBatis 的 Mapper 接口的实现类）-->
<bean class="org.mybatis.spring.mapper.MapperScannerConfigurer">
    <!-- mybatis-spring 组件的扫描器 -->
    <property name="basePackage" value="com.dao"/>
```

```
        <property name="sqlSessionFactoryBeanName" value="sqlSessionFactory"/>
    </bean>
    <!-- 指定需要扫描的包（包括子包），使注解生效。dao 包在 mybatis-spring 组件中已
    经扫描，这里不再需要扫描 -->
   <context:component-scan base-package="com.service"/>
</beans>
```

5）创建 MyBatis 的核心配置文件和 SQL 映射文件

在 src 目录下创建 com.mybatis 文件夹，并在该文件夹中创建 MyBatis 的核心配置文件 mybatis-config.xml。具体代码如下：

```
<?xml version="1.0" encoding="UTF-8" ?>
<!DOCTYPE configuration PUBLIC "-//mybatis.org//DTD Config 3.0//EN"
"http://mybatis.org/dtd/mybatis-3-config.dtd">
<configuration>
     <mappers><!-- 映射器告诉 MyBatis 到哪里去找映射文件 -->
       <mapper resource="com/mybatis/UserMapper.xml"/>
     </mappers>
</configuration>
```

在 com.mybatis 文件夹中创建 SQL 映射文件 UserMapper.xml。具体代码如下：

```
<!-- com.dao.UserDao 对应 Dao 接口 -->
<?xml version="1.0" encoding="UTF-8" ?>
<!DOCTYPE mapper PUBLIC "-//mybatis.org//DTD Mapper 3.0//EN"
"http://mybatis.org/dtd/mybatis-3-mapper.dtd">
<mapper namespace="com.dao.UserDao">
    <!-- 查询用户信息，id 的值对应 dao 的接口方法 -->
    <select id="selectUserByUname" resultType="com.po.MyUser"
    parameterType="com.po.MyUser">
        select * from user where 1=1
        <if test="uname !=null and uname!=''">
           and uname like concat('%',#{uname},'%')
        </if>
    </select>
</mapper>
```

❾ 发布并运行应用

首先将 ch19 应用发布到 Tomcat 服务器并启动 Tomcat 服务器，然后通过地址“http://localhost:8080/ch19/input.jsp”访问信息输入页面，运行结果如图 19.2 所示。

图 19.2　信息输入页面

在图 19.2 中输入用户名，例如“张”，然后单击“提交”按钮，显示如图 19.3 所示的

结果。

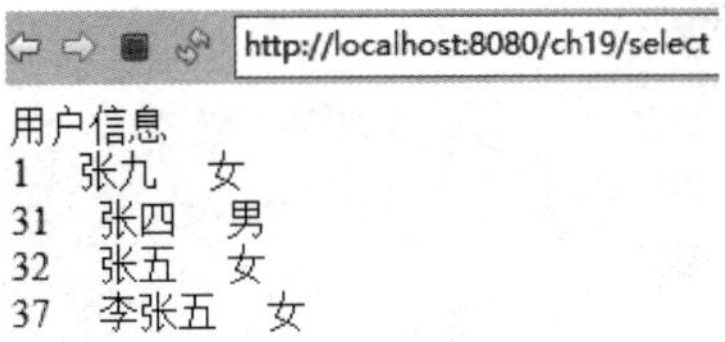

http://localhost:8080/ch19/select

用户信息
1　张九　女
31　张四　男
32　张五　女
37　李张五　女

图 19.3　用户列表页面

19.3　本章小结

本章首先简单介绍了 SSM 框架整合所需的 JAR 包，然后以 ch19 应用为例测试了 SSM 框架整合应用。

习题 19

1. 简述 SSM 框架整合所需的 JAR 包。
2. 简述 SSM 框架整合时 Spring MVC 核心配置文件和 web.xml 文件的配置信息。

电子商务平台的设计与实现

视频讲解

学习目的与要求

本章通过一个小型的电子商务平台讲述如何使用 SSM（Spring+Spring MVC+MyBatis）整合框架来开发一个 Web 应用。通过本章的学习，读者能够掌握 SSM 框架应用开发的流程、方法以及技术。

本章主要内容

- 系统设计;
- 数据库设计;
- 系统管理;
- 组件设计;
- 系统实现。

本章系统使用 SSM 框架实现各个模块，Web 服务器使用 Tomcat 9.0，数据库采用的是 MySQL 5.5，集成开发环境为 Eclipse IDE for Java EE Developers。

20.1 系统设计

电子商务平台分为两个子系统：一是后台管理子系统，一是电子商务子系统。下面分别说明这两个子系统的功能需求与模块划分。

20.1.1 系统功能需求

❶ 后台管理子系统

后台管理子系统要求管理员登录成功后才能对商品进行管理，包括添加商品、查询商品、修改商品以及删除商品。除商品管理以外，管理员还需要对商品类型、注册用户、用户的订单以及网站公告等进行管理。

❷ 电子商务子系统

1）非注册用户

非注册用户或未登录用户具有浏览首页、查看商品详情和查看公告的权限。

2）用户

成功登录的用户除具有未登录用户的权限以外，还具有购买商品、查看购物车、关注商品以及查看用户中心的权限。

20.1.2　系统模块划分

❶ 后台管理子系统

管理员登录成功后进入后台管理主页面（main.jsp），可以对商品及商品类型、注册用户、用户的订单以及网站公告进行管理。后台管理子系统的模块划分如图 20.1 所示。

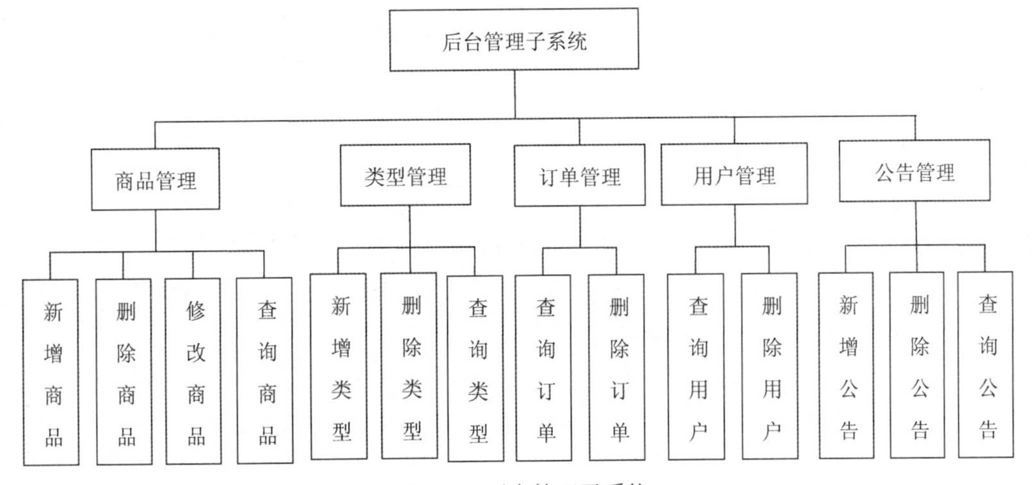

图 20.1　后台管理子系统

❷ 电子商务子系统

非注册用户只可以浏览商品、查看公告，不能购买商品、关注商品、查看购物车和查看用户中心。成功登录的用户可以完成电子商务子系统的所有功能，包括购买商品、支付等功能。电子商务子系统的模块划分如图 20.2 所示。

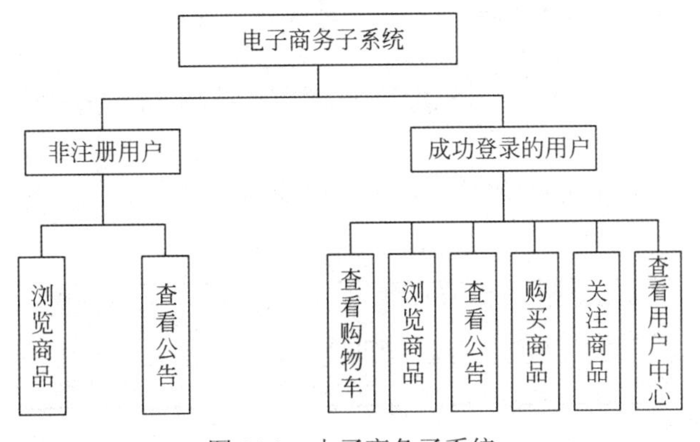

图 20.2　电子商务子系统

20.2 数据库设计

系统采用加载纯 Java 数据库驱动程序的方式连接 MySQL 5.5 数据库。在 MySQL 5.5 中创建数据库 shop，并在 shop 中创建 9 张与系统相关的数据表，即 ausertable、busertable、carttable、focustable、goodstable、goodstype、orderdetail、orderbasetable 和 noticetable。

20.2.1 数据库概念结构设计

根据系统设计与分析可以设计出如下数据结构。

（1）管理员：包括用户名和密码。管理员的用户名和密码由数据库管理员预设，不需要注册。

（2）用户：包括用户 ID、邮箱和密码。注册用户的邮箱不能相同，用户 ID 唯一。

（3）商品类型：包括类型 ID 和类型名称。商品类型由数据库管理员管理，包括新增和删除管理。

（4）商品：包括商品编号、名称、原价、现价、库存、图片以及类型。其中，商品编号唯一，类型与“商品类型”关联。

（5）购物车：包括购物车 ID、用户 ID、商品编号以及购买数量。其中，购物车 ID 唯一，用户 ID 与“用户”关联，商品编号与“商品”关联。

（6）关注商品：包括 ID、用户 ID、商品编号以及关注时间。其中，ID 唯一，用户 ID 与“用户”关联，商品编号与“商品”关联。

（7）订单基础信息：包括订单编号、用户 ID、订单金额、订单状态以及下单时间。其中，订单编号唯一，用户 ID 与“用户”关联。

（8）订单详情：包括订单编号、商品编号以及购买数量。其中，订单编号与“订单基础信息”关联，商品编号与“商品”关联。

（9）公告：包括 ID、标题、内容以及公告时间。其中，ID 唯一。

根据以上数据结构，结合数据库设计的特点，可以画出如图 20.3 所示的数据库概念结构图。

20.2.2 数据库逻辑结构设计

将数据库概念结构图转换为 MySQL 数据库所支持的实际数据模型，即数据库的逻辑结构。

管理员信息表（ausertable）的设计如表 20.1 所示。

表 20.1 管理员信息表

字　段	含　义	类　型	长　度	是否为空
aname	用户名（PK）	varchar	50	no
apwd	密码	varchar	50	no

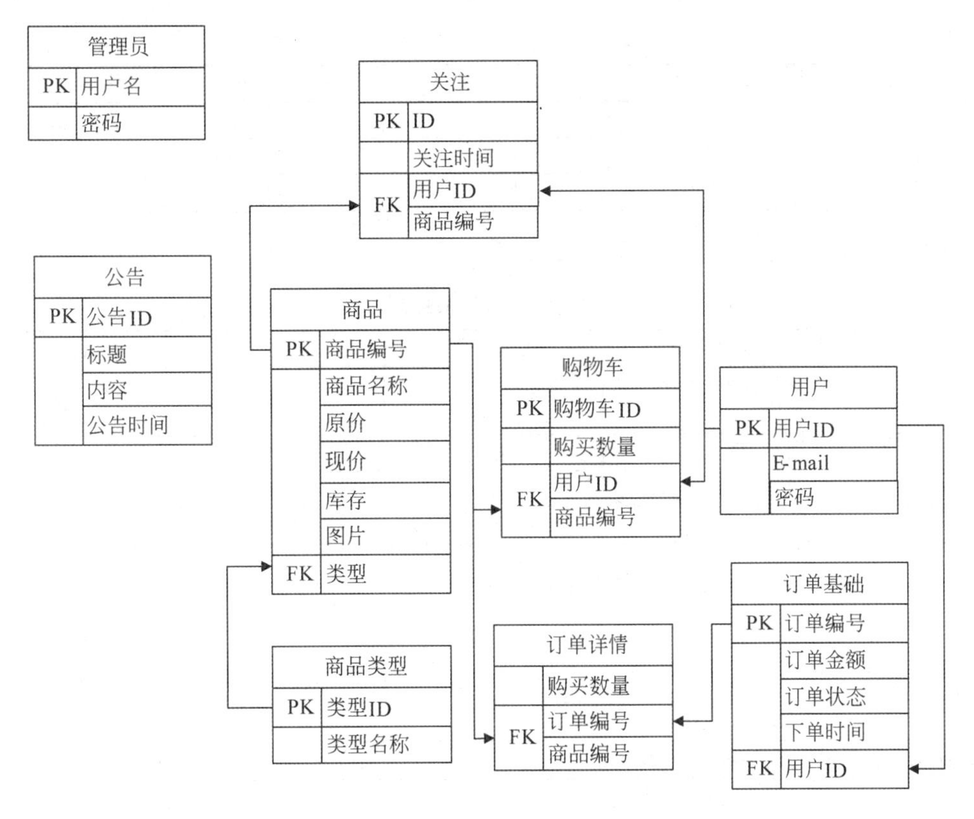

图 20.3　数据库概念结构图

用户信息表（busertable）的设计如表 20.2 所示。

表 20.2　用户信息表

字　段	含　义	类　型	长　度	是 否 为 空
id	用户 ID（PK 自增）	int	11	no
bemail	E-mail	varchar	50	no
bpwd	密码	varchar	50	no

商品类型表（goodstype）的设计如表 20.3 所示。

表 20.3　商品类型表

字　段	含　义	类　型	长　度	是 否 为 空
id	类型 ID（PK 自增）	int	11	no
typename	类型名称	varchar	50	no

商品信息表（goodstable）的设计如表 20.4 所示。

表 20.4　商品信息表

字　段	含　义	类　型	长　度	是 否 为 空
id	商品编号(PK 自增)	int	11	no
gname	商品名称	varchar	50	no
goprice	原价	double		no
grprice	现价	double		no

续表

字　段	含　义	类　型	长　度	是否为空
gstore	库存	int	11	no
gpicture	图片	varchar	50	
goodstype_id	类型（FK）	int	11	no

购物车表（carttable）的设计如表 20.5 所示。

表 20.5　购物车表

字　段	含　义	类　型	长　度	是否为空
id	购物车 ID（PK 自增）	int	11	no
busertable_id	用户 ID（FK）	int	11	no
goodstable_id	商品编号（FK）	int	11	no
shoppingnum	购买数量	int	11	no

商品关注表（focustable）的设计如表 20.6 所示。

表 20.6　关注表

字　段	含　义	类　型	长　度	是否为空
id	ID（PK 自增）	int	11	no
goodstable_id	商品编号（FK）	int	11	no
busertable_id	用户 ID（FK）	int	11	no
focustime	关注时间	datetime		no

订单基础表（orderbasetable）的设计如表 20.7 所示。

表 20.7　订单基础表

字　段	含　义	类　型	长　度	是否为空
id	订单编号(PK 自增)	int	11	no
busertable_id	用户 ID（FK）	int	11	no
amount	订单金额	double		no
status	订单状态	tinyint	4	no
orderdate	下单时间	datetime		no

订单详情表（orderdetail）的设计如表 20.8 所示。

表 20.8　订单详情表

字　段	含　义	类　型	长　度	是否为空
id	订单编号（PK 自增）	int	11	no
orderbasetable_id	订单编号（FK）	int	11	no
goodstable_id	商品编号（FK）	int	11	no
shoppingnum	购买数量	int	11	no

公告表（noticetable）的设计如表 20.9 所示。

表 20.9　公告表

字　段	含　义	类　型	长　度	是否为空
id	公告 ID（PK 自增）	int	11	no
ntitle	标题	varchar	100	no
ncontent	内容	varchar	500	no
ntime	公告时间	datetime		

20.2.3　创建数据表

根据 20.2.2 节的逻辑结构创建数据表。由于篇幅有限，创建数据表的代码请读者参考本书提供的源代码 shop.sql。

20.3　系统管理

20.3.1　导入相关的 JAR 包

新建一个 Web 应用 ch20，在 ch20 应用中开发本系统。系统的所有 JSP 页面尽量使用 EL 表达式和 JSTL 标签，采用纯 Java 数据库驱动程序连接 MySQL 5.5。除了将第 19 章中 ch19 应用的 JAR 包复制到 ch20/WebContent/WEB-INF/lib 目录下以外，还需要复制上传文件所需要的 JAR（commons-fileupload-1.3.1.jar 和 commons-io-2.4.jar）。

20.3.2　JSP 页面管理

系统由后台管理和电子商务两个子系统组成，为了方便管理，两个子系统的 JSP 页面、CSS 以及图片分开存放。在 WebContent/css/admin 目录下存放与后台管理子系统相关的 CSS；在 WebContent/images/admin 目录下存放与后台管理子系统相关的图片；在 WebContent/css/before 目录下存放与电子商务子系统相关的 CSS；在 WebContent/images/before 目录下存放与电子商务子系统相关的图片；在 WEB-INF/jsp/admin 目录下存放与后台管理子系统相关的 JSP 页面；在 WEB-INF/jsp/before 目录下存放与电子商务子系统相关的 JSP 页面。由于篇幅有限，本章仅附上 JSP 和 Java 文件的核心代码，具体代码请读者参考本书提供的源代码 ch20。

❶ 后台管理子系统

管理员在浏览器的地址栏中输入“http://localhost:8080/ch20/admin”访问登录页面，登录成功后进入后台管理主页面（main.jsp）。main.jsp 的运行效果如图 20.4 所示。

图 20.4　后台管理主页面

后台管理主页面 main.jsp 的核心代码如下：

```
…
<body>
    <div id="header">
        <br>
        <br>
        <h1>欢迎${auser.aname}进入后台管理系统！</h1>
    </div>
    <div id="navigator">
        <ul>
            <li><a>商品管理</a>
                <ul>
                    <li><a href="adminGoods/toAddGoods" target="center">
                    添加商品</a></li>
                    <li><a href="adminGoods/selectGoods?act=deleteSelect"
                    target="center">删除商品</a></li>
                    <li><a href="adminGoods/selectGoods?act=updateSelect"
                    target="center">修改商品</a></li>
                    <li><a href="adminGoods/selectGoods" target="center">
                    查询商品</a></li>
                </ul>
            </li>
```

```
            <li><a>类型管理</a>
                <ul>
                    <li><a href="adminType/toAddType" target="center">添加
                    类型</a></li>
                    <li><a href="adminType/toDeleteType" target="center">
                    删除类型</a></li>
                </ul>
            </li>
            <li><a>用户管理</a>
                <ul>
                    <li><a href="adminUser/userInfo" target="center">删除
                    用户</a></li>
                </ul>
            </li>
            <li><a>订单管理</a>
                <ul>
                    <li><a href="adminOrder/orderInfo" target="center">删
                    除订单</a></li>
                </ul>
            </li>
            <li><a>公告管理</a>
                <ul>
                    <li><a href="adminNotice/toAddNotice" target="center">
                    添加公告</a></li>
                    <li><a href="adminNotice/deleteNoticeSelect" target=
                    "center">删除公告</a></li>
                </ul>
            </li>
            <li><a href="exit">安全退出</a></li>
        </ul>

    </div>
    <div id="content">
        <iframe src="adminGoods/selectGoods" name="center" frameborder="0">
        </iframe>
    </div>
    <div id="footer">Copyright ©清华大学出版社</div>
</body>
…
```

❷ 电子商务子系统

注册用户或游客在浏览器的地址栏中输入“http://localhost:8080/ch20/before”可以访问电子商务子系统的首页（index.jsp）。index.jsp 的运行效果如图 20.5 所示。

图 20.5　电子商务子系统的首页

电子商务子系统的首页 index.jsp 的核心代码如下：

```
…
<div class="top10List clearfix">
    <c:forEach items="${salelist }" var="sg" varStatus="status">
        <ul class="clearfix">
            <img class="iteration" src="images/before/top_${status.index+
            1 }.gif" />
            <li class="topimg">
                <a href="goodsDetail?id=${sg.id }">
                <img class="samllimg" alt="" src="logos/${sg.gpicture}" />
                </a>
            </li>
            <li class="iteration1">
                <a href="goodsDetail?id=${sg.id }">${sg.gname }</a><br />
                售价：<font class="f1">￥${sg.grprice }元</font><br />
            </li>
        </ul>
    </c:forEach>
</div>
…
<div class="top10List clearfix">
        <c:forEach items="${focuslist }" var="sg" varStatus="status">
                <ul class="clearfix">
```

```
                        <img class="iteration" src="images/before/top_$
                        {status.index+1 }.gif" />
                    <li class="topimg">
                        <a href="goodsDetail?id=${sg.id }">
                        <img class="samllimg" alt="" src="logos/
                        ${sg.gpicture}" /></a>
                    </li>
                    <li class="iteration1">
                        <a href="goodsDetail?id=${sg.id }">${sg.gname }
                        </a><br />
                            售价：<font class="f1">￥${sg.grprice }元
                            </font><br />
                    </li>
                </ul>
        </c:forEach>
</div>
…
<div class="post_list ared">
    <ul>
    <c:forEach items="${noticelist}" var="nt">
    <li><a href="javascript:openNotice('/ch20/selectANotice?id=
    ${nt.id }');">${nt.ntitle }</a></li>
    </c:forEach>
</ul>
</div>
…
<div class="itemgood_nr clearfix">
    <ul>
    <c:forEach items="${lastedlist }" var="sg">
        <li>
            <div>
                <p class="pic">
                    <a href="goodsDetail?id=${sg.id }">
                    <img src="logos/${sg.gpicture}" /></a>
                </p>
                <p class="wz">
                   <strong><a href="goodsDetail?id=${sg.id }">${sg.gname }
                    </a></strong>
                    <em>现价:<span>￥${sg.grprice}</span></em>
                </p>
            </div>
        </li>
```

```
        </c:forEach>
    </ul>
</div>
...
```

20.3.3 应用的目录结构

ch20 应用的目录结构如图 20.6 所示。

```
src
  com.controller.admin
  com.controller.before
  com.dao
  com.exception
  com.mybatis
  com.mybatis.admin
  com.mybatis.before
  com.po
  com.service.admin
  com.service.before
  com.util
  applicationContext.xml
  log4j.properties
Libraries
JavaScript Resources
build
WebContent
  css
  images
  META-INF
  WEB-INF
    jsp
      admin
      before
      error
    lib
    springmvc-servlet.xml
    web.xml
```

图 20.6 目录结构

❶ controller 包

系统的控制器类都在该包中，与后台管理相关的控制器类在 admin 子包中，与电子商务相关的控制器类在 before 子包中。

❷ dao 包

dao 包中存放的 Java 接口程序用于实现数据库的持久化操作。每个 dao 的接口方法与 SQL 映射文件中的 id 相同。

❸ exception 包

该包中的异常类有 3 个，其中，AdminLoginNoException 处理管理员未登录异常；UserLoginNoException 处理前台用户未登录异常；MyExceptionHandler 对系统进行统一异常处理，包括管理员未登录异常、前台用户未登录异常以及一些未知异常。

❹ mybatis 包

MyBatis 的核心配置文件 mybatis-config.xml 和 SQL 映射文件在该包中。

❺ po 包

持久化类存放在该包中。

❻ service 包

service 包中有两个子包，即 admin 和 before，admin 子包存放后台管理相关业务层的接口与实现类；before 子包存放电子商务相关业务层的接口与实现类。

❼ util 包

util 包中存放的是系统的工具类，包括获取时间字符串方法以及获取前台用户登录 ID 方法。

20.3.4 配置文件管理

系统配置文件共分为五大类，即 MyBatis 的核心配置文件 mybatis-config.xml（在 mybatis 包中）、Spring 的核心配置文件 applicationContext.xml、MyBatis 的日志记录文件 log4j.properties、Spring MVC 的核心配置文件 springmvc-servlet.xml 以及 Web 应用的配置文件 web.xml。本节具体介绍如下：

❶ mybatis-config.xml

该配置文件配置了 PO 类的别名，并指定了 SQL 映射文件的位置。具体代码如下：

```
<?xml version="1.0" encoding="UTF-8" ?>
<!DOCTYPE configuration
PUBLIC "-//mybatis.org//DTD Config 3.0//EN"
"http://mybatis.org/dtd/mybatis-3-config.dtd">
<configuration>
    <typeAliases>
        <typeAlias alias="Auser" type="com.po.Auser"/>
        <typeAlias alias="Goods" type="com.po.Goods"/>
        <typeAlias alias="GoodsType" type="com.po.GoodsType"/>
        <typeAlias alias="Notice" type="com.po.Notice"/>
        <typeAlias alias="Buser" type="com.po.Buser"/>
        <typeAlias alias="Order" type="com.po.Order"/>
    </typeAliases>
    <mappers>
        <mapper resource="com/mybatis/admin/AdminGoodsMapper.xml"/>
        <mapper resource="com/mybatis/admin/AdminMapper.xml"/>
        <mapper resource="com/mybatis/admin/AdminTypeMapper.xml"/>
        <mapper resource="com/mybatis/admin/AdminNoticeMapper.xml"/>
        <mapper resource="com/mybatis/admin/AdminUserMapper.xml"/>
        <mapper resource="com/mybatis/admin/AdminOrderMapper.xml"/>
        <mapper resource="com/mybatis/before/IndexMapper.xml"/>
        <mapper resource="com/mybatis/before/UserMapper.xml"/>
        <mapper resource="com/mybatis/before/CartMapper.xml"/>
        <mapper resource="com/mybatis/before/OrderMapper.xml"/>
        <mapper resource="com/mybatis/before/UserCenterMapper.xml"/>
    </mappers>
</configuration>
```

❷ applicationContext.xml

该配置文件配置了数据源、事务管理、MyBatis 工厂（与 MyBatis 整合）、扫描包等内容。具体代码如下：

```
<?xml version="1.0" encoding="UTF-8"?>
<beans xmlns="http://www.springframework.org/schema/beans"
    xmlns:xsi="http://www.w3.org/2001/XMLSchema-instance"
    xmlns:context="http://www.springframework.org/schema/context"
    xmlns:tx="http://www.springframework.org/schema/tx"
    xsi:schemaLocation="http://www.springframework.org/schema/beans
        http://www.springframework.org/schema/beans/spring-beans.xsd
        http://www.springframework.org/schema/context
        http://www.springframework.org/schema/context/spring-context.xsd
```

```
    http://www.springframework.org/schema/tx
    http://www.springframework.org/schema/tx/spring-tx.xsd">
<!-- 配置数据源 -->
<bean id="dataSource" class="org.apache.commons.dbcp2.BasicDataSource">
        <property name="driverClassName" value="com.mysql.jdbc.Driver" />
    <property name="url" value="jdbc:mysql://localhost:3306/shop?
    characterEncoding=utf8" />
        <property name="username" value="root" />
        <property name="password" value="root" />
        <!-- 最大连接数 -->
        <property name="maxTotal" value="30"/>
        <!-- 最大空闲连接数 -->
        <property name="maxIdle" value="10"/>
        <!-- 初始化连接数 -->
        <property name="initialSize" value="5"/>
</bean>
<!-- 添加事务支持 -->
<bean id="txManager"
    class="org.springframework.jdbc.datasource.
    DataSourceTransactionManager">
    <property name="dataSource" ref="dataSource" />
</bean>
<!-- 开启事务注解 -->
<tx:annotation-driven transaction-manager="txManager" />
<!-- 配置 MyBatis 工厂，同时指定数据源，并与 MyBatis 完美整合 -->
<bean id="sqlSessionFactory" class="org.mybatis.spring.
SqlSessionFactoryBean">
    <property name="dataSource" ref="dataSource" />
    <!-- configLocation 的属性值为 MyBatis 的核心配置文件 -->
    <property name="configLocation" value="classpath:com/mybatis/
    mybatis-config.xml"/>
</bean>
<!-- Mapper 代理开发，使用 Spring 自动扫描 MyBatis 的接口并装配
（Spring 将指定包中所有被@Mapper 注解标注的接口自动装配为 MyBatis 的映射接口） -->
<bean class="org.mybatis.spring.mapper.MapperScannerConfigurer">
    <!-- mybatis-spring 组件的扫描器，必须写全 dao 的包名，且只能扫描一个 dao
    包 -->
    <property name="basePackage" value="com.dao"/>
    <property name="sqlSessionFactoryBeanName" value=
    "sqlSessionFactory"/>
</bean>
<!-- 指定需要扫描的包（包括子包），使注解生效。dao 包在 mybatis-spring 组件中已
```

```
    经扫描，这里不再需要扫描 -->
    <context:component-scan base-package="com.service"/>
</beans>
```

❸ **log4j.properties**

MyBatis 默认使用 log4j 输出日志信息，如果开发者需要查看控制台输出的 SQL 语句，那么需要在 classpath 路径下配置其日志文件。在应用的 src 目录下创建 log4j.properties 文件，其内容如下：

```
# Global logging configuration
log4j.rootLogger=ERROR, stdout
# MyBatis logging configuration...
log4j.logger.com.dao=DEBUG
# Console output...
log4j.appender.stdout=org.apache.log4j.ConsoleAppender
log4j.appender.stdout.layout=org.apache.log4j.PatternLayout
log4j.appender.stdout.layout.ConversionPattern=%5p [%t]-%m%n
```

❹ **springmvc-servlet.xml**

该配置文件配置了控制层的包扫描、静态资源处理、视图解析器、文件上传以及统一异常处理。具体代码如下：

```
<?xml version="1.0" encoding="UTF-8"?>
<beans xmlns="http://www.springframework.org/schema/beans"
    xmlns:xsi="http://www.w3.org/2001/XMLSchema-instance"
    xmlns:mvc="http://www.springframework.org/schema/mvc"
    xmlns:p="http://www.springframework.org/schema/p"
    xmlns:context="http://www.springframework.org/schema/context"
    xsi:schemaLocation="
    http://www.springframework.org/schema/beans
    http://www.springframework.org/schema/beans/spring-beans.xsd
    http://www.springframework.org/schema/context
    http://www.springframework.org/schema/context/spring-context.xsd
    http://www.springframework.org/schema/mvc
    http://www.springframework.org/schema/mvc/spring-mvc.xsd">
    <!-- 使用扫描机制扫描包 -->
    <context:component-scan base-package="com.controller" />
    <mvc:annotation-driven />
    <!-- 静态资源需要单独处理，不需要 dispatcher servlet -->
    <mvc:resources location="/css/" mapping="/css/**"></mvc:resources>
    <mvc:resources location="/images/" mapping="/images/**"></mvc:resources>
    <!-- 查看图片时 logos 文件夹不需要 dispatcher servlet -->
    <mvc:resources location="/logos/" mapping="/logos/**"></mvc:resources>
    <!-- 配置视图解析器 -->
```

```
    <bean
        class="org.springframework.web.servlet.view.InternalResourceViewResolver"
        id="internalResourceViewResolver">
        <!-- 前缀 -->
        <property name="prefix" value="/WEB-INF/jsp/" />
        <!-- 后缀 -->
        <property name="suffix" value=".jsp" />
    </bean>
    <!-- 配置 MultipartResolver 用于文件上传，使用 Spring 的
    CommonsMultipartResolver -->
    <bean id="multipartResolver" class="org.springframework.web.multipart.
    commons.CommonsMultipartResolver"
       p:defaultEncoding="UTF-8"
       p:maxUploadSize="5400000"
       p:uploadTempDir="fileUpload/temp" >
    </bean>
    <!-- 托管 MyExceptionHandler -->
    <bean class="com.exception.MyExceptionHandler"/>
</beans>
```

❺ **web.xml**

该配置文件配置了 ApplicationContext 容器、Spring MVC 的 DispatcherServlet 以及字符编码过滤器。具体代码如下：

```
<?xml version="1.0" encoding="UTF-8"?>
<web-app xmlns:xsi="http://www.w3.org/2001/XMLSchema-instance"
    xmlns="http://xmlns.jcp.org/xml/ns/javaee"
    xsi:schemaLocation="http://xmlns.jcp.org/xml/ns/javaee
http://xmlns.jcp.org/xml/ns/javaee/web-app_3_1.xsd"
    id="WebApp_ID" version="3.1">
    <!-- 实例化 ApplicationContext 容器 -->
    <context-param>
        <!-- 加载 src 目录下的 applicationContext.xml 文件 -->
        <param-name>contextConfigLocation</param-name>
        <param-value>
            classpath:applicationContext.xml
        </param-value>
    </context-param>
    <!-- 指定以 ContextLoaderListener 方式启动 Spring 容器 -->
    <listener>
        <listener-class>
            org.springframework.web.context.ContextLoaderListener
        </listener-class>
```

```
    </listener>
    <!-- 配置 DispatcherServlet -->
    <servlet>
        <servlet-name>springmvc</servlet-name>
        <servlet-class>org.springframework.web.servlet.DispatcherServlet
        </servlet-class>
        <load-on-startup>1</load-on-startup>
    </servlet>
    <servlet-mapping>
        <servlet-name>springmvc</servlet-name>
        <url-pattern>/</url-pattern>
    </servlet-mapping>
    <!-- 避免中文乱码 -->
    <filter>
    <filter-name>characterEncodingFilter</filter-name>
    <filter-class>org.springframework.web.filter.CharacterEncodingFilter
    </filter-class>
    <init-param>
            <param-name>encoding</param-name>
            <param-value>UTF-8</param-value>
    </init-param>
    <init-param>
            <param-name>forceEncoding</param-name>
            <param-value>true</param-value>
    </init-param>
  </filter>
  <filter-mapping>
    <filter-name>characterEncodingFilter</filter-name>
    <url-pattern>/*</url-pattern>
  </filter-mapping>
</web-app>
```

20.4　组件设计

本系统的组件包括管理员登录权限验证控制器、前台用户登录权限验证控制器、验证码、统一异常处理以及工具类。

20.4.1　管理员登录权限验证

从系统分析得知，管理员成功登录后才能管理商品、商品类型、用户、订单以及公告等功能模块，因此本系统需要对这些功能模块的操作进行管理员登录权限控制。在com.controller.admin 包中创建了 BaseController 控制器类，在该类中有一个@ModelAttribute 注解的方法 isLogin，isLogin 方法的功能是判断管理员是否已成功登录，需要进行管理员登录权限控制的控制器类继承 BaseController 类，因为@ModelAttribute 注解的方法首先被控制器执行。BaseController 控制器类的代码如下：

```
package com.controller.admin;
import javax.servlet.http.HttpServletRequest;
import javax.servlet.http.HttpSession;
import org.springframework.stereotype.Controller;
import org.springframework.web.bind.annotation.ModelAttribute;
import com.exception.AdminLoginNoException;
@Controller
public class BaseController {
    /**
     * 登录权限控制，在处理方法执行前执行该方法
     * @throws AdminLoginNoException
     */
    @ModelAttribute
    public void isLogin(HttpSession session, HttpServletRequest request)
    throws AdminLoginNoException{
      if(session.getAttribute("auser")==null){
          throw new AdminLoginNoException("没有登录");
      }
    }
}
```

20.4.2 前台用户登录权限验证

从系统分析得知，用户成功登录后才能购买商品、关注商品、查看购物车以及用户中心。与管理员登录权限验证同理，在 com.controller.before 包中创建了 BaseBeforeController 控制器类，在该类中有一个@ModelAttribute 注解的方法 isLogin，isLogin 方法的功能是判断前台用户是否已成功登录，需要进行前台用户登录权限控制的控制器类继承 BaseBeforeController 类，因为@ModelAttribute 注解的方法首先被控制器执行。BaseBeforeController 控制器类的代码如下：

```
package com.controller.before;
import javax.servlet.http.HttpServletRequest;
import javax.servlet.http.HttpSession;
import org.springframework.stereotype.Controller;
import org.springframework.web.bind.annotation.ModelAttribute;
```

```
import com.exception.UserLoginNoException;
@Controller
public class BaseBeforeController {
    /**
     * 前台用户登录权限控制，在处理方法执行前执行该方法
     * @throws UserLoginNoException
     */
    @ModelAttribute
    public void isLogin(HttpSession session, HttpServletRequest request)
    throws UserLoginNoException {
      if(session.getAttribute("bruser")==null){
          throw new UserLoginNoException("没有登录");
      }
    }
}
```

20.4.3　验证码

本系统验证码的使用步骤如下：

❶ 创建产生验证码的控制器类

在 com.controller.before 包中创建产生验证码的控制器类 ValidateCodeController，具体代码如下：

```
package com.controller.before;
import java.awt.Color;
import java.awt.Font;
import java.awt.Graphics;
import java.awt.image.BufferedImage;
import java.io.IOException;
import java.io.OutputStream;
import java.util.Random;
import javax.imageio.ImageIO;
import javax.servlet.ServletException;
import javax.servlet.http.HttpServletRequest;
import javax.servlet.http.HttpServletResponse;
import javax.servlet.http.HttpSession;
import org.springframework.stereotype.Controller;
import org.springframework.web.bind.annotation.RequestMapping;
/**
 * 验证码
 */
@Controller
public class ValidateCodeController {
```

```
private char code[]={ 'a', 'b', 'c', 'd', 'e', 'f', 'g', 'h', 'i', 'j',
        'k', 'm', 'n', 'p', 'q', 'r', 's', 't', 'u', 'v', 'w', 'x', 'y',
        'z', 'A', 'B', 'C', 'D', 'E', 'F', 'G', 'H', 'J', 'K', 'L', 'M',
        'N', 'P', 'Q', 'R', 'S', 'T', 'U', 'V', 'W', 'X', 'Y', 'Z', '2',
        '3', '4', '5', '6', '7', '8', '9' };
private static final int WIDTH=50;
private static final int HEIGHT=20;
private static final int LENGTH=4;
@RequestMapping("/validateCode")
public void validateCode(HttpServletRequest request,
        HttpServletResponse response) throws ServletException, IOException {
    //TODO Auto-generated method stub
    //设置响应报头信息
    response.setHeader("Pragma", "No-cache");
    response.setHeader("Cache-Control", "no-cache");
    response.setDateHeader("Expires", 0);
    //设置响应的MIME类型
    response.setContentType("image/jpeg");
    BufferedImage image=new BufferedImage(WIDTH, HEIGHT,
            BufferedImage.TYPE_INT_RGB);
    Font mFont=new Font("Arial", Font.TRUETYPE_FONT, 18);
    Graphics g=image.getGraphics();
    Random rd=new Random();
    //设置背景颜色
    g.setColor(new Color(rd.nextInt(55)+200, rd.nextInt(55)+200,
    rd.nextInt(55)+200));
    g.fillRect(0, 0, WIDTH, HEIGHT);
    //设置字体
    g.setFont(mFont);
    //画边框
    g.setColor(Color.black);
    g.drawRect(0, 0, WIDTH-1, HEIGHT-1);
    //随机产生的验证码
    String result="";
    for (int i=0; i < LENGTH; ++i) {
        result+=code[rd.nextInt(code.length)];
    }
    HttpSession se=request.getSession();
    se.setAttribute("code", result);
    //画验证码
    for (int i=0; i<result.length(); i++) {
        g.setColor(new Color(rd.nextInt(200), rd.nextInt(200),
                rd.nextInt(200)));
        g.drawString(result.charAt(i)+"", 12 * i+1, 16);
    }
```

```
            //随机产生两个干扰线
            for (int i=0; i<2; i++) {
                g.setColor(new Color(rd.nextInt(200), rd.nextInt(200),
                        rd.nextInt(200)));
                int x1=rd.nextInt(WIDTH);
                int x2=rd.nextInt(WIDTH);
                int y1=rd.nextInt(HEIGHT);
                int y2=rd.nextInt(HEIGHT);
                g.drawLine(x1, y1, x2, y2);
            }
            //释放图形资源
            g.dispose();
            try {
                OutputStream os=response.getOutputStream();
                //输出图像到页面
                ImageIO.write(image, "JPEG", os);
            } catch (IOException e) {
                e.printStackTrace();
            }
        }
    }
```

❷ 使用验证码

在需要验证码的 JSP 页面中调用产生验证码的控制器显示验证码，示例代码片段如下：

```
<tr>
    <td>
        <img id="code" src="validateCode"/>
    </td>
    <td class="ared">
        <a href="javascript:refreshCode();"><font color="blue">看不清，换一
        个！</font></a>
    </td>
</tr>
```

20.4.4　统一异常处理

系统对管理员未登录异常、前台用户未登录异常以及程序未知异常进行了统一异常处理。具体步骤如下：

❶ 创建未登录自定义异常

创建管理员未登录异常 AdminLoginNoException 和前台用户未登录异常 UserLoginNoException，代码略。

❷ 创建 HandlerExceptionResolver 的实现类

ch20 使用实现 HandlerExceptionResolver 接口的方式进行统一异常处理，具体代码

如下：

```
package com.exception;
import java.util.HashMap;
import java.util.Map;
import javax.servlet.http.HttpServletRequest;
import javax.servlet.http.HttpServletResponse;
import org.springframework.web.servlet.HandlerExceptionResolver;
import org.springframework.web.servlet.ModelAndView;
import com.po.Auser;
import com.po.Buser;
public class MyExceptionHandler implements HandlerExceptionResolver {
    @Override
public ModelAndView resolveException(HttpServletRequest arg0,
HttpServletResponse arg1, Object arg2,
           Exception arg3) {
        Map<String, Object> model=new HashMap<String, Object>();
       model.put("ex", arg3);
       //根据不同错误转向不同页面
       if(arg3 instanceof AdminLoginNoException){
       //登录页面需要 auser 对象
        arg0.setAttribute("auser", new Auser());
        arg0.setAttribute("msg", "没有登录，请登录！");
        return new ModelAndView("/admin/login", model);
       } else if(arg3 instanceof UserLoginNoException){
        arg0.setAttribute("buser", new Buser());
        arg0.setAttribute("msg", "没有登录，请登录！");
        return new ModelAndView("/before/login", model);
      }else{
        return new ModelAndView("/error/error", model);
      }
    }
}
```

❸ 托管 MyExceptionHandler

在 Spring MVC 的配置文件中使用<bean>元素将 MyExceptionHandler 托管，具体代码如下：

```
<bean class="com.exception.MyExceptionHandler"/>
```

20.4.5 工具类

本系统使用的工具类是 MyUtil，在该类中有两个工具方法：一是获得时间字符串，一是获得前台登录用户的 ID。具体代码如下：

```
package com.util;
import java.text.SimpleDateFormat;
import java.util.Date;
```

```
import javax.servlet.http.HttpSession;
import com.po.Buser;
public class MyUtil {
    /**
     * 获得时间字符串
     */
    public static String getStringID(){
        String id=null;
        Date date=new Date();
        SimpleDateFormat sdf=new SimpleDateFormat("yyyyMMddHHmmssSSS");
        id=sdf.format(date);
        return id;
    }
    /**
     * 获得用户 ID
     */
    public static Integer getUserId(HttpSession session) {
        Buser ruser=(Buser)session.getAttribute("bruser");
        return ruser.getId();
    }
}
```

20.5　后台管理子系统的实现

管理员登录成功后可以对商品及商品类型、注册用户、用户的订单以及网站公告进行管理。本节将详细讲解管理员的功能实现。

20.5.1　管理员登录

在管理员输入用户名和密码后，系统将对管理员的用户名和密码进行验证。如果用户名和密码都正确，则成功登录，进入系统管理主页面（main.jsp）；如果用户名或密码有误，则提示错误。实现步骤如下：

❶ 编写视图

login.jsp 页面提供登录信息输入的界面，如图 20.7 所示。

图 20.7　管理员登录界面

在 WEB-INF/jsp/admin 目录下创建 login.jsp。该 JSP 页面的核心代码如下：

```
<body>
    <form:form action="admin/login" modelAttribute="auser" method="post">
    <table>
        <tr>
            <td colspan="2"><img src="images/admin/login.gif"></td>
        </tr>
        <tr>
            <td>姓名：</td>
            <td>
                <form:input path="aname" cssClass="textSize"/>
            </td>
        </tr>
        <tr>
            <td>密码：</td>
            <td>
                <form:password path="apwd" cssClass="textSize" maxlength="20"/>
            </td>
        </tr>
        <tr>
            <td colspan="2">
                <input type="image" src="images/admin/ok.gif" onclick=
                "gogo()" >
                <input type="image" src="images/admin/cancel.gif" onclick=
                "cancel()" >
            </td>
        </tr>
    </table>
    </form:form>
    ${msg }
</body>
```

❷ **编写控制器层**

视图 Action 的请求路径为“admin/login”，系统根据请求路径和@RequestMapping 注解找到对应控制器类 com.controller.admin.AdminController 的 login 方法处理登录，在控制器类的 login 方法中调用 com.service.admin.AdminService 接口的 login 方法处理登录。登录成功后，首先将登录人信息存入 session；然后获得商品的类型并存入 session；最后进入 main.jsp 页面。若登录失败回到本页面。控制器层的相关代码如下：

```
@Autowired
private AdminService adminService;
@RequestMapping("/admin")
public String toLogin(@ModelAttribute Auser auser) {
    return "admin/login";
}
@RequestMapping("/admin/login")
public String login(@ModelAttribute Auser auser, Model model, HttpSession
```

```
session) {
    return adminService.login(auser, model, session);
}
```

❸ 编写 Service 层

Service 层由接口 com.service.admin.AdminService 和接口的实现类 com.service.admin.AdminServiceImpl 组成。Service 层是功能模块实现的核心，Service 层调用 Dao 层进行数据库操作。管理员登录的业务处理方法 login 的代码如下：

```
public String login(Auser auser, Model model, HttpSession session) {
        if(adminDao.login(auser) != null && adminDao.login(auser).size() > 0) {
            session.setAttribute("auser", auser);
            //添加商品与修改商品页面使用
            session.setAttribute("goodsType", adminTypeDao.
            selectGoodsType());
            return "admin/main";
        }
        model.addAttribute("msg", "用户名或密码错误！");
        return "admin/login";
}
```

❹ 编写 SQL 映射文件

在 SSM 框架中 Dao 层仅由@Mapper 注解的接口实现。Dao 层接口方法与 SQL 映射文件中 SQL 语句的 id 相同，因为 Dao 层只有接口方法，这里不再赘述。管理员登录的 SQL 映射文件为 AdminMapper.xml，实现的 SQL 语句如下：

```
<select id="login" resultType="Auser" parameterType="Auser">
    select * from ausertable where aname=#{aname} and apwd=#{apwd}
</select>
```

20.5.2　类型管理

类型管理分为添加类型和删除类型，如图 20.8 所示。

欢迎admin进入名片管理系统！

商品管理　类型管理　用户管理　订单管理　公告管理　安全退出

添加类型
删除类型

ID	名称	价格	库存	详情
12	衣X	100.0	500	详情
13	果	8.0	500	详情
14	樱桃	5.0	800	详情
15	微波炉	300.0	500	详情
16	电视99	9000.0	600	详情
17	孕装55	800.0	1000	详情
18	牛奶11	50.0	6000	详情
19	红酒99	5000.0	30000	详情
20	化妆品22	3600.0	5989	详情
21	aa	12.0	11	详情

共11条记录　共2页　第1页　下一页

图 20.8　类型管理

❶ 添加类型

添加类型的实现步骤如下：

1）编写视图

单击图 20.8 中的“添加类型”超链接（adminType/toAddType），打开如图 20.9 所示的添加页面。

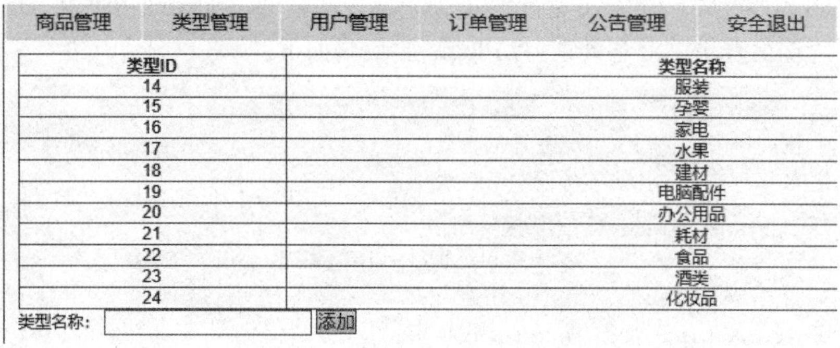

商品管理 类型管理 用户管理 订单管理 公告管理 安全退出

类型ID	类型名称
14	服装
15	孕婴
16	家电
17	水果
18	建材
19	电脑配件
20	办公用品
21	耗材
22	食品
23	酒类
24	化妆品

类型名称： 添加

图 20.9　添加类型

在打开“添加类型”页面时将查询出所有类型并显示在添加页面中。在 WEB-INF/jsp/admin 目录下创建添加类型页面 addType.jsp，该 JSP 页面的核心代码如下：

```
<body>
    <c:if test="${allTypes.size() == 0 }">
        您还没有类型。
    </c:if>
    <c:if test="${allTypes.size() != 0 }">
        <table border="1" bordercolor="PaleGreen" >
            <tr>
                <th width="200px">类型 ID</th>
                <th width="600px">类型名称</th>
            </tr>
            <c:forEach items="${allTypes }" var="goodsType">
                <tr>
                    <td>${goodsType.id }</td>
                    <td>${goodsType.typename }</td>
                </tr>
            </c:forEach>
        </table>
    </c:if>
    <form action="adminType/addType" method="post">
        类型名称：
        <input type="text" name="typename"/>
        <input type="submit" value="添加"/>
    </form>
</body>
```

2）编写控制器层

此功能共有两个处理请求，即“添加类型”超链接 adminType/toAddType 与视图 Action 的请求路径 adminType/addType。系统根据@RequestMapping 注解找到对应控制器类 com.controller.admin.AdminTypeController 的 toAddType 和 addType 方法处理请求。在控制器类的处理方法中调用 com.service.admin.AdminTypeService 接口的 toAddType 和 addType 方法处理业务。控制器层的相关代码如下：

```
@Autowired
private AdminTypeService adminTypeService;
/**
 * 到添加类型页面
 */
@RequestMapping("/toAddType")
public String toAddType(Model model) {
    return adminTypeService.toAddType(model);
}
/**
 * 添加类型
 */
@RequestMapping("/addType")
public String addType(String typename,Model model,HttpSession session) {
    return adminTypeService.addType(typename, model, session);
}
```

3）编写 Service 层

超链接 adminType/toAddType 的业务处理方法 toAddType 的代码如下：

```
public String toAddType(Model model) {
    model.addAttribute("allTypes", adminTypeDao.selectGoodsType());
    return "admin/addType";
}
```

添加类型 adminType/addType 的业务处理方法 addType 的代码如下：

```
public String addType(String typename, Model model, HttpSession session) {
    adminTypeDao.addType(typename);
    //添加商品与修改商品页面使用
    session.setAttribute("goodsType", adminTypeDao.selectGoodsType());
    return "forward:/adminType/toAddType";
}
```

4）编写 SQL 映射文件

实现超链接 adminType/toAddType 的 SQL 语句如下：

```
<select id="selectGoodsType" resultType="GoodsType" >
    select * from goodstype
</select>
```

实现添加类型 adminType/addType 的 SQL 语句如下：

```
<insert id="addType" parameterType="String">
    insert into goodstype (id, typename) values (null, #{typename})
</insert>
```

❷ 删除类型

删除类型的实现步骤如下：

1）编写视图

单击图 20.8 中的“删除类型”超链接（adminType/toDeleteType），打开如图 20.10 所示的删除页面。

商品管理　类型管理　用户管理　订单管理　公告管理　安全退出

类型ID	类型名称	删除操作
14	服装	删除
15	孕婴	删除
16	家电	删除
17	水果	删除
18	建材	删除
19	电脑配件	删除
20	办公用品	删除
21	耗材	删除
22	食品	删除
23	酒类	删除
24	化妆品	删除

图 20.10　删除类型

在 WEB-INF/jsp/admin 目录下创建删除类型页面 deleteType.jsp，该 JSP 页面的核心代码如下：

```
<body>
    <c:if test="${allTypes.size()==0 }">
        您还没有类型。
    </c:if>
    <c:if test="${allTypes.size()!=0 }">
        <table border="1" bordercolor="PaleGreen" >
            <tr>
                <th width="200px">类型 ID</th>
                <th width="300px">类型名称</th>
                <th width="300px">删除操作</th>
            </tr>
            <c:forEach items="${allTypes }" var="goodsType">
                <tr>
                    <td>${goodsType.id }</td>
                    <td>${goodsType.typename }</td>
                    <td><a href="javascript:checkDel
                    ('${goodsType.id }')">删除</a></td>
                </tr>
            </c:forEach>
            <tr>
```

```
                <td colspan="3">${msg }</td>
            </tr>
        </table>
    </c:if>
</body>
```

2）编写控制器层

此功能模块共有两个处理请求，即“删除类型”超链接 adminType/toDeleteType 与视图“删除”的请求路径 adminType/deleteType。系统根据@RequestMapping 注解找到对应控制器类 com.controller.admin.AdminTypeController 的 toDeleteType 和 deleteType 方法处理请求。在控制器类的处理方法中调用 com.service.admin.AdminTypeService 接口的 toDeleteType 和 deleteType 方法处理业务。控制器层的相关代码如下：

```
/**
 * 到删除页面
 */
@RequestMapping("/toDeleteType")
public String toDeleteType(Model model) {
    return adminTypeService.toDeleteType(model);
}
/**
 * 删除类型
 */
@RequestMapping("/deleteType")
public String deleteType(Integer id,Model model) {
    return adminTypeService.deleteType(id, model);
}
```

3）编写 Service 层

超链接 adminType/toDeleteType 的业务处理方法 toDeleteType 的代码如下：

```
public String toDeleteType(Model model) {
        model.addAttribute("allTypes", adminTypeDao.selectGoodsType());
        return "admin/deleteType";
}
```

删除 adminType/deleteType 的业务处理方法 deleteType 的代码如下：

```
public String deleteType(Integer id, Model model) {
    //类型有关联
    if(adminTypeDao.selectGoodsByType(id).size()>0) {
        model.addAttribute("msg", "类型有关联，不允许删除！");
        return "forward:/adminType/toDeleteType";
    }
    if(adminTypeDao.deleteType(id)>0)
        model.addAttribute("msg", "类型成功删除！");
    //回到删除页面
```

```
    return "forward:/adminType/toDeleteType";
}
```

4）编写 SQL 映射文件

实现超链接 adminType/toDeleteType 的 SQL 语句如下：

```
<select id="selectGoodsType"  resultType="GoodsType" >
    select * from goodstype
</select>
```

实现删除 adminType/deleteType 的 SQL 语句如下：

```
<delete id="deleteType" parameterType="Integer">
    delete from goodstype where id=#{id}
</delete>
<select id="selectGoodsByType" resultType="Goods" parameterType="Integer">
    select * from goodstable where goodstype_id=#{id}
</select>
```

20.5.3 添加商品

单击图 20.11 中的“添加商品”超链接，打开如图 20.12 所示的“添加商品”页面。

图 20.11　商品管理

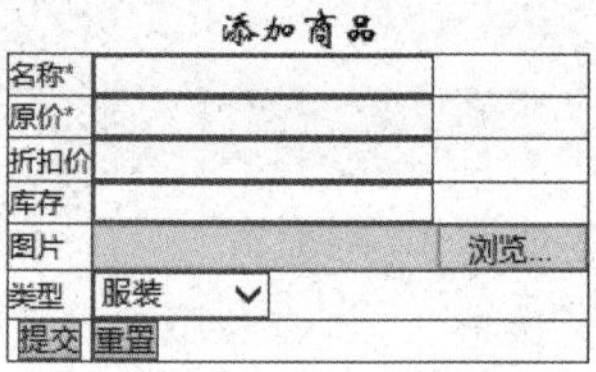

图 20.12　添加商品

添加商品的实现步骤如下：

❶ 编写视图

在 WEB-INF/jsp/admin 目录下创建添加商品页面 addGoods.jsp，该 JSP 页面的核心代码如下：

```
<body>
    <form:form action="adminGoods/addGoods" method="post" modelAttribute=
    "goods" enctype="multipart/form-data">
        <table border=1 style="border-collapse: collapse">
            <caption>
                <font size=4 face=华文新魏>添加商品</font>
            </caption>
            <tr>
                <td>名称<font color="red">*</font></td>
                <td>
```

```
                <form:input path="gname"/>
            </td>
        </tr>
        <tr>
            <td>原价<font color="red">*</font></td>
            <td>
                <form:input path="goprice"/>
            </td>
        </tr>
        <tr>
            <td>折扣价</td>
            <td>
                <form:input path="grprice"/>
            </td>
        </tr>
        <tr>
            <td>库存</td>
            <td>
                <form:input path="gstore"/>
            </td>
        </tr>
        <tr>
            <td>图片</td>
            <td>
                <input type="file" name="logoImage"/>
            </td>
        </tr>
        <tr>
            <td>类型</td>
            <td>
                <form:select path="goodstype_id">
            <form:options items="${goodsType }" itemLabel="typename"
            itemValue="id"/>
                </form:select>
            </td>
        </tr>
        <tr>
            <td align="center">
                <input type="submit" value="提交"/>
            </td>
            <td align="left">
                <input type="reset" value="重置"/>
            </td>
        </tr>
</table>
```

```
    </form:form>
</body>
```

❷ 编写控制器层

此功能模块共有两个处理请求，即“添加商品”超链接 adminGoods/toAddGoods 与视图“添加”的请求路径 adminGoods/addGoods。系统根据@RequestMapping 注解找到对应控制器类 com.controller.admin.AdminGoodsController 的 toAddGoods 和 addGoods 方法处理请求。在控制器类的处理方法中调用 com.service.admin.AdminGoodsService 接口的 toAddGoods 和 addGoods 方法处理业务。控制器层的相关代码如下：

```
/**
 * add 页面初始化
 */
@RequestMapping("/toAddGoods")
public String toAddGoods(Model model){
    model.addAttribute("goods", new Goods());
    return "admin/addGoods";
}
/**
 * 添加与修改
 */
@RequestMapping("/addGoods")
public String addGoods(@ModelAttribute Goods goods, HttpServletRequest
request, String updateAct){
    return adminGoodsService.addOrUpdateGoods(goods, request, updateAct);
}
```

❸ 编写 Service 层

添加商品的 Service 层的相关代码如下：

```
/**
 * 添加或更新
 */
@Override
public String addOrUpdateGoods(Goods goods, HttpServletRequest request,
String updateAct) {
    /*上传文件的保存位置"/logos"，该位置是指
    workspace\.metadata\.plugins\org.eclipse.wst.server.core\tmp0\wtpweb
    apps,
    发布后使用*/
    //防止文件名重名
    String newFileName="";
    String fileName=goods.getLogoImage().getOriginalFilename();
    //选择了文件
    if(fileName.length()>0){
        String realpath=request.getServletContext().getRealPath("logos");
```

```
            //实现文件上传
            String fileType=fileName.substring(fileName.lastIndexOf('.'));
            //防止文件名重名
            newFileName=MyUtil.getStringID()+fileType;
            goods.setGpicture(newFileName);
            File targetFile=new File(realpath, newFileName);
            if(!targetFile.exists()){
                targetFile.mkdirs();
            }
            //上传
            try {
            goods.getLogoImage().transferTo(targetFile);
            } catch (Exception e) {
                e.printStackTrace();
            }
        }
        //修改
        if("update".equals(updateAct)){  //updateAct 不能与 act 重名，因为使用了转发
            //修改到数据库
          if(adminGoodsDao.updateGoodsById(goods) > 0){
            return "forward:/adminGoods/selectGoods?act=updateSelect";
          }else{
            return "/adminGoods/updateAgoods";
          }
        }else{
            //保存到数据库
            if(adminGoodsDao.addGoods(goods) > 0){
                //转发到查询的 controller
                return "forward:/adminGoods/selectGoods";
            }else{
                return "card/addCard";
            }
        }
    }
```

❹ 编写 SQL 映射文件

添加商品的 SQL 语句如下：

```
<!-- 添加商品 -->
<insert id="addGoods" parameterType="Goods">
    insert into goodstable (id,gname,goprice,grprice,gstore,gpicture,
    goodstype_id)
    values (null, #{gname}, #{goprice}, #{grprice}, #{gstore}, #{gpicture},
    #{goodstype_id})
</insert>
```

20.5.4 查询商品

管理员登录成功后进入后台管理子系统的主页面，在主页面中初始显示查询商品页面。用户也可以通过单击图 20.11 中的“查询商品”超链接显示查询商品页面。查询页面的运行效果如图 20.13 所示。单击图 20.13 中的“详情”链接，显示如图 20.14 所示的详情页面。

商品管理 类型管理 用户管理 订单管理 公告管理 安全退出

ID	名称	价格	库存	详情
12	上衣X	100.0	500	详情
13	苹果	8.0	500	详情
14	樱桃	5.0	800	详情
15	微波炉	300.0	500	详情
16	电视99	9000.0	600	详情
17	孕装55	800.0	1000	详情
18	牛奶11	50.0	6000	详情
19	红酒99	5000.0	30000	详情
20	化妆品22	3600.0	5989	详情
21	aa	12.0	11	详情

共11条记录 共2页 第1页 下一页

图 20.13 查询商品

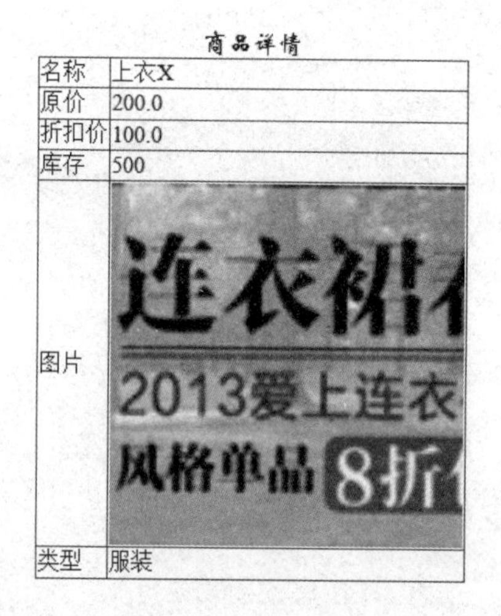

商品详情

名称	上衣X
原价	200.0
折扣价	100.0
库存	500
图片	
类型	服装

图 20.14 商品详情

查询商品的实现步骤如下：

❶ 编写视图

在 WEB-INF/jsp/admin 目录下创建查询商品页面 selectGoods.jsp，该 JSP 页面的核心代码如下：

```
<body>
    <c:if test="${allGoods.size()==0 }">
        您还没有商品。
    </c:if>
    <c:if test="${allGoods.size()!=0 }">
```

```
        <table border="1" bordercolor="PaleGreen">
            <tr>
                <th width="100px">ID</th>
                <th width="200px">名称</th>
                <th width="200px">价格</th>
                <th width="100px">库存</th>
                <th width="200px">详情</th>
            </tr>
            <c:forEach items="${allGoods }" var="goods">
                <tr onmousemove="changeColor(this)" onmouseout=
                "changeColor1(this)">
                    <td>${goods.id }</td>
                    <td>${goods.gname }</td>
                    <td>${goods.grprice }</td>
                    <td>${goods.gstore }</td>
        <td><a href="adminGoods/selectAGoods?id=${goods.id }" target=
        "_blank">详情</a></td>
                </tr>
            </c:forEach>
            <tr>
                <td colspan="5" align="right">
                       
           共${totalCount}条记录  共${totalPage}
        页  
                    第${pageCur}页  
                    <c:url var="url_pre" value="adminGoods/selectGoods">
                        <c:param name="pageCur" value="${pageCur-1 }"/>
                    </c:url>
                    <c:url var="url_next" value="adminGoods/selectGoods">
                        <c:param name="pageCur" value="${pageCur+1 }"/>
                    </c:url>
                    <!-- 第一页，没有上一页 -->
                    <c:if test="${pageCur != 1 }">
                        <a href="${url_pre}">上一页</a>   
                         
                    </c:if>
                    <!-- 最后一页，没有下一页 -->
                    <c:if test="${pageCur != totalPage && totalPage != 0}">
                        <a href="${url_next}">下一页</a>
                    </c:if>
                </td>
            </tr>
        </table>
    </c:if>
</body>
```

在 WEB-INF/jsp/admin 目录下创建商品详情页面 goodsDetail.jsp，该 JSP 页面的核心代码如下：

```
<body>
    <table border=1 style="border-collapse: collapse">
        <caption>
            <font size=4 face=华文新魏>商品详情</font>
        </caption>
        <tr>
            <td>名称</td>
            <td>
                ${goods.gname }
            </td>
        </tr>
        <tr>
            <td>原价</td>
            <td>
                ${goods.goprice }
            </td>
        </tr>
        <tr>
            <td>折扣价</td>
            <td>
                ${goods.grprice }
            </td>
        </tr>
        <tr>
            <td>库存</td>
            <td>
                ${goods.gstore }
            </td>
        </tr>
        <tr>
            <td>图片</td>
            <td>
                <c:if test="${goods.gpicture != '' }">
                    <img alt="" width="250" height="250"
                    src="logos/${goods.gpicture}"/>
                </c:if>
            </td>
        </tr>
        <tr>
            <td>类型</td>
            <td>
                ${goods.typename }
```

```
            </td>
        </tr>
    </table>
</body>
```

❷ 编写控制器层

此功能模块共有两个处理请求，即 adminGoods/selectGoods 和 adminGoods/selectAGoods。系统根据@RequestMapping 注解找到对应控制器类 com.controller.admin.AdminGoodsController 的 selectGoods 和 selectAGoods 方法处理请求。在控制器类的处理方法中调用 com.service.admin.AdminGoodsService 接口的 selectGoods 和 selectAGoods 方法处理业务。控制器层的相关代码如下：

```
@RequestMapping("/selectGoods")
public String selectGoods(Model model, Integer pageCur, String act) {
    return adminGoodsService.selectGoods(model, pageCur, act);
}
/**
 * 查询一个名片
 */
@RequestMapping("/selectAGoods")
public String selectAGoods(Model model, Integer id, String act){
    return adminGoodsService.selectAGoods(model, id, act);
}
```

❸ 编写 Service 层

查询商品和查看详情的 Service 层的相关代码如下：

```
/**
 * 查询商品
 */
@Override
public String selectGoods(Model model, Integer pageCur, String act) {
    List<Goods> allGoods=adminGoodsDao.selectGoods();
    int temp=allGoods.size();
    model.addAttribute("totalCount", temp);
    int totalPage=0;
    if (temp==0) {
        totalPage=0;  //总页数
    } else {
        //返回大于或者等于指定表达式的最小整数
        totalPage=(int) Math.ceil((double) temp/10);
    }
    if (pageCur==null) {
        pageCur=1;
    }
    if ((pageCur-1)*10>temp) {
```

```
            pageCur=pageCur-1;
        }
        //分页查询
        Map<String, Object> map=new HashMap<String, Object>();
        map.put("startIndex", (pageCur-1)*10);  //起始位置
        map.put("perPageSize", 10);  //每页10个
        allGoods=adminGoodsDao.selectAllGoodsByPage(map);
        model.addAttribute("allGoods", allGoods);
        model.addAttribute("totalPage", totalPage);
        model.addAttribute("pageCur", pageCur);
        //删除查询
        if("deleteSelect".equals(act)){
            return "admin/deleteSelectGoods";
        }
        //修改查询
        else if("updateSelect".equals(act)){
            return "admin/updateSelectGoods";
        }else{
            return "admin/selectGoods";
        }
    }
    /**
     * 查询一个商品
     */
    @Override
    public String selectAGoods(Model model, Integer id, String act) {
        Goods agoods=adminGoodsDao.selectGoodsById(id);
        model.addAttribute("goods", agoods);
        //修改页面
        if("updateAgoods".equals(act)){
            return "admin/updateAgoods";
        }
        //详情页面
        return "admin/goodsDetail";
    }
```

❹ 编写 SQL 映射文件

查询商品和查看详情的 SQL 语句如下：

```
<!-- 查询商品 -->
<select id="selectGoods" resultType="Goods">
    select * from goodstable
</select>
<!-- 分页查询商品 -->
<select id="selectAllGoodsByPage" resultType="Goods" parameterType=
"map">
```

```
    select * from goodstable order by id limit #{startIndex},
#{perPageSize}
</select>
<!-- 根据 id 查询一个商品 -->
<select id="selectGoodsById" resultType="Goods" parameterType="Integer">
select gt.*,gy.typename from goodstable gt,goodstype gy where gt.id=#{id}
and gt.goodstype_id=gy.id
</select>
```

20.5.5　修改商品

单击图 20.11 中的“修改商品”超链接（adminGoods/selectGoods?act=updateSelect），打开修改查询页面 updateSelectGoods.jsp，如图 20.15 所示。

单击图 20.15 中的“修改”超链接（adminGoods/selectAGoods?id=${goods.id}&act=updateAgoods），打开修改商品信息页面 updateAgoods.jsp，如图 20.16 所示。在图 20.16 中输入要修改的信息后单击“提交”按钮，将商品信息提交给 adminGoods/addGoods?updateAct=update 处理。

商品管理　类型管理　用户管理　订单管理　公告管理　安全退出

ID	名称	价格	库存	详情	操作
12	上衣X	100.0	500	详情	修改
13	苹果	8.0	500	详情	修改
14	樱桃	5.0	800	详情	修改
15	微波炉	300.0	500	详情	修改
16	电视99	9000.0	600	详情	修改
17	孕装55	800.0	1000	详情	修改
18	牛奶11	50.0	6000	详情	修改
19	红酒99	5000.0	30000	详情	修改
20	化妆品22	3600.0	5989	详情	修改
21	aa	12.0	11	详情	修改
				共11条记录 共2页	第1页 下一页

图 20.15　修改查询页面

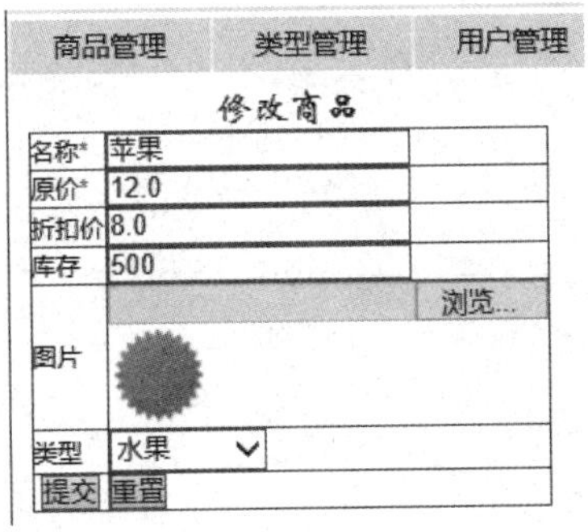

商品管理　类型管理　用户管理

修改商品

名称*	苹果
原价*	12.0
折扣价	8.0
库存	500
图片	浏览...
类型	水果
提交	重置

图 20.16　修改商品页面

修改商品的实现步骤如下：

❶ 编写视图

在 WEB-INF/jsp/admin 目录下创建修改查询页面 updateSelectGoods.jsp 和修改商品信

息页面 updateAgoods.jsp。updateSelectGoods.jsp 与查询商品页面的内容基本一样，updateAgoods.jsp 与添加商品页面的内容基本一样，这里不再赘述。

❷ 编写控制器层

此功能模块共有 3 个处理请求，即 adminGoods/selectGoods?act=updateSelect、adminGoods/selectAGoods?id=${goods.id}&act=updateAgoods 和 adminGoods/addGoods?updateAct=update。adminGoods/selectGoods?act=updateSelect 和 adminGoods/selectAGoods?id=${goods.id}&act=updateAgoods 请求已在 20.5.4 节介绍，adminGoods/addGoods?updateAct=update 请求已在 20.5.3 节介绍。

❸ 编写 Service 层

同理，Service 层请参考 20.5.3 节与 20.5.4 节。

❹ 编写 SQL 映射文件

修改商品的 SQL 语句如下：

```
<!-- 修改一个商品 -->
<update id="updateGoodsById" parameterType="Goods">
update goodstable
<set>
   <if test="gname!=null">
     gname=#{gname},
   </if>
   <if test="goprice!=null">
     goprice=#{goprice},
   </if>
   <if test="grprice!=null">
     grprice=#{grprice},
   </if>
   <if test="gstore!=null">
     gstore=#{gstore},
   </if>
   <if test="gpicture!=null">
     gpicture=#{gpicture},
   </if>
   <if test="goodstype_id!=null">
     goodstype_id=#{goodstype_id},
   </if>
 </set>
 where id=#{id}
</update>
```

20.5.6 删除商品

单击图 20.11 中的“删除商品”超链接（adminGoods/selectGoods?act=deleteSelect），打开删除查询页面 deleteSelectGoods.jsp，如图 20.17 所示。

单击图 20.17 中的“删除”超链接（adminGoods/deleteAGoods?id=）可实现单个商品的删除；在图 20.17 中选择商品 ID 并单击“删除”按钮（adminGoods/deleteGoods）可实现多个商品的删除。成功删除（关联商品不允许删除）后返回删除查询页面。

商品管理　类型管理　用户管理　订单管理　公告管理　安全退出

ID	名称	价格	库存	详情	操作
□12	上衣X	100.0	500	详情	删除
□13	苹果	8.0	500	详情	删除
□14	樱桃	5.0	800	详情	删除
□15	微波炉	300.0	500	详情	删除
□16	电视99	9000.0	600	详情	删除
□17	孕装55	800.0	1000	详情	删除
□18	牛奶11	50.0	6000	详情	删除
□19	红酒99	5000.0	30000	详情	删除
□20	化妆品22	3600.0	5989	详情	删除
□21	aa	12.0	11	详情	删除

删除

共11条记录 共2页 第1页 下一页

图 20.17　删除查询页面

❶ 编写视图

在 WEB-INF/jsp/admin 目录下创建删除查询页面 deleteSelectGoods.jsp。该页面与查询商品页面的内容基本一样，这里不再赘述。

❷ 编写控制器层

此功能模块共有 3 个处理请求，即 adminGoods/selectGoods?act=deleteSelect、adminGoods/deleteAGoods?id=和 adminGoods/deleteGoods。adminGoods/selectGoods?act=deleteSelect 请求已在 20.5.4 节介绍，这里不再赘述。adminGoods/deleteAGoods?id=和 adminGoods/deleteGoods 请求的相关控制器层代码如下：

```
/**
 * 删除多个商品
 */
@RequestMapping("/deleteGoods")
public String deleteGoods(Integer ids[], Model model) {
    return adminGoodsService.deleteGoods(ids, model);
}
/**
 * 删除单个商品
 */
@RequestMapping("/deleteAGoods")
public String deleteAGoods(Integer id, Model model) {
    return adminGoodsService.deleteAGoods(id, model);
}
```

❸ 编写 Service 层

删除单个商品和删除多个商品的相关业务处理代码如下：

```
/**
 * 删除多个商品
```

```
 */
@Override
public String deleteGoods(Integer[] ids, Model model) {
    List<Integer> list=new ArrayList<Integer>();
    for (int i=0; i<ids.length; i++) {
        //商品有关联
        if(adminGoodsDao.selectCartGoods(ids[i]).size() > 0 ||
            adminGoodsDao.selectFocusGoods(ids[i]).size() > 0 ||
            adminGoodsDao.selectOrderdetailGoods(ids[i]).size() > 0) {
                model.addAttribute("msg", "商品有关联，不允许删除！");
                return "forward:/adminGoods/selectGoods?act=deleteSelect";
            }
        list.add(ids[i]);
    }
    adminGoodsDao.deleteGoods(list);
    model.addAttribute("msg", "成功删除商品！");
    return "forward:/adminGoods/selectGoods?act=deleteSelect";
}
/**
 * 删除一个商品
 */
@Override
public String deleteAGoods(Integer id, Model model) {
    //商品有关联
    if(adminGoodsDao.selectCartGoods(id).size() > 0 ||
        adminGoodsDao.selectFocusGoods(id).size() > 0 ||
        adminGoodsDao.selectOrderdetailGoods(id).size() > 0) {
            model.addAttribute("msg", "商品有关联，不允许删除！");
            return "forward:/adminGoods/selectGoods?act=deleteSelect";
        }
    adminGoodsDao.deleteAGoods(id);
    model.addAttribute("msg", "成功删除商品！");
    return "forward:/adminGoods/selectGoods?act=deleteSelect";
}
```

❹ 编写 SQL 映射文件

“删除商品”功能模块的相关 SQL 语句如下：

```
<!-- 删除多个商品 -->
<delete id="deleteGoods" parameterType="List">
    delete from goodstable where id in
    <foreach item="item" index="index" collection="list"
    open="(" separator="," close=")">
        #{item}
    </foreach>
</delete>
```

```
<!-- 删除单个商品 -->
<delete id="deleteAGoods" parameterType="Integer">
    delete from goodstable where id=#{id}
</delete>
<!-- 查询关联商品 -->
<select id="selectCartGoods" parameterType="Integer" resultType="map">
    select * from carttable where goodstable_id=#{id}
</select>
<select id="selectFocusGoods" parameterType="Integer" resultType="map">
    select * from focustable where goodstable_id=#{id}
</select>
<select id="selectOrderdetailGoods" parameterType="Integer" resultType=
"map">
    select * from orderdetail where goodstable_id=#{id}
</select>
```

20.5.7　订单管理

单击后台管理主页面中“订单管理”的“删除订单”超链接（adminOrder/orderInfo），打开订单管理页面 orderManager.jsp，如图 20.18 所示。

单击图 20.18 中的“删除”超链接（adminOrder/deleteorderManager?id=）可删除未付款的订单。

商品管理　类型管理　用户管理　订单管理　公告管理　安全退出

订单编号	用户E-mail	订单金额	订单状态	订单日期	操作
8	1@126.com	39600.0	已付款	2018-02-12 05:04:33.0	
9	1@126.com	3600.0	未付款	2018-02-13 07:25:36.0	删除

图 20.18　订单管理页面

❶ 编写视图

在 WEB-INF/jsp/admin 目录下创建订单管理页面 orderManager.jsp，该页面的核心代码如下：

```
<body>
    <br>
    <table border="1" bordercolor="PaleGreen">
        <tr>
            <th width="150px">订单编号</th>
            <th width="150px">用户 E-mail</th>
            <th width="100px">订单金额</th>
            <th width="100px">订单状态</th>
            <th width="150px">订单日期</th>
            <th width="100px">操作</th>
        </tr>
```

```
        <c:forEach var="n" items="${orderList}">
        <tr onmousemove="changeColor(this)" onmouseout="changeColor1(this)">
            <td>${n.id}</td>
            <td>${n.bemail}</td>
            <td>${n.amount}</td>
    <td><c:if test="${n.status==0}" >未付款</c:if><c:if test="${n.status
    ==1}" >已付款</c:if></td>
            <td>${n.orderdate}</td>
            <td>
                <c:if test="${n.status==0}" >
                    <a href="javaScript:checkDel('${n.id}')">删除</a>
                </c:if>
                 
            </td>
        </tr>
        </c:forEach>
    </table>
</body>
```

❷ **编写控制器层**

此功能模块共有两个处理请求，即 adminOrder/orderInfo 和 adminOrder/deleteorderManager?id=。两个请求的相关控制器层代码如下：

```
@Autowired
private AdminOrderService adminOrderService;
@RequestMapping("/orderInfo")
public String orderInfo(Model model) {
    return adminOrderService.orderInfo(model);
}
@RequestMapping("/deleteorderManager")
public String deleteorderManager(Integer id) {
    return adminOrderService.deleteorderManager(id);
}
```

❸ **编写 Service 层**

“订单管理”功能模块的相关 Service 层代码如下：

```
@Autowired
private AdminOrderDao adminOrderDao;
@Override
public String orderInfo(Model model) {
    List<Map<String,Object>> list=adminOrderDao.orderInfo();
    model.addAttribute("orderList", list);
    return "admin/orderManager";
}
@Override
```

```
public String deleteorderManager(Integer id) {
    //先删除明细
    adminOrderDao.deleteOrderDetail(id);
    //再删除订单基础
    adminOrderDao.deleteOrderBase(id);
    return "forward:/adminOrder/orderInfo";
}
```

❹ 编写 SQL 映射文件

“订单管理”功能模块的相关 SQL 语句如下：

```
<select id="orderInfo" resultType="map" >
    select ot.id, ot.amount, ot.status, orderdate, bt.bemail,
    ot.busertable_id
    from ORDERBASETABLE ot, BUSERTABLE bt where ot.busertable_id=bt.id
</select>
<delete id="deleteOrderDetail" parameterType="Integer">
    delete from orderdetail where orderbasetable_id=#{id}
</delete>
<delete id="deleteOrderBase" parameterType="Integer">
    delete from orderbasetable where id=#{id}
</delete>
```

20.5.8　用户管理

单击后台管理主页面中“用户管理”的“删除用户”超链接（adminUser/userInfo），打开用户管理页面 userManager.jsp，如图 20.19 所示。

单击图 20.19 中的“删除”超链接（adminUser/deleteuserManager?id=）可删除未关联的用户。

商品管理　类型管理　用户管理　订单管理　公告管理　安全退出

用户ID	用户E-mail	用户密码	删除
5	1@126.com	••••••	删除

图 20.19　用户管理页面

“用户管理”与 20.5.7 节中“订单管理”的实现方式基本一样，这里不再赘述。

20.5.9　公告管理

单击后台管理主页面中“公告管理”的“添加公告”超链接（adminNotice/toAddNotice），打开“添加公告”页面 addNotice.jsp，如图 20.20 所示。单击后台管理主页面中“公告管理”的“删除公告”超链接（adminNotice/deleteNoticeSelect），打开“删除公告”页面 deleteNoticeSelect.jsp，如图 20.21 所示。

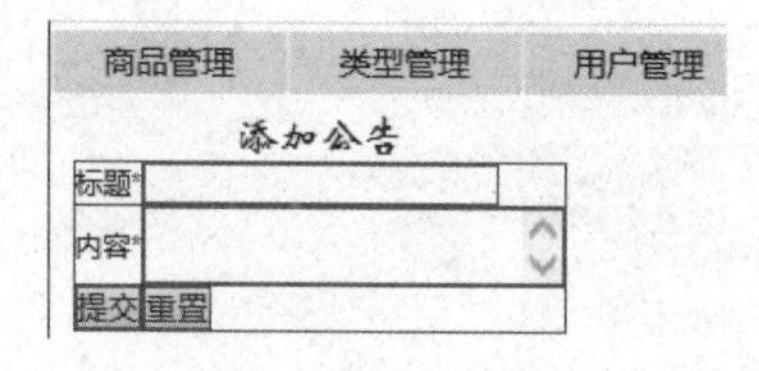

图 20.20　添加公告页面

商品管理　类型管理　用户管理　订单管理　公告管理　安全退出

ID	标题	时间	详情	操作
4	2018不忘初心 砥志前行	2018-02-11 07:05:19.0	详情	删除
5	2018年大减价	2018-02-11 07:06:25.0	详情	删除
6	大减价	2018-02-11 07:11:41.0	详情	删除
7	春节不打烊	2018-02-11 07:11:59.0	详情	删除
8	春节不打烊春节不打烊	2018-02-11 07:12:07.0	详情	删除
9	欢度春节	2018-02-11 07:12:27.0	详情	删除
10	欢度春节欢度春节	2018-02-11 07:12:39.0	详情	删除
11	低价大放送	2018-02-11 07:13:10.0	详情	删除
12	低价大放送低价大放送	2018-02-11 07:13:19.0	详情	删除
13	勇往直前	2018-02-11 07:14:29.0	详情	删除
14	勇往直前勇往直前	2018-02-11 07:14:38.0	详情	删除
15	春节不打烊春节不打烊春节不打烊	2018-02-11 07:15:00.0	详情	删除

图 20.21　删除公告页面

单击图 20.21 中的“删除”超链接（adminNotice/deleteNotice?id=）可实现删除公告的功能，单击“详情”超链接（adminNotice/selectANotice?id=${notice.id }）可打开详情页面 noticeDetail.jsp，如图 20.22 所示。

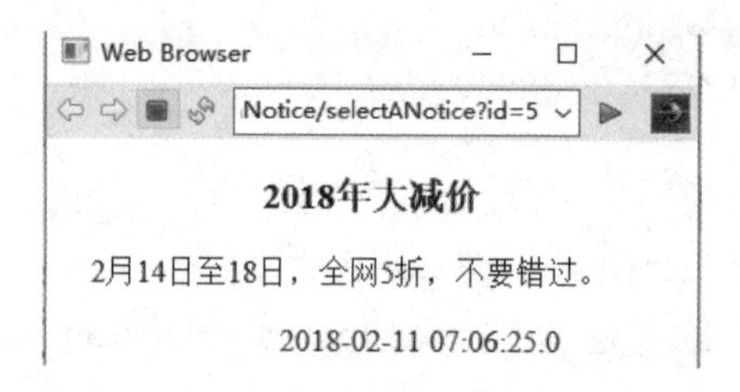

图 20.22　公告详情页面

“公告管理”与“商品管理”的实现方式基本一样，这里不再赘述。

20.5.10　退出系统

在后台管理主页面中单击“退出系统”超链接（exit）将返回后台登录页面。系统根据@RequestMapping 注解找到对应控制器类 com.controller.admin.AdminController 的 exit 方法处理请求，在 exit 方法中执行 session.invalidate()将 session 失效，并返回后台登录页面。具体代码如下：

```
@RequestMapping("/exit")
public String exit(@ModelAttribute Auser auser,HttpSession session) {
    session.invalidate();
    return "admin/login";
}
```

20.6　前台电子商务子系统的实现

游客具有浏览首页、查看商品详情和查看公告等权限。成功登录的用户除了具有游客的权限外，还具有购买商品、查看购物车、关注商品以及查看用户中心的权限。本节将详细讲解前台电子商务子系统的实现。

20.6.1　导航栏

在前台的每个 JSP 页面中都引入了一个名为 head.jsp 的页面，引入代码如下：

```
<jsp:include page="head.jsp"></jsp:include>
```

其中，head.jsp 中的商品类型是从 session 对象获得的，该 session 对象是用户请求电子商务子系统首页时初始化的。在请求首页时，系统根据@RequestMapping 注解找到 com.controller.before.IndexController 类的 before 方法处理请求。处理请求后转发到 before/index.jsp 页面。index.jsp 页面的实现过程见 20.3.2 节。head.jsp 页面的运行效果如图 20.23 所示。

图 20.23　导航栏

在导航栏的搜索框中输入信息，单击“搜索”按钮，将搜索信息提交给 search 请求处理，系统根据@RequestMapping 注解找到 com.controller.before.IndexController 控制器类的 search 方法处理请求，并将搜索到的商品信息转发给 searchResult.jsp。searchResult.jsp 页面的运行效果如图 20.24 所示。

图 20.24　搜索结果

❶ 编写视图

该模块的视图涉及 WEB-INF/jsp/before 目录下的两个 JSP 页面，即 head.jsp 和

searchResult.jsp。

head.jsp 的核心代码如下：

```
<body>
    <div class="all_zong">
        <!-- 最上面 灰色条部分 -->
        <div class="all_zong_top">
            <div class="all_zong_top_right a8c">
                <table border="0" cellspacing="0" cellpadding="0" class=
                "main_login">
                    <tr>
                        <td>
                            <p id="content">
                               <c:if test="${bruser!=null}">欢迎 ${bruser.
                                bemail }</c:if>
                                <c:if test="${bruser==null}"><a href=
                                "toLogin">登录</a></c:if>
                            </p>
                        </td>
                        <td>
                            <p>
                                <a href="toRegister">注册</a>
                            </p>
                        </td>
                        <td><span class="xx">|</span><a href="userCenter">
                        用户中心</a><span class="xx">|</span></td>
                        <!-- 没有登录 -->
                        <c:if test="${bruser!= null}">
                        <td><a href="user/exit">退出</a><span
                            class="xx">|</span></td>
                        </c:if>
                    </tr>
                </table>
            </div>
        </div>
        <!-- end -->
        <!-- logo 搜索 -->
        <div class="all_zong_logo">
            <div class="all_zong_logo2">
                <img src="images/before/mylogo.png" />
            </div>
            <div class="back_search">
                <div class="back_search_red">
                    <form action="search" name="myForm" method="post">
                        <div class="div2">
```

```
<input type="text" name="mykey" class="txt" value="请输入您要查询的内容"
onfocus="clearValue()" />
                        </div>
                        <div class="div1">
                            <input type="submit" class="an" value="搜索" />
                        </div>
                    </form>
                </div>
            </div>
            <!-- end -->
        </div>
        <!-- 红色 导航 -->
        <div class="skin_a">
            <div class="front_daohangbj">
                <div class="all_zong">
                    <div class="front_daohang">
                        <ul>
                            <li class="backbj"><a href="before?id=0">首页
                            </a></li>
                            <!-- 显示商品类型 -->
                            <c:forEach items="${goodsType}" var="g">
                                <li><a href="before?id=${g.id }">
                                ${g.typename }</a>
                                </li>
                            </c:forEach>
                            <li class="buy">
                                <p class="car">
                                    <a href="cart/selectCart">购物车</a>
                                </p>
                            </li>
                        </ul>
                    </div>
                </div>
            </div>
        </div>
        <! --红色 导航 end -->
    </div>
</body>
```

searchResult.jsp 的核心代码如下：

```
<c:forEach items="${searchlist }" var="mf">
    <tr>
        <td bgcolor="#ffffff" align="center">${mf.id }</td>
        <td bgcolor="#ffffff" align="center">${mf.gname }</td>
        <td align="center" bgcolor="#ffffff" height="60px"><img
```

```
            style="width: 50px; height: 50px;"
            src="logos/${mf.gpicture }" border="0" title=
            "${mf.gname }" />
        </td>
        <td align="center" bgcolor="#ffffff">${mf.grprice }</td>
        <td align="center" bgcolor="#ffffff"><a
            style="text-decoration: none;" class="f6"
            href="goodsDetail?id=${mf.id}">去看看</a>
        </td>
    </tr>
</c:forEach>
```

❷ **编写控制器层**

该功能模块的控制器层涉及 com.controller.before.IndexController 控制器类的两个处理方法，具体代码如下：

```
/**
 * 首页
 */
@RequestMapping("/before")
public String before(Model model,HttpSession session, Goods goods) {
    return indexService.before(model, session, goods);
}
/**
 * 首页搜索
 */
@RequestMapping("/search")
public String search(Model model,String mykey) {
    return indexService.search(model, mykey);
}
```

❸ **编写 Service 层**

该功能模块的 Service 层的代码如下：

```
@Override
public String before(Model model, HttpSession session, Goods goods) {
    session.setAttribute("goodsType", adminTypeDao.selectGoodsType());
    model.addAttribute("salelist", indexDao.getSaleOrder());
    model.addAttribute("focuslist", indexDao.getFocusOrder());
    model.addAttribute("noticelist", indexDao.selectNotice());
    if(goods.getId()==null)
        goods.setId(0);
    model.addAttribute("lastedlist", indexDao.getLastedGoods(goods));
    return "before/index";
}
@Override
```

```
public String search(Model model, String mykey) {
    List<Goods> list=indexDao.search(mykey);
    model.addAttribute("searchlist", list);
    return "before/searchResult";
}
```

❹ 编写 SQL 映射文件

该功能模块涉及的 SQL 语句如下：

```
<!-- 查询销售排行 -->
    <select id="getSaleOrder"  resultType="map">
        select sum(od.shopnumber) shopnumber,
                gd.id id,
                gd.gname gname,
                gd.goprice goprice,
                gd.grprice grprice,
                gd.gpicture gpicture
        from GOODSTABLE gd  LEFT JOIN ORDERDETAIL  od  ON od.goodstable_
        id=gd.id
                group by
                gd.id,
                gd.gname,
                gd.goprice,
                gd.grprice,
                gd.gpicture
        order by shopnumber desc limit 10
    </select>
    <!-- 人气排行 -->
    <select id="getFocusOrder" resultType="map">
        select count(ft.goodstable_id) fn, gt.id id, gt.gname gname,
        gt.grprice grprice, gt.gpicture gpicture
        from GOODSTABLE gt
        LEFT JOIN FOCUSTABLE ft  ON ft.goodstable_id=gt.id
        group by gt.id, gt.gname, gt.grprice, gt.gpicture
        order by fn desc limit 10
    </select>
    <!-- 公告 -->
    <select id="selectNotice" resultType="Notice">
        select * from noticetable order by ntime desc
    </select>
    <!-- 最新商品 -->
    <select id="getLastedGoods" resultType="Goods" parameterType="Goods">
        select gt.*, gy.typename from GOODSTABLE gt,GOODSTYPE gy where
        gt.goodstype_id=gy.id
        <if test="id!=0">
            and gy.id=#{id}
```

```
        </if>
        order by  gt.id desc limit 15
    </select>
    <!-- 首页搜索 -->
    <select id="search" resultType="Goods" parameterType="String">
        select gt.*, gy.typename from GOODSTABLE gt,GOODSTYPE gy where
        gt.goodstype_id=gy.id
        and gt.gname like concat('%',#{mykey},'%')
    </select>
```

20.6.2 销售排行

销售排行是以订单详情表中每种商品的销量总和排序的，具体实现请参考 20.6.1 节。

20.6.3 人气排行

人气排行是以关注表中每种商品的关注次数总和排序的，具体实现请参考 20.6.1 节。

20.6.4 最新商品

最新商品是以商品 ID 排序的，因为商品 ID 是自增的，具体实现请参考 20.6.1 节。

20.6.5 公告栏

公告栏的具体实现请参考 20.6.1 节。

20.6.6 用户注册

单击导航栏中的“注册”超链接（toRegister），打开注册页面 register.jsp，如图 20.25 所示。

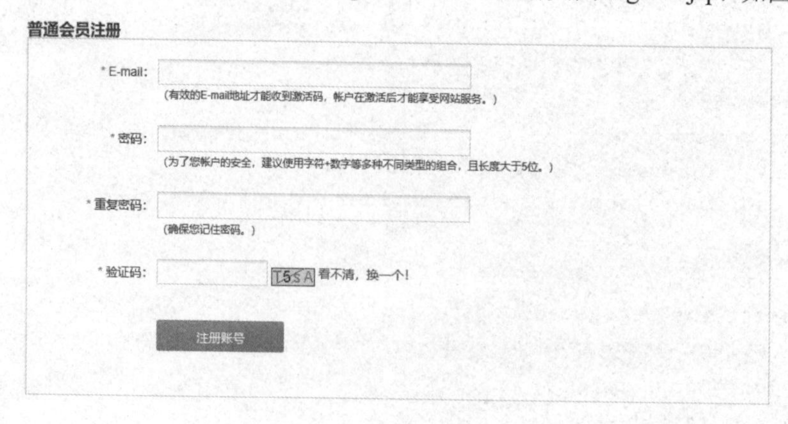

图 20.25　注册页面

输入用户信息，单击“注册账号”按钮，将用户信息提交给 user/register 处理请求，系统根据@RequestMapping 注解找到 com.controller.before.UserController 控制器类的 register 方法处理请求。注册模块的实现步骤如下：

❶ 编写视图

该模块的视图涉及 WEB-INF/jsp/before 目录下的 register.jsp。register.jsp 的代码略。

❷ 编写控制器层

该功能模块涉及 com.controller.before.IndexController 控制器类的 toRegister 方法和 com.controller.before.UserController 控制器类的 register 方法。具体代码如下：

```
/**
 * 转到注册页面
 */
@RequestMapping("/toRegister")
public String toRegister(Model model) {
    return indexService.toRegister(model);
}
@RequestMapping("/register")
public String register(@ModelAttribute Buser buser,Model model, HttpSession
session, String code) {
    return userService.register(buser, model, session, code);
}
```

❸ 编写 Service 层

该功能模块的 Service 层的代码如下：

```
@Override
public String toRegister(Model model) {
    model.addAttribute("rbuser", new Buser());
    return "before/register";
}
@Override
public String register(Buser buser, Model model, HttpSession session, String
code) {
    if(!code.equalsIgnoreCase(session.getAttribute("code").toString())) {
        model.addAttribute("codeError", "验证码错误！");
        return "before/register";
    }
    int n=userDao.register(buser);
    if(n>0) {
        return "before/login";
    }else {
        model.addAttribute("msg", "注册失败！");
        return "before/register";
    }
}
```

❹ **编写 SQL 映射文件**

该功能模块涉及的 SQL 语句如下：

```
<insert id="register" parameterType="Buser">
    insert into busertable (id, bemail, bpwd) values (null, #{bemail}, #{bpwd})
</insert>
```

20.6.7 用户登录

用户注册成功后跳转到登录页面 login.jsp，如图 20.26 所示。

图 20.26 登录页面

在图 20.26 中输入信息后单击“确定”按钮，将用户输入的 E-mail、密码以及验证码提交给 user/login 请求处理。系统根据@RequestMapping 注解找到 com.controller.before.UserController 控制器类的 login 方法处理请求。登录成功后将用户的登录信息保存在 session 对象中，然后回到网站首页。具体实现步骤如下：

❶ **编写视图**

该模块的视图涉及 WEB-INF/jsp/before 目录下的 login.jsp。login.jsp 的代码略。

❷ **编写控制器层**

该功能模块涉及 com.controller.before.UserController 控制器类的 login 方法。具体代码如下：

```
@RequestMapping("/login")
public String login(@ModelAttribute Buser buser,Model model, HttpSession
session, String code) {
    return userService.login(buser, model, session, code);
}
```

❸ **编写 Service 层**

该功能模块的 Service 层的代码如下：

```
@Override
public String login(Buser buser, Model model, HttpSession session, String
code) {
```

```
    if(!code.equalsIgnoreCase(session.getAttribute("code").toString())) {
        model.addAttribute("msg", "验证码错误！");
        return "before/login";
    }
    Buser ruser=null;
    List<Buser> list=userDao.login(buser);
    if(list.size() > 0) {
        ruser=list.get(0);
    }
    if(ruser!=null) {
        session.setAttribute("bruser", ruser);
        return "forward:/before";
    }else {
        model.addAttribute("msg", "用户名或密码错误！");
        return "before/login";
    }
}
```

❹ 编写 SQL 映射文件

该功能模块的 SQL 语句如下：

```
<select id="login" resultType="Buser" parameterType="Buser">
    select * from busertable where bemail=#{bemail} and bpwd=#{bpwd}
</select>
```

20.6.8　商品详情

用户可以从销售排行、人气排行以及最新商品等位置单击商品图片或商品名称进入商品详情页面 goodsdetail.jsp，如图 20.27 所示。

图 20.27　商品详情页面

商品详情的具体实现步骤如下：

❶ 编写视图

该模块的视图涉及 WEB-INF/jsp/before 目录下的 goodsDetail.jsp，goodsDetail.jsp 的核心代码如下：

```
<body>
    <form action="cart/putCart" name="putcartform" method="post">
        <div class="blank"></div>
        <div class="block clearfix">
        <div class="location ared">
            当前位置：<a href="#">首页</a> <a href="#">商品详情</a>
        </div>
        <div class="blank"></div>
        <div id="goodsInfo">
            <div class="imgInfo">
                <input type="hidden" name="id"
                    value="${goods.id }"/><img
                    src="logos/${goods.gpicture}"
                    width="230px" height="230px" />
            </div>
        </div>
        <!-- 商品描述 -->
        <div class="goods_desc">
            <div class="bt">
                ${goods.gname }
            </div>
            <div class="goods_show">
                <ul>
            <li><span>价格:</span> <strong class="yj">${goods.goprice }元
            </strong></li>
                    <li><span>折扣价:</span><strong
                        class="xj">${goods.grprice }元</strong>
                    </li>
                    <li><span>类型:</span> ${goods.typename }</li>
                    <li><span>购买数量:</span><input type="text" name=
                    "shoppingnum"
                        class="good_txt" value="1" /> (库存${goods.gstore }
                        件)</li>
                </ul>
            </div>
            <p class="bottom10 top5">
                <img src="images/before/goods_ann2.gif" style="cursor: pointer"
                    onclick="goCart()" />  <input type="button"
                    style="cursor: pointer" class="sh_bnt2"
                    onclick="gofocus('${goods.id }')"
```

```
                    value="关注" />
              </p>${msg }
          </div>
          <!--end-->
          </div>
      </form>
</body>
```

❷ **编写控制器层**

该功能模块涉及 com.controller.before.IndexController 控制器类的 goodsDetail 方法，具体代码如下：

```
/**
 * 转到商品详情页
 */
@RequestMapping("/goodsDetail")
public String goodsDetail(Model model,Integer id) {
    return indexService.goodsDetail(model, id);
}
```

❸ **编写 Service 层**

该功能模块的 Service 层的代码如下：

```
@Override
public String goodsDetail(Model model, Integer id) {
    Goods goods=indexDao.selectGoodsById(id);
    model.addAttribute("goods", goods);
    return "before/goodsdetail";
}
```

❹ **编写 SQL 映射文件**

该功能模块的 SQL 语句如下：

```
<!-- 根据 id 查询一个商品 -->
<select id="selectGoodsById"  resultType="Goods" parameterType="Integer">
    select gt.*,gy.typename from goodstable gt,goodstype gy where gt.id=#{id} and
    gt.goodstype_id=gy.id
</select>
```

20.6.9 关注商品

登录成功的用户可以在商品详情页面中单击“关注”按钮关注该商品，此时请求路径为 cart/focus?id=。系统根据@RequestMapping 注解找到 com.controller.before.CartController 控制器类的 focus 方法处理请求。具体实现步骤如下：

❶ **编写控制器层**

该功能模块涉及 com.controller.before.CartController 控制器类的 focus 方法，具体代码

如下：

```
/**
 * 关注商品
 */
@RequestMapping("/focus")
public String focus(Model model,Integer id, HttpSession session) {
    return cartService.focus(model, id, session);
}
```

❷ 编写 Service 层

该功能模块的 Service 层的代码如下：

```
@Override
public String focus(Model model, Integer id, HttpSession session) {
    Map<String, Object> map=new HashMap<String, Object>();
    map.put("uid", MyUtil.getUserId(session));
    map.put("gid", id);
    List<Map<String, Object>> list=cartDao.isFocus(map);
    if(list.size()>0) {
        model.addAttribute("msg", "已关注该商品！");
    }else {
        int n=cartDao.focus(map);
        if(n>0)
            model.addAttribute("msg", "成功关注该商品！");
        else
            model.addAttribute("msg", "关注失败！");
    }
    return "forward:/goodsDetail?id="+id;
}
```

❸ 编写 SQL 映射文件

该功能模块的 SQL 语句如下：

```
<!-- 关注商品 -->
<insert id="focus" parameterType="map">
    insert into focustable(id, goodstable_id, busertable_id,focustime)
    values (null, #{gid}, #{uid},now())
</insert>
<!-- 是否已关注 -->
<select id="isFocus" parameterType="map" resultType="map">
    select * from focustable where goodstable_id=#{gid} and busertable_
    id=#{uid}
</select>
```

20.6.10 购物车

单击商品详情页面中的“加入购物车”按钮或导航栏中的“购物车”超链接，打开购物车页面 cart.jsp，如图 20.28 所示。

图 20.28　购物车

与购物车有关的处理请求有 cart/putCart（加入购物车）、cart/clear（清空购物车）、cart/selectCart（查询购物车）和 cart/deleteAgoods?id=（删除购物车）。系统根据@RequestMapping 注解分别找到 com.controller.before.CartController 控制器类的 putCart、clear、selectCart、deleteAgoods 等方法处理请求。具体实现步骤如下：

❶ 编写视图

该模块的视图涉及 WEB-INF/jsp/before 目录下的 cart.jsp，cart.jsp 的核心代码如下：

```
<body>
    <div class="blank"></div>
    <div class="block clearfix">
        <!-- 当前位置 -->
        <div class="location ared">
            当前位置： <a href="before?id=0">首页</a> > 购物流程 > 购物车
        </div>
        <div class="blank"></div>
        <div>
            <div class="nFlowBox">
                <table width="99%" align="center" border="0" cellpadding="5"
                    cellspacing="1" bgcolor="#dddddd">
                    <tr>
                        <th>商品信息</th>
                        <th>单价（元）</th>
                        <th>数量</th>
                        <th>小计</th>
```

```
        <th>操作</th>
    </tr>
    <tr>
        <td colspan="5" height="15px"
            style="border: 0 none; background: #FFF"></td>
    </tr>
    <!-- 这里使用了 JSTL 标签 -->
    <c:forEach var="ce" items="${cartlist}">
    <tr>
        <td bgcolor="#ffffff" align="center"><a href=
      "goodsDetail?id=${ce.id}"> <imgstyle="width: 100px;
        height: 100px;"src="logos/${ce.gpicture}" border=
        "0"title="${ce.gname}" />
            </a><br/><a style="text-decoration: none;"
            href="goodsDetail?id=${ce.id}" class="f6">
            ${ce.gname}</a></td>
            <td bgcolor="#ffffff" width="110px" align=
            "center"><span>${ce.grprice}</span></td>
            <td align="center" bgcolor="#ffffff" width=
            "115px"valign="middle"><input type="text" name=
            "goods_number"value="${ce.shoppingnum}"
            size="4"class="inputBg"style="text-align:
            center; width: 36px; color: #999999" /></td>
            <td align="center" bgcolor="#ffffff" width=
            "115px">￥ <span>${ce.smallsum}</span> 
            </td>
            <td align="center" bgcolor="#ffffff" width=
            "185px"><a style="text-decoration: none;" href=
            "javaScript:deleteAgoods('${ce.id}')"class=
            "f6" title="删除"><img src="images/before/
            sc.png" />
            </a></td>
    </tr>
    </c:forEach>
    <tr>
        <td align="right" bgcolor="#ffffff" colspan="4"
        height="41px;"style="border-left: 0 none;"><font
        style="color: #a60401; font-size: 13px; font-
        weight: bold; letter-spacing: 0px;">
        购物金额总计(不含运费) ￥ <span id="stotal">
        </span>
            ${total}元
```

```
                    </font></td>
                    <td align="center" bgcolor="#ffffff"><input
                    type="button"value="清空购物车" onclick="godelete()"
                    class="bnt_blue_1" id="bnt11" />
                    </td>
                </tr>
                <tr>
                    <td bgcolor="#ffffff" colspan="4" align="right"
                        style="padding: 5px; padding-left: 2px;
                        border-right: 0 none">
                        <a href="before"> <img src="images/before/
                        jxgw.jpg" alt="continue" />
                        </a>
                    </td>
                    <td bgcolor="#ffffff" align="center"
                        style="border-left: 0 none; padding: 5px;
                        padding-right: 2px;">
                        <a style="cursor: pointer;" href="javascript:
                        goOrderConfirm()">
                            <img src="images/before/qjs.jpg" alt=
                            "checkout" />
                        </a>
                    </td>
                </tr>
            </table>
        </div>
    </div>
</div>
</body>
```

❷ 编写控制器层

该功能模块涉及 com.controller.before.CartController 控制器类的 putCart、clear、selectCart、deleteAgoods 等方法，具体代码如下：

```
/**
 * 添加购物车
 */
@RequestMapping("/putCart")
public  String  putCart(Model  model,Integer  shoppingnum,  Integer  id,
HttpSession session) {
    return cartService.putCart(model, shoppingnum, id, session);
}
/**
 * 查询购物车
```

```
 */
@RequestMapping("/selectCart")
public String selectCart(Model model, HttpSession session) {
    return cartService.selectCart(model, session);
}
/**
 * 删除购物车
 */
@RequestMapping("/deleteAgoods")
public String deleteAgoods(Integer id,HttpSession session) {
    return cartService.deleteAgoods(id, session);
}
/**
 * 清空购物车
 */
@RequestMapping("/clear")
public String clear(HttpSession session) {
    return cartService.clear(session);
}
```

❸ 编写 Service 层

该功能模块的 Service 层的代码如下：

```
@Override
public String putCart(Model model, Integer shoppingnum, Integer id,
HttpSession session) {
    Map<String, Object> map=new HashMap<String, Object>();
    map.put("uid", MyUtil.getUserId(session));
    map.put("gid", id);
    map.put("shoppingnum", shoppingnum);
    //是否已添加购物车
    List<Map<String, Object>> list=cartDao.isPutCart(map);
    if(list.size() > 0)
        cartDao.updateCart(map);
    else
        cartDao.putCart(map);
    return "forward:/cart/selectCart";
}
@Override
public String selectCart(Model model, HttpSession session) {
    List<Map<String, Object>> list=cartDao.selectCart(MyUtil.getUserId
    (session));
    double sum=0;
    for(Map<String, Object> map : list) {
        sum=sum+(Double)map.get("smallsum");
    }
```

```
        model.addAttribute("total", sum);
        model.addAttribute("cartlist", list);
        return "before/cart";
    }
    @Override
    public String deleteAgoods(Integer id, HttpSession session) {
        Map<String, Object> map=new HashMap<String, Object>();
        map.put("uid", MyUtil.getUserId(session));
        map.put("gid", id);
        cartDao.deleteAgoods(map);
        return "forward:/cart/selectCart";
    }
    @Override
    public String clear(HttpSession session) {
        cartDao.clear(MyUtil.getUserId(session));
        return "forward:/cart/selectCart";
    }
```

❹ 编写 SQL 映射文件

该功能模块的 SQL 语句如下：

```
<!-- 是否已添加购物车 -->
<select id="isPutCart" parameterType="map" resultType="map">
    select * from carttable where goodstable_id=#{gid} and busertable_id=
    #{uid}
</select>
<!-- 添加购物车 -->
<insert id="putCart" parameterType="map">
    insert into carttable(id, busertable_id,goodstable_id,shoppingnum)
    values (null, #{uid}, #{gid}, #{shoppingnum})
</insert>
<!-- 更新购物车 -->
<update id="updateCart" parameterType="map">
    update carttable set shoppingnum=shoppingnum+#{shoppingnum} where
    busertable_id=#{uid} and goodstable_id=#{gid}
</update>
<!-- 查询购物车 -->
<select id="selectCart" parameterType="Integer" resultType="map">
    select gt.id, gt.gname, gt.gpicture, gt.grprice, ct.shoppingnum,
    ct.shoppingnum*gt.grprice smallsum
    from GOODSTABLE gt, CARTTABLE ct where gt.id=ct.goodstable_id and
    ct.busertable_id=#{id}
</select>
<!-- 删除购物车 -->
<delete id="deleteAgoods" parameterType="map">
    delete from carttable where busertable_id=#{uid} and goodstable_id=
```

```
    #{gid}
</delete>
    <!-- 清空购物车 -->
<delete id="clear" parameterType="Integer">
    delete from carttable where busertable_id=#{uid}
</delete>
```

20.6.11 下单

在购物车页面中单击“去结算”按钮，进入订单确认页面 orderconfirm.jsp，如图 20.29 所示。

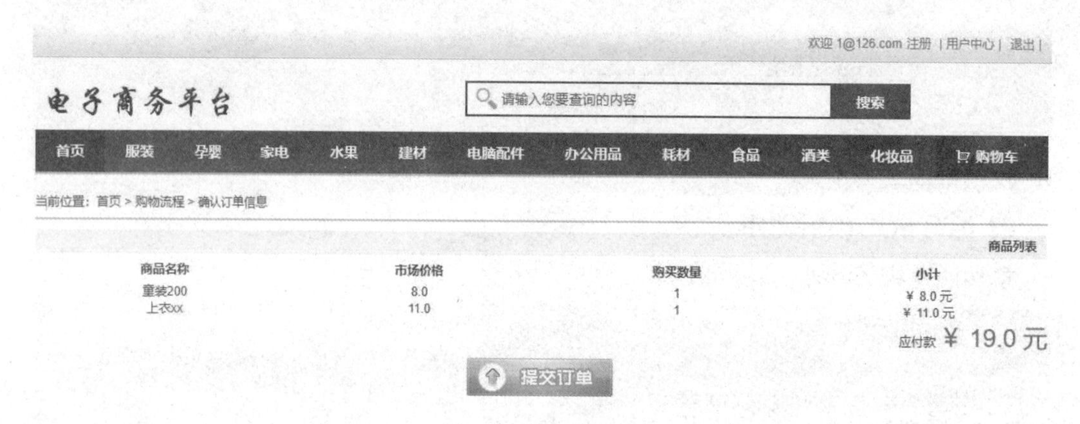

图 20.29　订单确认

在订单确认页面中单击“提交订单”按钮，完成订单。订单完成时页面效果如图 20.30 所示。

图 20.30　订单完成页面

单击图 20.30 中的“银联”图片完成订单支付。具体实现步骤如下：

❶ **编写视图**

该模块的视图涉及 WEB-INF/jsp/before 目录下的 orderconfirm.jsp，orderconfirm.jsp 的代码与购物车页面的代码基本一样，这里不再赘述。

❷ 编写控制器层

该功能模块涉及 com.controller.before.CartController 控制器类的 orderConfirm 方法和 com.controller.before.OrderController 控制器类的 orderSubmit 与 pay 方法。具体代码如下：

```
/**
 * 去结算
 */
@RequestMapping("/orderConfirm")
public String orderConfirm(Model model, HttpSession session) {
    return cartService.orderConfirm(model, session);
}
/**
 * 提交订单
 */
@RequestMapping("/orderSubmit")
public String orderSubmit(Model model, HttpSession session,Double amount) {
    return orderService.orderSubmit(model, session, amount);
}
/**
 * 支付订单
 */
@RequestMapping("/pay")
public String pay(Integer ordersn) {
    return orderService.pay(ordersn);
}
```

❸ 编写 Service 层

该功能模块的 Service 层的代码如下：

```
@Override
public String orderConfirm(Model model, HttpSession session) {
    List<Map<String, Object>> list=cartDao.selectCart(MyUtil.getUserId
    (session));
    double sum=0;
    for (Map<String, Object> map : list) {
        sum=sum+(Double)map.get("smallsum");
    }
    model.addAttribute("total", sum);
    model.addAttribute("cartlist", list);
    return "before/orderconfirm";
}
/**
 * 订单提交，连续的事务处理
 */
@Override
public String orderSubmit(Model model, HttpSession session, Double amount) {
```

```
    Order order=new Order();
    order.setAmount(amount);
    order.setBusertable_id(MyUtil.getUserId(session));
    //生成订单，并将主键返回 order
    orderDao.addOrder(order);
    //生成订单详情
    Map<String, Object> map=new HashMap<String, Object>();
    map.put("ordersn", order.getId());
    map.put("uid", MyUtil.getUserId(session));
    orderDao.addOrderDetail(map);
    //更新商品库存
    //更新商品库存 1.查询商品购买量，以便更新库存使用
    List<Map<String, Object>> list=orderDao.selectGoodsShop(MyUtil.
    getUserId(session));
    //更新商品库存 2.根据商品购买量更新库存
    for (Map<String, Object> map2 : list) {
        orderDao.updateStore(map2);
    }
    //清空购物车
    orderDao.clear(MyUtil.getUserId(session));
    model.addAttribute("ordersn", order.getId());
    return "before/orderdone";
}
@Override
public String pay(Integer ordersn) {
    orderDao.pay(ordersn);
    return "before/paydone";
}
```

❹ 编写 SQL 映射文件

该功能模块涉及的 SQL 语句如下：

```
<!-- 添加一个订单，成功后将主键值回填给 id（po 类的属性）-->
<insert id="addOrder" parameterType="Order" keyProperty="id"
useGeneratedKeys="true">
    insert into orderbasetable (busertable_id, amount, status, orderdate)
    values (#{busertable_id}, #{amount}, 0, now())
</insert>
<!-- 生成订单详情 -->
<insert id="addOrderDetail" parameterType="map">
    insert into ORDERDETAIL (orderbasetable_id, goodstable_id, SHOPPINGNUM)
    select #{ordersn}, goodstable_id, SHOPPINGNUM from CARTTABLE where
    busertable_id=#{uid}
</insert>
<!-- 查询商品购买量，以便更新库存使用 -->
<select id="selectGoodsShop" parameterType="Integer" resultType="map">
```

```
        select shoppingnum gshoppingnum, goodstable_id gid from carttable where
        busertable_id=#{uid}
    </select>
    <!-- 更新商品库存 -->
    <update id="updateStore" parameterType="map">
        update GOODSTABLE set GSTORE=GSTORE-#{gshoppingnum} where id=#{gid}
    </update>
    <!-- 清空购物车 -->
    <delete id="clear" parameterType="Integer">
        delete from carttable where busertable_id=#{uid}
    </delete>
    <!-- 支付订单 -->
    <update id="pay" parameterType="Integer">
        update orderbasetable set status=1 where id=#{ordersn}
    </update>
```

20.6.12　用户中心

成功登录的用户在导航栏的上方单击“用户中心”超链接（userCenter），进入用户中心页面 userCenter.jsp，如图 20.31 所示。

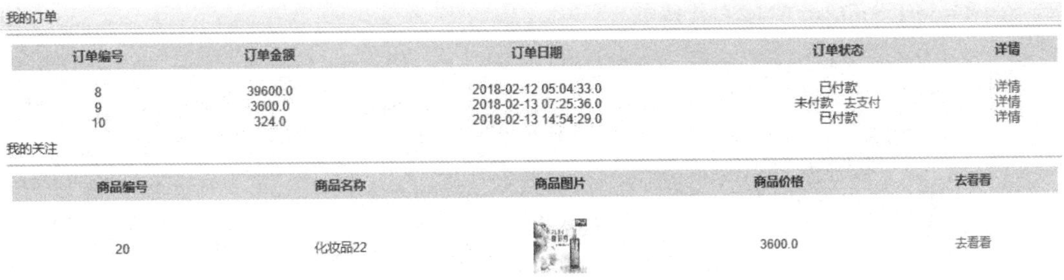

我的订单

订单编号	订单金额	订单日期	订单状态	详情
8	39600.0	2018-02-12 05:04:33.0	已付款	详情
9	3600.0	2018-02-13 07:25:36.0	未付款　去支付	详情
10	324.0	2018-02-13 14:54:29.0	已付款	详情

我的关注

商品编号	商品名称	商品图片	商品价格	去看看
20	化妆品22		3600.0	去看看

图 20.31　用户中心

在用户中心页面单击“去看看”超链接（goodsDetail?id=${mf.id}），打开该行商品的详情页面；单击“详情”超链接（orderDetail?ordersn=${mo.id}），打开该行订单的详情页面 userOrderDetail.jsp。userOrderDetail.jsp 页面的运行效果如图 20.32 所示。

订单详情

商品编号	商品名称	商品图片	商品价格	购买数量
15	微波炉		300.0	1
13	苹果		8.0	3

图 20.32　订单详情

具体实现步骤如下：

❶ 编写视图

该模块的视图涉及 WEB-INF/jsp/before 目录下的 userCenter.jsp 和 userOrderDetail.jsp。userCenter.jsp 的核心代码如下：

```
<c:forEach var="mo" items="${myOrder}">
    <tr>
        <td bgcolor="#ffffff" align="center">${mo.id}</td>
        <td bgcolor="#ffffff" align="center">${mo.amount}</td>
        <td align="center" bgcolor="#ffffff">${mo.orderdate}</td>
        <td bgcolor="#ffffff" align="center">
        <c:if test="${mo.status==0}" >
            未付款  
        <a style="text-decoration: none;" class="f6" href="order/pay?
        ordersn=${mo.id}">去支付</a>
        </c:if>
        <c:if test="${mo.status==1}" >已付款</c:if>
        </td>
        <td align="center" bgcolor="#ffffff">
        <a style="text-decoration: none;" class="f6" href="orderDetail?
        ordersn=${mo.id}" target="_blank">详情</a>
        </td>
    </tr>
</c:forEach>
```

userOrderDetail.jsp 的核心代码如下：

```
<c:forEach var="mf" items="${myFocus}">
    <tr>
        <td bgcolor="#ffffff" align="center">${mf.id}</td>
        <td bgcolor="#ffffff"  align="center">${mf.gname}</td>
        <td align="center" bgcolor="#ffffff" height="60px"> <img
            style="width: 50px; height: 50px;"
            src="logos/${mf.gpicture}" border="0"
            title="${mf.gname}" /> </td>
        <td align="center" bgcolor="#ffffff">${mf.grprice}</td>
        <td align="center" bgcolor="#ffffff">
        <a style="text-decoration: none;" class="f6" href="goodsDetail?id=
        ${mf.id}">去看看</a>
        </td>
    </tr>
</c:forEach>
```

❷ 编写控制器层

该功能模块涉及 com.controller.before.UserCenterController 控制器类的 userCenter 方法和 orderDetail 方法。具体代码如下：

```
@RequestMapping("/userCenter")
public String userCenter(HttpSession session, Model model) {
    return userCenterService.userCenter(session, model);
}
@RequestMapping("/orderDetail")
```

```
public String orderDetail(Model model, Integer ordersn) {
    return userCenterService.orderDetail(model, ordersn);
}
```

❸ **编写 Service 层**

该功能模块的 Service 层的代码如下：

```
@Override
public String userCenter(HttpSession session, Model model) {
    model.addAttribute("myOrder", userCenterDao.myOrder(MyUtil.getUserId
    (session)));
    model.addAttribute("myFocus", userCenterDao.myFocus(MyUtil.getUserId
    (session)));
    return "before/userCenter";
}
@Override
public String orderDetail(Model model, Integer ordersn) {
    model.addAttribute("myOrderDetail", userCenterDao.orderDetail(ordersn));
    return "before/userOrderDetail";
}
```

❹ **编写 SQL 映射文件**

该功能模块的 SQL 语句如下：

```
<select id="myOrder"  resultType="map" parameterType="Integer">
    select id, amount, busertable_id, status, orderdate from ORDERBASETABLE
where busertable_id=#{bid}
</select>
<select id="myFocus" resultType="map"  parameterType="Integer">
    select gt.id, gt.gname, gt.goprice, gt.grprice, gt.gpicture from
    FOCUSTABLE ft, GOODSTABLE gt
    where ft.goodstable_id=gt.id and  ft.busertable_id=#{bid}
</select>
<select id="orderDetail" resultType="map"  parameterType="Integer">
    select gt.id, gt.gname, gt.goprice, gt.grprice, gt.gpicture,
    odt.shoppingnum from
    GOODSTABLE gt, ORDERDETAIL odt
    where  odt.orderbasetable_id=#{ordersn} and gt.id=odt.goodstable_id
</select>
```

20.7　本章小结

本章讲述了电子商务平台通用功能的设计与实现。通过本章的学习，读者不仅应该掌握 SSM 框架整合应用开发的流程、方法和技术，还应该熟悉电子商务平台的业务需求、设计以及实现。

项目案例——基于SSM的邮件管理系统

该系统仅实现某内部邮件系统的基本功能，不考虑其安全性。用户分为两大类：系统管理员和注册用户。

1．系统管理员

系统管理员具有锁定非法用户、删除非法邮件等功能。

2．注册用户

注册用户具有写信、收信、已发送、用户中心等基本功能，具体如下。

（1）写信：写信时，只可以自动获取收件人地址，不可以手动录入收件人地址，这是因为该系统是内部邮件系统。信件包括标题、内容、附件等信息。

（2）收信：收信时，未读邮件的标题以粗体标注。收件信息包括发件人地址、标题、内容、附件等信息。该模块具有回复、查看、删除收到邮件的功能。

（3）已发送：该模块具有查看、删除已发送邮件的功能。

（4）用户中心：该模块具有修改用户密码的功能。

读者扫描二维码，可阅读本系统详细的设计文档、观看视频讲解以及下载源代码。

项目案例

项目案例——基于SSM的人事管理系统

系统总体目标是搭建某单位的人事信息管理平台，不仅满足目前的业务需要，还要满足公司未来的发展，而且具备良好的可扩展性，形成公司未来人力资源管理信息化平台。系统包括部门管理、岗位管理、入职管理、试用期管理、岗位调动管理、员工离职管理、员工信息中心、报表管理等功能模块。具体功能如下。

（1）部门管理：主要用于描述组织的部门信息，以及部门的上下级关系。包括新建部门、修改部门、查询部门下的员工等功能。

（2）岗位管理：主要用于对组织内各岗位进行管理，包括增加、修改、删除岗位，以及查询岗位下的在职人员等功能。

（3）入职管理：主要用于员工基本信息进行录入，包括员工部门、岗位、试用期及其他信息的录入。

（4）试用期管理：主要对试用期员工进行管理，包括试用期转正、试用期延期、试用期不通过、已转正员工信息查询等功能。

（5）岗位调动管理：主要对员工岗位调动进行管理，包括部门内岗位调动、部门间岗位调动、调动员工查询等功能。

（6）员工离职管理：主要对员工离职进行管理，包括确定离职员工、已离职员工信息查询等功能。离职的类型包括主动辞职、辞退、退休、开除、试用期未通过。

（7）员工信息中心：主要对员工的信息进行归类管理，包括职业生涯信息、外语能力信息、家庭成员及社会关系等信息。

（8）报表管理：包括给定时间段新聘员工报表、给定时间段离职员工报表、给定时间段岗位调动员工报表、人事月报等报表查询。

读者扫描二维码，可阅读本系统详细的设计文档、观看视频讲解以及下载源代码。

项目案例

在 Eclipse 中使用 Maven 整合 SSM 框架

Apache Maven 是一个软件项目管理工具，基于项目对象模型（Project Object Model，POM）的理念，通过一段核心描述信息来管理项目构建、报告和文档信息。在 Java 项目中，Maven 主要完成两件工作：① 统一开发规范与工具；② 统一管理 jar 包。

Maven 统一管理项目开发所需要的 jar 包，但这些 jar 包将不再包含在项目内（即不在 lib 目录下），而是存放于仓库当中。仓库主要包括：

（1）中央仓库：存放开发过程中的所有 jar 包，例如 JUnit，用户可以从中央仓库中下载这些包，仓库地址：http://mvnrepository.com。

（2）本地仓库：本地计算机中的仓库。

官方下载 Maven 的本地仓库，配置在“%MAVEN_HOME%\conf\settings.xml”文件中，找到“localRepository”即可；Eclipse 中自带 Maven 的默认本地仓库地址在“{user.home}/.m2/repository/settings.xml”文件中，同样找到“localRepository”即可。

Maven 项目首先会从本地仓库中获取所需要的 jar 包，当无法获取指定 jar 包时，本地仓库会从远程仓库（中央仓库）中下载 jar 包，并放入本地仓库以备将来使用。

如何在 Eclipse 中使用 Maven 整合 SSM 框架，请扫描图 C.1 所示二维码，查看配置说明文档。

图 C.1　二维码

参考文献

[1] 史胜辉，王春明，陆培军．Java EE 轻量级框架 Struts 2+Spring+Hibernate 整合开发[M]．北京：清华大学出版社，2014．

[2] 范新灿．基于 Struts、Hibernate、Spring 架构的 Web 应用开发[M]．2 版．北京：电子工业出版社，2014．

[3] （美）Paul Deck．Spring MVC 学习指南[M]．林仪明，崔毅，译．北京：人民邮电出版社，2015．

[4] 韩路彪．看透 Spring MVC：源代码分析与实践[M]．北京：机械工业出版社，2015．

[5] （加）Budi Kurniawan，（美）Paul Deck．Servlet JSP 和 Spring MVC 初学指南[M]．北京：人民邮电出版社，2016．

[6] 杨开振，等．Java EE 互联网轻量级框架整合开发——SSM 框架（Spring MVC+Spring+MyBatis）和 Redis 实现[M]．北京：电子工业出版社，2017．

[7] 黑马程序员．Java EE 企业级应用开发教程（Spring+Spring MVC+MyBatis）[M]．北京：人民邮电出版社，2017．

图书资源支持

感谢您一直以来对清华版图书的支持和爱护。为了配合本书的使用，本书提供配套的资源，有需求的读者请扫描下方的“书圈”微信公众号二维码，在图书专区下载，也可以拨打电话或发送电子邮件咨询。

如果您在使用本书的过程中遇到了什么问题，或者有相关图书出版计划，也请您发邮件告诉我们，以便我们更好地为您服务。

我们的联系方式：

地　　址：北京海淀区双清路学研大厦 A 座 707

邮　　编：100084

电　　话：010－62770175－4604

资源下载：http://www.tup.com.cn

电子邮件：weijj@tup.tsinghua.edu.cn

QQ：883604(请写明您的单位和姓名)

书圈

用微信扫一扫右边的二维码，即可关注清华大学出版社公众号“书圈”。